Bluejacket
Admiral

Bluejacket Admiral

The Navy Career of Chick Hayward

JOHN T. HAYWARD AND C. W. BORKLUND

Naval Institute Press • Annapolis, Maryland
Naval War College Foundation • Newport, Rhode Island

Naval Institute Press
291 Wood Road
Annapolis, MD 21402

© 2000 by John T. Hayward and C. W. Borklund

All rights reserved. No part of this book may be reproduced or utilized in any form or by any means, electronic or mechanical, including photocopying and recording, or by any information storage and retrieval system, without permission in writing from the publisher.

Library of Congress Cataloging-in-Publication Data
Hayward, John T., 1908–
   Bluejacket Admiral : the Navy career of Chick Hayward / John T. Hayward and C. W. Borklund.
   p.   cm.
   ISBN 1-55750-189-0 (alk. paper)
   1. Hayward, John T., 1908–   2. United States. Navy—Aviation—Biography.
3. Admirals—United States—Biography.   I. Borklund, Carl W.   II. Title.

V64.U6 H39 2000
359'.0092—dc21
[B]   00-023159

Printed in the United States of America on acid-free paper ∞
07 06 05 04 03 02 01 00   9 8 7 6 5 4 3 2
First printing

With special thanks to the Naval War College Foundation in the publication of this book.

Contents

Preface   vii

1. The Dropout Years   1
2. Up-Checks and Down-Checks   12
3. The Yellow Peril   27
4. Life on the Covered Wagon   41
5. The Mid-1930s: Getting on the Step   55
6. Peace in Our Time?   69
7. Preparing to Fight while Fighting   86
8. From Air Combat to Atomic Bombs   104
9. A Sailor's Start into the Nuclear Age   118
10. A-Bombs and Turf Wars   134
11. When in Doubt, Reorganize   150
12. Pushing the Envelope   166
13. This Is Not a Boat, Hayward   182
14. Turf Wars: Some Angry, Some Deadly   198
15. The George Cruz Ascom Story   212
16. On Making Admiral   227
17. The Pentagon: A Place to Pass By   241

18 From R&D to Cuba   256
19 From Cuba to the Naval War College   272
20 Where Are the Leaders?   290

Appendix:
Thumbnail Biography of Marine Corps Brigadier General Richard W. Hayward, 1906–1989   301

Notes   305

Bibliography   317

Index   319

Preface

It has been said, "The most useful history is that which records the human mind's progress." This autobiography records the progress and career of a particularly unique mind. "Jack" to his family, "Chick" to the U.S. Navy, John T. Hayward enlisted in the navy in 1925 by claiming he was seventeen, the minimum age, which he wasn't. The "bluejacket" job he liked best was as a battleship "powder monkey," third class, in 1926. Though he was a high school dropout, he graduated from the Naval Academy in 1930, ranking fifty-first in his 406-man class. Of sixty-seven students who started navy flight training at Pensacola, Florida, in October 1931, he was one of only seventeen to receive his gold pilot's wings.

Ensign Hayward's flying career began on the navy's first aircraft carrier, an old *Orion*-class collier with a flat, barren 610-foot wooden flight deck superimposed on it—the origin of the "flattop" nickname for aircraft carriers. Following a personal routine started in 1937, of graduate-level study at night and flying every kind of aircraft he could commandeer during the day, by 1950 he was both a nuclear physicist and one of naval aviation's first jet-qualified pilots. He ended his career as president of the Naval War College, starting its growth into one of the best institutions of its kind anywhere. By then, he had amassed more pilot hours than any other flag officer ever, and had earned dozens of combat and distinguished-service medals and two science awards.

But this book is as much military history as memoir, a record of amazing growth in science and engineering technology; of dramatic shifts in military threats; of wars (for him, World War II, Korea, the Cuban missile crisis, Vietnam); of creating nuclear-powered ships, antisubmarine defenses, "smart" bombs, ballistic and cruise missiles, satellites, and supersonic aircraft. Chick was involved in all of these events, many of which had a major impact on politico-military policies and plans.

Prior to World War II, at the Naval Aircraft Factory in Philadelphia, Lieutenant Hayward was head of a program to create new aircraft instruments, altimeters, compasses, radios, and navigation aids for both the U.S. Navy and the British air forces. In 1944 at the China Lake Naval Weapons Test Center, Commander Hayward ran programs to develop solid-fueled rockets, the ancestors of many of today's air- and ground-launched missiles—Sidewinder, Harpoon, and others. He and mathematics wizard Johnny von Neumann also devised a way to implode a plutonium bomb (the Fat Man dropped on Nagasaki to end World War II), a task necessary since the gun-type detonation used for the uranium in Little Boy, dropped on Hiroshima, couldn't make Fat Man's plutonium fission.

Later, he oversaw creation of an atomic depth charge and of the navy's present-day mines. Then he set up the first carrier-based atomic-bomb squadron. As the Atomic Energy Commission's head of weapons research, he helped found the Lawrence Livermore Laboratory. After retirement, he sold General Dynamics on making the first Tomahawk, a torpedo then. As an example of those gains, his first navy plane cruised at 110 miles an hour; one he flew (blind in one eye) just thirty-two years later *climbed* at sixteen hundred miles an hour. "That kind of advance won't continue unless it's encouraged and well-managed," he said, "which it isn't in today's climate."

In mid-1959, early in my thirty-year career (1957–1987) as a magazine editor-publisher in Washington, D.C., the then-CNO (Chief of Naval Operations) Arleigh Burke split the office of Assistant CNO for Research and Development into two divisions: Chief of Research, and Deputy CNO for Development. The use of "deputy" in the second title subtly said he wanted a faster, more focused conversion into useful military hardware of the rush of scientific knowledge then being discovered. (In the arcane job-title business, an "assistant" advises the CNO what to do, but a "deputy" can do the deciding on the CNO's behalf.) When Chick Hayward was picked to be the "deputy," I rushed to his Pentagon office, research and development being my magazine's editorial meat-and-potatoes at the time. After that first two-hour meeting, I saw him often over the years, for lunch, dinner, and golf. Then in 1994, his nephew, Steve Madey, Jr., sent me a three-hundred-page single-spaced tome Chick had written, mostly in diary form. The result of that, plus some Naval Institute Oral Histories and a dozen sessions with him myself, is *Bluejacket Admiral,* all read and approved by Chick in its final draft. During its long gestation, he was fighting cancer but assured me, "Don't worry. These navy doctors are going to keep me alive until they can *prove* the reason I have cancer is because I watched our first A-bomb test at Alamogordo in 1945."

One interesting quirk about his navy career was that often, by the time one of his innovative programs began to gain public attention, he'd left, assigned to launch a new one. Similarly, on 20 May 1999, I called to tell him, "The Naval Institute Press has confirmed it will publish our book." He was delighted. Three days later, he died, gone to join his wife (who'd died in 1998) and start some new project.

C. W. Borklund

Bluejacket Admiral

1

The Dropout Years

On 15 May 1925 at the Madison Square recruiting office in New York City, John Tucker Hayward enlisted in the United States Navy. It took a lie, a bribe, and a lieutenant commander more worried about meeting a quota than about a law saying recruits must be at least seventeen years old. That apprentice seaman, myself, retired from active duty as Vice Admiral Hayward on 1 September 1968. Did I enlist to serve my country, to be part of the military drama and awesome advances in science and engineering that occurred during those forty-three years? Hardly.

At the time, I thought I was exactly fifteen-and-a-half years old. Fifty years later, the Social Security folks told me I'd been born in 1947. By then, official records of my birth date either had burned up in a fire or been lost, but a baptismal entry in our family Bible implied that I was born in 1908, not 1909 as I'd thought for years. In any case, when I enlisted I was underage, a dead loss in school with a crazy notion to escape, to sail off to romantic adventures in new places where life would be better. It was a juvenile idea, of course. My family already was fairly well-off, my life already a record of exciting events.

Born in New York City on 15 November, whatever the year (the navy chose 1908), one of Charles Brian and Rosa Valdetarro Hayward's eight children, I was one of only two boys—Richard and myself—and two girls—Eleanor (who insisted it be spelled Elinor) and Marjorie—who survived. The others died in childhood of the kinds of illnesses (pneumonia, diphtheria, spinal meningitis) that rarely kill today in our country—one proof of medical science's amazing gains, especially in the past forty years. My father wanted me named after a family hero, Royal Navy Lt. Benjamin Tucker. My mother rarely denied him anything but this time she did. "No," she said. "I'll not have a son of mine being called 'Benjie!'" Dad settled for John Tucker Hayward.

My maternal grandfather Valdetarro was from La Spezia, a small port city in northwest Italy. In the 1800s many English notables, including the Shelleys from Mother's side of my family, treated Italy as a second home. One, English author Mary Shelley, née Godwin, had written the grim classic, *Frankenstein*. Her husband, British poet Percy Bysshe Shelley, an avid yachtsman but a lousy sailor who couldn't swim, got himself drowned in 1822 at age twenty-nine, sailing his yacht from Livorno, Italy, into a raging storm. (When his fish-bitten, decomposed body washed ashore, it was cremated, the ashes interred in Rome—except for his heart that eerily refused to burn. So Mary Shelley preserved it until she died in 1851, and it was buried with her.)

My father's heritage was even more richly historic. In 1663, King Charles II gave eight English lords a chunk of America they named Carolina (now North and South Carolina). "Our illiterate cousins," Father called them for spelling it "Heywoerd," went there in the seventeenth century. Our own Tucker clan, which settled in Bermuda, included Dr. Samuel Tucker, Chief Justice John Tucker, and Bermuda's first governor, Daniel Tucker. They owned St. David's island for a time, and most of my paternal grandfather's ancestors are buried in a churchyard on St. George's. Born in Bermuda in 1876, my dad moved in the 1890s to New York where he met and married my mother.

We grew up in a fine house on Grace Avenue in tiny Great Neck, Long Island, and had two maids, Margaret and Katherine, imported by my parents from Ireland. A very happy part of my early years, these women fussed over us as if we were little angels. So did my strict but caring mother, Rosa. A vivacious brunette, only five feet tall, she ran the household as most women do and regularly bought us fine books, urging us to read. Early on, I was irked at grandfather Hayward for favoring his namesake, my elder brother, Richard Wright Hayward, V. The peeve died off, however, after my grandfather growled at me, "English custom is to give the eldest boy first priority!"

In any case, my father was the boss, made the decisions, set and enforced the rules. He frequently told me firmly, "You can't be a boy, forever. Grow up! Accept responsibility even when tempted to avoid it," as I often was. His "tough love," as it's now called, riled me then, but I now know it was the bedrock for most of what I achieved in my career.

At six feet, he was tall for that era, dark-haired, slim, very talented, worked hard, and studied all the time. As a partner in a small advertising firm, he helped it prosper. He was a lawyer, and was adept in mechanical, electronic, automotive, and aeronautical engineering as well. That led to our having one of the first refrigerators and first automatic dishwashers in Great Neck. He also built in our attic a radio with a spark-gap transmitter. An extra-special treat for me was being invited up to listen to Pittsburgh's KDKA when it became, in November 1920, the country's first radio station to do regularly scheduled broadcasts.

My father also wrote several engineering books and was one of America's first aviators. His *Practical Aeronautics*, published in 1912, is a "how to build an airplane in your garage and fly it" book. Its introduction, written by a far more famous aviation pioneer, says in part, "[It is] remarkably free [of] errors usually found in aeronautical works of this character.... The chapter on patent litigation the best [I have seen].... [Your book is] a valuable addition to the literature of Aeronautics. [Signed] Orville Wright."

My father and his brother, my Uncle Billy, volunteers in New York's Naval Reserve, had fought in the Spanish-American War on the *Yankee*, the auxiliary gunboat that captured the Guantánamo Bay fortress.[1] Predictably, when our tiny Atlantic Fleet visited New York once, my father took me down to the Hudson River to board the battleship *Connecticut*. I was impressed, awestruck, really. Uncle Billy, a lifelong bachelor, added in the fun of sailing on Long Island Sound during our summer vacations in his sleek, gaff-rigged, seventy-five-foot sloop, the *Weonah*. I heard a lot about the navy on those trips and acquired a love of sailing.

My father first took me flying when I was not yet six years old. We used Long Island's Flying Field #2, renamed Hazelhurst Field when the Army "Rainbow" Division at nearby Camp Mills began pilot training there prior to World War I. So I saw "Jennies" overhead regularly from 1916 to 1918.[2] By then, I'd read or had read to me everything I could find on aviation, and ached to fly an airplane myself. Flying had become my first and would remain my lifelong love.

After the war, first as Hazelhurst, then Mitchell (for New York Mayor John Jerry Mitchell), then Roosevelt Field, it was the takeoff or landing site for many record-setting dirigible and aircraft flights, including Charles A. Lindbergh's historic solo flight to Paris in 1927. The field also hosted aerobatic shows, including one I watched in which an Italian World War I air ace got himself killed. One morning, I saw the British dirigible, R-34, float over our yard on its historic flight from England.[3] I did a lot of "hangar flying" there myself, listening to Jimmy Doolittle, Casey Jones, and other aviation pioneers.[4] Back then, of all the ways to become a pilot, military training was said to be the best. My father told me that learning what pilots had to know to fly well meant knowing engineering and mathematics—in other words, go to school.

I paid his advice little mind. In fact, by high-school age, I had no use for school at all. My service record says, "Education: Oakdale Military Academy, Long Island," but it was my brother who went to Oakdale. I went to Clason Point Military Academy, a Christian Brothers high school in the Bronx. I hated it, unlike my brother who was "born military." But it was the Roaring Twenties, the Jazz Age, the era of Prohibition, the "flapper," bathtub gin, and bootleg whiskey. The nation's economy was booming, John Calvin "Silent Cal" Coolidge was president, and I was wrapped up in baseball.

Most New York baseball fans rooted for the New York Giants, owned by John McGraw; and called their National League rival, the Brooklyn Dodgers, "Bums." But our neighbor, Thomas A. "Tad" Dorgan, was a New York Yankees fan. He drew a popular "Inside Sports" cartoon for Hearst newspapers; had named a "hot dog" the sausage in a bun sold at the ball park; and coined the phrase, "Yes, we have no bananas."[5] And he had dozens of celebrity friends. Through him I met cowboy–movie actor Tom Mix, heavyweight boxing champion Jack Dempsey, and many other stars called the "Golden People" by author Paul Gallico. I met Gallico at Dorgan's too, and over the years, got to know him very well. Eventually, he dedicated one of his many best-selling books, *The Poseidon Adventure*, to me, I guess because I showed him how to capsize a ship—the main event in the book.

Dorgan was a heavy drinker, some said alcoholic—a distinction too fine for me to draw. Not the booze but a bad heart, he said, prevented his going to Yankee games. So he persuaded Col. Jacob Ruppert to hire me as assistant bat boy, just a "gofer," really, to Eddie Bennett, a tiny hunchback who tended to Yankees equipment a little but mostly played team mascot. I joined the team in 1923 as it quit playing at the Polo Grounds, owned by the Giants, and moved to newly built Yankee Stadium. My job, Tad said, was to report in detail after games what his beloved team had done. I think all he really wanted was to stop this screaming kid from organizing ball games in his backyard.

In any case, afternoons at the ballpark didn't help my record at Clason. They cut me, finally, for "Unauthorized Absence." By then, while wondering what I wanted to do in life besides not go to school, I was and still am a Yankee fan. In 1923, they were rated the epitome of excellence in their unique pinstriped uniforms. Quietly proud of their record, they behaved like champions off the field, too—kind, polite, generous (most of them), always in coat and tie in public. I still can recall clearly my going to see those fine men after I'd enlisted, and their greeting this gofer like a proverbial prodigal son. And they hated to lose, an attitude they rubbed off onto me, one I've nursed ever since.

They won the American League pennant in 1921 and 1922, the World Series in 1923. But in 1924, they lost the pennant to the lowly Washington

Senators, an awful upset that inspired a marvelous musical comedy, *Damn Yankees*. They fell on even darker days in 1925—"Babe" Ruth suspended, the team in seventh place, manager Miller Huggins in a rage. This just put vinegar on the whole business for me, the maverick teenager who idolized the Yankees.

To be near the ballpark, I was by then living at my grandmother's in the Bronx where, incidentally, she gave me the only lectures on sex I ever received. (In my own home, the subject was not allowed—very unlike the rule in most homes today.) And I had been enrolled in Loyola, a Jesuit-run school at Eighty-third and Park Avenue. Its headmaster, kindly old Father John Farley, taught me weekends at Hudson River Country Club to play golf, and tried tactfully to fix my mixed-up brain while stern Jesuit teachers tried to pound math, Latin, and Greek into me. But I just hated school and my grades were as poor as my chances for a college engineering degree.

Still, standing by Metropolitan Tower at Madison Square that sunny day in May 1925, I had no long-range plans and only a vague idea why I entered that recruiting shop. The officer there gave me a form to fill out, telling me I had to be seventeen to enlist. He glared at me when I said I was. Obviously, I had to find a way to finesse this age thing. But his problem in a robust economy was finding anyone at all who'd work for only $21 a month. So underage or not, I was given the physical. Noting I was under the 118-pound minimum acceptable weight, the doctor, staring at bare-ass me, short even for my real age, growled, "Besides, this boy can't possibly be seventeen."

"Well," the recruiter snorted, "if his folks sign the papers, it's not our argument."

Still, maybe hoping to scare me off, he recited a "minority enlistment" rule. It said that recruits under legal adult age, twenty-one at the time, had to agree to serve until that age. If the lie about my age held up, it meant having to stay in the navy five years, a prospect that didn't bother me a bit. Indeed, as I left, I was smugly pleased with myself and quickly found two strangers in the park ("winos" they were called then, "homeless" they're called now) thirstily eager for a small fee easily paid to forge my parent's names on a form saying I had their permission to enlist. A month later, on 29 June, I shipped out for boot camp in Newport, Rhode Island, my first "romantic adventure."

The navy had given me a ticket on the *Commonwealth*, a Fall River steamer, New York to Providence, with a 3:30 A.M. stop in Newport. The navy, I learned, doesn't buy a first-class bunk, let alone a private cabin, for apprentice seamen. I was sent to the fo'c'sle, to the top bunk in a six-tier stack. The instant I lay down, bedbugs began scurrying over me as if I were a playground. This was not the kind of romantic adventure I was after. So I climbed down and went out to walk the boat's bleak, empty decks. Crossing Long Island Sound, we passed Execution Light, Port Jefferson, Fisher's and Duck Islands, and other places I had known all my life, ports of call in my uncle's sailboat. Gazing at lights on shore and the ship's wake trailing off behind us made me a bit homesick. Worse, I knew my parents did not know exactly where I was.

At our Newport dock, the place seemed even more dreary than the trip had been. From there, seven of us were driven to Barracks A at the Training Center where we were lined up. A big old bosun's mate walked the line, inspecting what he'd been sent. When he got to me, the smallest, skinniest recruit there, he bellowed, "And how in hell did a little chick like you get in here amongst all these grown men?"

By sunset, "Chick" was and would remain my navy nickname. My wife has never liked it. Nor have my children. My mother hated it. In my family, I'm "Jack." But from then on, most people in the navy, even those I never met, called me "Chick" Hayward. Even on my service record at the Naval Historical Center, someone penned in brackets, "Nickname is Chick."

In 1925, some people were arguing the merit of a "League of Nations," or "how the next world war will start." Yet none seemed worried about it. For sure, bluejacket Hayward didn't foresee us fighting the Japanese or anyone else, nor did I expect to be in a couple more wars after that—nor expect to become an officer, let alone an admiral. The navy's image was much lower then than it is now. There'd be no more war, people said, so why do we need a navy at all? Most mothers forbade their daughters a date with a sailor. The public's general opinion was that we were mainly ne'er-do-wells, drunks, carousers after women, not much use to anybody, really. Many of my own mother's friends were horrified to learn that her son had become "one of those." Most Newport natives had an especially low opinion of us. Some even had signs on their lawns reading, "Sailors and Dogs, Keep Off the Grass!"

Boot training was dismal. The navy in those days was wrapped in a rigid caste system, beyond imagining today, of elites and have-nots. An enlisted "bluejacket" sailor was the lowest of the lowly. They cut all the hair off our heads, made us sleep in hammocks, gave us beans and prunes for breakfast. We had to scrub our own laundry, tie all kinds of knots, drill all morning on what we called "the grinder," the parade ground. We were harangued to lash our hammocks, store our seabags "just so;" and, on Fridays, parade in front of the old Naval War College, singing "Nancy Lee," "Strike up the Band," "Here Comes a Sailor," and other such sailor songs as we marched. Then some of us had to man the yards on the old *Constellation,* moored at what still is called Constellation Pier on Coasters Harbor Island.

Besides that tedium, we second-class seamen always got the drudge jobs. Once a year, for three months, a select group of us were assigned "mess cooking." It meant running the errands necessary to get chow together for everyone else, serving the meals, and washing the dishes. For this toil, we received a munificent bonus of $5 a month. Still, the day-to-day life was fairly easy. We didn't have to think too much. I was young and had "a strong back and a weak mind," as the bosun's mate frequently reminded me.

And I couldn't stand it. I was miserable, praying my father would hurry up and find me, would get me out of there, tell them that my enlistment paper was a forgery and that I'd lied about my age. He didn't catch up to me right away but did put out an "all points," as maybe only he could. It was one of his dozens of friends, Navy Assistant Secretary T. Douglas Robinson, who caught me during one of the endless inspections we had to endure. There he asked me, as had lots of people by then, "Son, how old are you?" I answered, as I had for quite some time, "Seventeen, sir." Asked twice more, I repeated the lie.

He knew my name, of course, and a couple of days later, I was told to report to the office of Capt. J. P. Orton Jackson, commander of bluejacket training and as friendly to us as a tiger with an abscessed tooth. Not being guilty of anything lately, I guessed the only explanation for my being summoned to a room otherwise off-limits to ordinary seamen was that Robinson had called my father. Jackson's conference room was a drab place with rigid, wooden chairs, the only kind the navy seemed to buy in those days. I was told to sit on one of them and wait. Minutes later, an eternity to me, Jackson bulled in with my father. *Saved!,* I thought, but all my father said, bluntly, was, "Son, you are in the navy, now. You can stay in the navy." And he left.

And that was the end of it. I was stunned, angry, snarling to myself, "I'll show this guy." It was the first time I'd ever felt that way. My mother was very unhappy over all of it but my father was the family boss, the ruler. Maybe that's why none of us ever again mentioned the crushing decision he made that day. Feeling like a whipped puppy, I went back to the barracks and back to my training.

We ordinary seaman seldom saw an officer. All training was done by chief petty officers who ran the navy then and I suspect still do. Our CPO was Chief Quartermaster Harlow, a first-rate teacher, expert in navy lore. His primary training of us lasted four weeks, on-base, never allowed "ashore" (off the base). Next came six weeks of second-level training, a less confining stage. Then we went to sea. Navy strength in 1925 was only 95,230 people including 8,430 officers and 488 women, mostly nurses. Seagoing days were very restricted. Our Atlantic Navy was just a small fleet of mainly scouting ships, our European Squadron only the light cruiser *Detroit*. But the ships did ply Narragansett Bay, making Newport, whether the townsfolk liked it or not, a big navy town.

My first ship was the famed coal-burning battleship *Wyoming*. I was on her to shovel coal as a rated "fireman-engineer," just as yesterday's "janitor" is called a "building maintenance engineer" today. Because of my small size, the job wore me down in a hurry. I ended up shoving coal down off the pile and let the big guys do the heavy lifting into the fires. I was on that grimy job only a short time when a "draft," a soft-soap word for a call for compulsory, temporary-duty "volunteers," came in for a crew to man a ship's tender serving the navy's and America's first rigid airship, the *Shenandoah*. I jumped at it.

Then in September in a storm over Ohio, a lightning bolt exploded the *Shenandoah*, killing several people. It provoked Navy Secretary Dwight Wilbur to say that enemy aircraft will never attack America because flying "is simply too hazardous," and it got us reassigned to runs between Texas and the East Coast, hauling fuel. But not for long. Late that month in Newport Barracks (now Sims Hall), a guy in the next berth belched a crude stream of profanity. I didn't know why he was irritated but suddenly a big broad-shouldered priest, a commander, spun me around and belted me one, upside the head. "Why'd you do that?" I yelped. "I wasn't doing anything."

After a stunned second, he barked as he left, "Well, get the guy who did and smack him!"

So little "Chick" did as ordered—and got flattened by the cusser.

The priest was Father John J. Brady. Eventually promoted to captain, he had been with our marines at Belleau Wood and the Marne in World War I where he earned the Navy Cross and a Silver Star.[6] He sort of adopted me after our dustup, even making me an altar boy at Mass. And he shoved and shaped this high-school dropout into a student who developed a lifelong desire to learn. Father Brady could be a tough cookie if he thought you needed it, but he was a good man who believed that most of us sailors were just decent young men cast adrift. He wanted all of us to get ourselves educated enough to become officers. A fabulous man, he knew how to study and did, and, as a result, his grasp of the world and what's in it amazed me. The effort to learn, to gain knowledge, must be unending, he said. By his example more than his sermons, he opened up the world of textbooks, the rewards of study, to me. He opened up everything for me.

Early on, he told me that Navy Secretary Josephus Daniels had persuaded Congress during World War I to let the president appoint one hundred sailors a year to the Naval Academy—if we passed the entrance exams. That quota never had been filled nor had the navy done much to put us on the road to Annapolis. I was in worse shape than most since I had only maybe two years of high school. But Father Brady didn't just say I could pass the entrance exams. He convinced me that I could survive the academy's four hard years of study, test, study, test, and graduate.

That same September 1925, an excursion steamer from Pawtucket, Rhode Island, heading for Newport Beach, blew up right offshore of our base. They were lucky that a bunch of us were there on a boat drill. We became instant lifesavers. I also was lucky, not from the glowing local praise we got, including Pawtucket making us honorary citizens—pretty heady stuff for a teenager—but because the rescue made headlines in Washington. That inspired President Coolidge to offer his "aid and assistance" to any of us who wanted to enter the academy. The aid I got was from Father Brady and it was double-barreled. First, since I didn't have a high school diploma, he had to recommend, which he did, that I be sent to the Naval Academy Prep School in Norfolk, Virginia. If he hadn't done that, I'd have been shipped out to sea.

Next, he had the Bureau of Navigation send him a copy of every Naval Academy entrance exam ever given, going back (I think) to 1898. He told me to copy down every question in all of them, then find the answers. Using textbooks he or I gathered up on each subject, I worked harder than ever before in my life. For plane geometry, his order was simple: Write on the blackboard all its theorems, then stare at them until I can recite every one, at random, from memory.

Newport was icy that winter—lots of snow, our barracks always cold. That and the unfriendly natives kept most of us on base. Father Brady didn't let me go "ashore" anyway, even if I'd had time left after my studies, which I didn't. In spite of Prohibition, the bars on Thames Street were wide open. Besides, he said, "There are girls out there!"

But the Naval Academy was a blank to me and when I first heard "Annapolis," I thought it was Indiana's state capitol. I did know I'd not have to wash clothes, cook meals, or sleep in a hammock there. Best of all, I learned, an academy degree could qualify me for pilot training. The naval base at Norfolk already had all the schools such as electrical and radio where sailors earned one of the many specialty ratings of that era. But in 1925, the navy also had started two academy prep schools, one in Norfolk, and one in San Diego, California. Norfolk's was in an abandoned World War I building. Father Brady had the five of us who had signed up sent there in December. It was bleak, unlike the prep school facilities the navy has now, but it was a start.

We had not quite four months to study for the exams. Starting the third Wednesday in April 1926 at the school, we would take two tests a day for three days, plus of course the ever-present physical. The exams would cover a full high school education: algebra, geometry, physics, American and ancient history, and English literature; and to be accepted into the academy, we had to pass not most of but *all* of them.

Early on at Norfolk, my newfound urge to go to college was almost derailed. As I was walking down the street one day, reading a textbook, my peripheral vision spotted a commander—for bluejackets the next thing to God—with a marine orderly across the street, coming my way. I didn't salute. In quickstep time, the marine was at my side, and seconds later I was facing the commander who snarled, "Well, son, why didn't you salute? Aren't we in the same navy?"

A thorough student of the rules with pretty good recall and, at the time, a guardhouse-lawyer attitude toward such things, I said, "Yes, sir, I am, but Article 1-330 of the Landing Force Manual says one is not required to salute unless within thirty paces. I do not believe I was within thirty paces but I will be delighted to measure it, sir."

Instantly, I was "On Report," etching into me a new law: Don't argue with commanders over what the rules may say, a "law" I followed all my career—well, most of the time. My shipmates teased me but warned me not to be so quick to quote rules. (Four years later, when that commander was my senior watch officer, he seemed to have forgotten our run-in.) Our instructors at school were Lt. Lawrence Schulten and Ensigns Rose and Woods, but actually the study was mostly up to us.[7] We hadn't much time. Father Brady's intense program, which I continued, made that third Wednesday in April arrive too quickly, it seemed to me.

I still remember those tests. Even now, I can find the length of the side of a pentagon with a diagonal length A; I know when and how every piece of territory was acquired since the United States began. English was, still is, the subject hardest for me to handle. Here is a typical question: "Identify the source of 'Sleep, beloved sleep, from Pole to Pole; Mary, Queen of Heaven, save my soul.'" Since then, I've asked many college-English majors to name it. Few could. It's from a balladlike poem, "The Rhyme of the Ancient Mariner," by Samuel Coleridge, a tale of the terrors a sailor has to endure for senselessly killing an albatross.[8]

After the tests, back home, I guessed my chances were fifty-fifty. But, I told my parents, since I can keep taking the exams until I'm twenty, if I fail, I'd take them again, next year. Then I got a wire: perfect in geometry, algebra; tops in physics, history; a "pass" in English, probably because of my high marks on the first three, at the time an academy priority. Only 19 of 119 sailors who took the exams passed, I the only one of the Newport five. The wire said I was to be at the Washington (D.C.) Navy Yard by 1 July for transfer to the academy. I'd made it, a tribute to Father Brady as much as to me. I was full of equal parts elation and fear of classroom failure.

2

Up-Checks and Down-Checks

In large organizations, the foul-up is ever present. When I arrived at the Washington (D.C.) Navy Yard as ordered, the bureaucracy, though small in 1926, had lost my records—service record, medical record, everything. Officially, I was not of this Earth. I was sent to Annapolis anyway, to the station ship *Reina Mercedes* as a motor-launch coxswain, my sole task being to ferry my should-be classmates to the rifle range and back. "Chick," they said daily, "living the easy life while we must toil and suffer."

I ran that boat most of a frustrating month until the "system" finally found my records. It was a fine, warm day, 2 August 1926, when I raised my right hand to be sworn in, midshipman, United States Navy, class of 1930.[1] It was a marvelous class. Those academy years were among the happiest of my life, the kind of experience I'd urge on every youngster in America—a sentiment I'm sure most West Point and Air Force Academy graduates share. But it was no four-year frolic. I'd hauled to school a batch of burdens: high school dropout, youngest in my class, late arrival, maybe the worst-prepared ever to enter the academy, competing with at least some in my class who already had been to a college somewhere before coming to Annapolis. At 606, we were the largest plebe class in some time. Still, though we were sworn to serve four years, we were told that by 1930 a third of us would flunk out or be dismissed for "bad conduct," whatever that meant.

Our only classwork in August was solid geometry. I did well in daily recitations, worth 60 percent toward a grade, but poorly in the final exam, a barely passing 2.5 overall—a hint to how tough plebe year was to be. They'd nicked my wallet, too. My sailor's pay had been $54 a month. I'd felt well-off, even rich when I, on my own, paid the $100 academy entry fee. Now I was being paid $65 a month—but given only $3. Sure, they banked the rest for us and, sure, we had no time to buy much. (Beyond ten days at Christmas, plebes could "go ashore" [off the academy grounds] only on Saturdays,

2 to 5 P.M.—except on football game days, attendance required.) The whole business peeved me.

Still, puffed up as I was by my new uniform, plebe summer, the month of August, was for me just a unique vacation. Then in September the upper classmen returned, adding new woes to our now-full load of classroom, military, and athletic training. That year, plebes who'd been bluejackets were the upper classmen's favored targets to "run," a euphemism for mindless harassment. If we didn't answer a first classman's every—often dumb—question and obey every order, we got punished, usually a beating with a broom. My hazing ragged on until the day a first classman, the equivalent of a senior in civilian universities, told me to throw a pie at another first classman. It put me astride a dilemma. If I don't, he beats me. If I do, the other guy beats me. So I threw the pie. And striding into its arc at exactly that moment came the academy's dour, stoic Superintendent Thompson, "T. V." to his pals. When the pie hit, chaos erupted, with upset people everywhere.

The navy has no rule that explicitly forbids throwing pies at officers, but it does issue rules written vaguely enough to cover unanticipated events. The navy is very sharp this way. I was charged with "conduct prejudicial to good order and discipline," throwing the pie; and "willful destruction of government property," the pie itself. After the furor faded, first classmen avoided me and I them, excepting a few who became my friends.

Anyway, we all were immersed in competition, a cultural trait really at the academy: competition in academics, in athletics, in military training. College freshmen could not compete in intercollegiate varsity athletics, back then, but plebes had to join a sports program—our choice, but do it. So I signed on for swimming and softball, then water polo and baseball in the spring. But mostly I studied. And studied. And studied.

That year's highlight was having lovely Evelyn Lyons of Port Washington, Long Island, come to our Christmas Ball. I guess we all remember our first love and she had been mine since high school, at least according to me. But I think she accepted my invitation only to see the academy. Many girls do that, just for the thrill. It is a beautiful place but that ended it for Evelyn. I wasn't romantic enough, she said. I never did learn how I failed in that department.

But the academy allowed no time for psychoanalysis of lost loves. All plebes had to take the same courses, no electives. The pressure was constant, the curriculum heavy on engineering subjects. So being first in my English class part of a semester, while it pleased me no end, didn't boost my ranking much. I realized then how lucky I was that Father Brady had taught me how to study. Duty and love of country helped, too. It's embedded in the academy, hard to describe, but it's everywhere. Often, it lured me to John Paul "I have not yet begun to fight!" Jones' softly lit tomb, or Memorial Hall to gaze at Perry's flag, "Don't Give Up the Ship!" To me, they were pep talks, not symbols.[2] I ended the year ranked 279th, a poor but fixable start since fourth-year class standing is worth four times plebe-year rank. However narrowly, I'd jumped a huge hurdle. No promotion in a navy career is as rewarding as the one from plebe to "youngster." I itched to flash it around, especially in front of my father.

In June 1927, we formed up for my first summer cruise. Usually it was a three-battleship squadron, but we had only two. Mine was the twenty-nine-thousand-ton *Nevada*, an oil-burner, crew of 1,301, ten-knot cruise speed. On board was one of my prep school teachers, Ensign Rose, a future admiral. As we sailed to Panama, San Francisco, and Guantánamo Bay, we youngsters did able-bodied seamen's work, swabbing decks, serving mess, hearing for the first time the jarring roar of 14-inch guns, and enjoying it all immensely. At Guantánamo, "Gitmo," we fired main and secondary batteries in a short-range drill, using not the remote fire-control system but spotters pointing the guns at targets fifteen hundred yards away. This was a test of a gun crew's ability to hold on target in spite of the ship's motion, and competition was fierce between ships and among crews. We ended with a "self-brag" to tell my father, an "E" for excellence awarded my gun crew.

On that sail, our ship's three floatplanes were set up only for scouting. Navy doctrine back then said that's all they were good for, to help battleships square up against an enemy as they had in the Battle of Jutland.[3] I also recall our Capt. "Pluvie" Kemp, constantly smoking his pipe, snarling about "turncoat" Adm. William Sims, who had said in an editorial, "The aircraft ship, not the battleship, is the capital ship of the future."[4]

It was a weird era. Midshipmen studied the Battle of Jutland while admirals in navy bureaus and the Executive Office Building in Washington, D.C., debated the value, if any, of aircraft. The army's Gen. Billy Mitchell had been court-martialed in 1925 for ignoring an order to stop his public promotion of "airpower." His bomber dropped a bomb on a ship in 1921 and again in 1923, inflaming the battleship admirals. Sims was the only naval officer to testify in Mitchell's behalf at the court-martial, horrifying other navy senior officers. By 1930, the controversy these two had provoked in 1925 was intense.[5]

That wasn't the navy's only headache at the time. Its top priority was the care and feeding of our fifteen battleships with their 14- and 16-inch guns, and our ten light (seventy-five-hundred-ton) cruisers with 6-inch guns. There was controversy over gun size. In 1916, Congress had authorized the building of the *Lexington* and *Saratoga* as thirty-five-thousand-ton battle cruisers with 16-inch guns, and four ten-thousand-ton cruisers with 8-inch guns, fueling a fierce debate over which had better firepower. The fact is that they chose ten thousand tons and 8-inch guns simply because that's what the Washington Naval Conference and Treaty of 1921–22 allowed.

Set up by the White House, that nine-nation meeting (the United States, Great Britain, France, Japan, Italy, China, the Netherlands, Belgium, and Portugal) was "to prevent war" (as were a 1925 Protocol and 1927 Disarmament Conference in Geneva), and to halt a Japan-Britain-United States "shipbuilding race" in the Pacific. The "major naval powers," the United States, Britain, France, Japan, and Italy, agreed to halt building of combat ships for ten years and to limit to ten thousand tons the size of what they built after 1932. Typical of "disarmament" treaties ever since, it was unenforceable. And in fact, nobody paid any attention to that treaty at the time except the United States.

We weren't exactly pure as Ivory soap, however. Since the pact didn't mention aircraft carriers, the navy in 1922 put a wooden top deck on the collier *Jupiter* to make her our first carrier, renamed the *Langley* to honor aviation pioneer Samuel Pierpont Langley.[6] And in 1925, work began to convert the *Lexington* and *Saratoga* into carriers. But it was done simply because it was allowed. The battleship admirals still had first call in the ship-buying business, which told me that naval aviation might always be just "a little chicken amongst all these grown men."

I didn't dwell on that in 1927. I had too many more immediate things to do—like summer leave. My father was delighted to see me and I, him. I think that's why he took us to Bermuda, which had been a happy part of my many childhood vacations in an historically Hayward playground. Then the return to school in the fall of 1928 was the start of differential calculus and other scholastic bugbears. I also found, amazingly, that I had become a good-enough swimmer to make the water polo team. It's a rough sport, very physical, a lot like doing isometrics nonstop. But I was up to 155 pounds by then and, anyway, talent on the rest of the team was Grade A. We went undefeated, won the intercollegiate championship, and I earned my block-letter "N."

History says the second year is when an academy class usually suffers its highest flunk-out rate. In June 1928, I learned how well a year's hard work in study hall had paid off. My class shrank to 490 students, nearly a third fewer than a year earlier. I ranked 59th that second year. That and the water polo success pleased me no end. Our cruise the second summer was on the sixteen-year-old, just-refitted thiry-one-thousand-ton (full-loaded), turbine-powered battleship *Arkansas*. She had a 1,330-man ship's complement, twenty-knot speed, twelve 12-inch and sixteen 5-inch guns, and carried three aircraft. Maybe it was us soon-to-be second classmen getting to navigate, and to stand all deck watches as junior officer of the deck; or maybe it was the cruise itself, to Maine, Venezuela, Panama, and Guantánamo (where the 5-inch gun I captained won an "E"). Whatever the reason, I reveled in every minute of that cruise.

Summer leave the family spent in Bermuda again, land of the sky-blue sea, seeing family friends. To a just-caught second classman, it was especially exciting. I floated back to the academy—to be hit by the classroom grind. But study had become a habit natural as breathing. I could handle it. And did. At the end of that third year, my undefeated water polo team had won the intercollegiate championship for the second year in a row and I stood thirtieth in my class. We've had graduates who've said, "I could have stood first if I'd worked harder." I'm one sailor who can't say that. I worked as hard as I could.

But in early 1929, signs of an economic sag were on the horizon, as was an impending London Naval Conference of 1930 that brought a new scheme to restrict the size of, as it turned out, only the U.S. Navy. The Kellogg-Briand Pact, also named the Pact of Paris, had been signed in 1928 by fifteen nations. Another forty-five agreed to abide by its Article I, to "condemn recourse to war for the solution of international controversies, and renounce it as an instrument of national policy." Added up, it said the lights had dimmed on a navy career. At the academy, we talked about it some, but mostly we just wanted to graduate.

That third year's summer cruise was in the same squadron as the year before, with the *Florida, Utah,* and *Arkansas*—my ship again. We went to Paris, Rome, London, Madrid, and Barcelona, a heady trip for an impressionable young man seeing a kaleidoscope of culture. Ever since, whenever my ship has passed the Rock of Gibraltar, I've recalled that cruise to the beautiful Mediterranean, the ports of Spain, the Riviera, Sardinia, Italy, and Greece. Probably the birthplace of Western civilization, the "Med" from then on had a romantic, almost hypnotic appeal to me.

I recall, on a train from Naples to Rome, hearing people jabber, "Isn't it great! Mussolini has the trains running on time!" (I rode third class and do not recommend it.) Adolf Hitler had written *Mein Kampf,* "My Struggle," but few people believed what it said. Reparation imposed on Germany after World War I was a tax so high their economy couldn't pay it, and didn't. And our allies, saying the two were linked, didn't pay what was owed us for military supplies. The French even said they'd never considered us an ally, anyway. In the streets, they snarled, "We fought and bled for democracy while you stayed home and got rich off the war." How easily history can be rewritten when truth becomes inconvenient. Strolling Paris in my uniform, I was tailed by French children yapping, "N'pas une sou pour l'American," roughly translated, "I'd not give one penny for an American." (A year later, after a decade of coldly ignoring the debt, France called it officially "deferred," lawyer talk for, "We're not going to pay, ever!")

And in 1929, the country's economy, the whole world's, in fact, sank into the Great Depression, but my first class year was, well, first class. I ended up ranked ninth in class, we were water polo intercollegiate champions again, and I was elected to the national All-American Water Polo team. I was hit by one jarring event. Advised to put my real age on record to avoid problems later, I wrote a paper saying that a 1908 entry in the family Bible, noting my eldest sister's birth, implied that I'd been born in 1909. I went to see Academy Commandant C. P. "Peck" Snyder to request the change. As I entered his office, he said, "Young man, if I accept that paper, I'll have to charge you with fraudulent enlistment in the navy and fraudulent enrollment in the academy. If I were you, I'd turn right around and get out of here. And take that paper with you!" So the navy said, still does, that I was born in 1908. But I'm not sure.

In May, we heard that Congress, to show us it was "fighting the Depression," was going to let us all get a diploma, an engineering degree, but half of us would be "invited out," the navy's way of saying we'd not receive a commission. Like most rumors, it died off (though that was done to Tom Moorer's class of 1933.) Elated at graduating, I invited my family and my, I thought, "forever" girl friend to the 6 June 1930 rite. At its end, per tradition, we went to Lover's Lane where my mother pinned on me one ensign's epaulet and my "forever" girl the other.

Multiplying and adding together my four class standings, my overall standing was fifty-first in the 406-man class of 1930. In 1926, I hadn't thought that final rating even remotely possible. From day one at the academy, we'd been told, "Do your best to stand as high in your class as possible. It's important." No one was much impressed by that at the time. Thanks mainly to Father Brady, my aim all along had been to acquire the knowledge for its own sake. I soon learned that where you stand in your class and where it stands in seniority decides where you stand in line for "perks," housing and other allowances, the "fun" missions, and all the rest.

But that awareness came later. Meantime, a thirty-day leave was enough time to learn that my "eternal-love" girl was not interested. I guess most everyone takes that kind of hit some time. For another year or so, I thought of her fondly, unhappily, off and on, even dated her a few times, but my ship remained dead in the water. My long fight to change her mind was mostly, I think, because of what the Yankees taught me. I just hated to lose. At anything. But I did accept it, finally, helped by the fact that I had a career to chase.

First-year ensigns usually did their first duty at sea. "The navy is ships," dogma intoned. "You will become a qualified ship's officer first; fool with this flying stuff later if you must." But with the *Lexington* and *Saratoga* as converted aircraft carriers, they needed pilots, so mine was the first class they let go directly to aviation. Well, for me, sort of. Signing up was no problem. Class standing put me fifty-first in line to pick the navy service I wanted to join. Given a Pensacola flight-training class size at the time, all fifty ahead of me could pick aviation and they'd still have room for me. Then, if we passed the physical, a navy drill as routine as meals, we'd be sent to either Hampton Roads Naval Air Station, Virginia, or North Island at San Diego, California.

There, an instructor would give us up to fifteen flying hours to prove we'd be able to cope with the stress of Pensacola, which was then and still is father- and mother-in-law of naval aviation recruits. The tryout was a short copy of Pensacola's curriculum: fly half a day, aviation engineering study in class the other half. It echoed my father: "To be a good pilot, you must know engineering." As Mark Twain said, "It was amazing how much smarter my father got, the older I got."

We flew a two-seat, single-float NY-1 seaplane driven by the same Curtiss Wright 225-horsepower engine Lindbergh used to fly the Atlantic. After flying with an instructor for the time it took to satisfy him, beyond a required four hours, the student was put in for a "check." That meant going up to do turns, landings, and such for another instructor. If the student got an "up-check" grade, he took a three-hour solo "final exam." A "down-check" meant going back up twice with two other different instructors. The student had to get an up-check from both. If just one gave him a down-check, it meant no more naval aviation, a defeat that surely would have left me a psychological mess.

Parts of my Hampton Roads diary sum up the torments: "(July 13) Fair on the turns, today. Lord, I love to fly. Assigned me to Lt. Brum Nichols, Class of '24, one of the best, they say." (Nichols became a much-praised aide to "Bull" Halsey, commander of the Allied Naval Forces in the South Pacific in World War II.) "(July 15) Letter from Mother. Did spins, today. (July 17) Must do better on turns. (July 21) Black despair, those damn spirals. (July 22) Brum is a peach. (July 23) Improving. Letter from Mother. (July 24) Need work on landings. Have to get it. (July 28) Terrible on emergency landings. Brum certainly has patience."

After seven flying hours, Brum put me in for "solo check," giving me a chance to fulfill a dream I'd nurtured for fifteen years. As I went to the aircraft, I told myself, "You've got to deliver. You can't let Brum down." Fact is, as I taxied out alone, gunned the seaplane's engine, took off, flew the triangular course, did my spirals, turns, touched down, and lifted off again, I felt a thrill beyond my ability to describe. I passed easily, maybe because I wanted those wings so badly. I know the fun of it never faded. In my career, I'd fly 150 different aircraft types, log more navy pilot hours (13,023) than any other flag officer, and 900 more hours in commercial, foreign-made, and private aircraft. And I was as pumped up on my last flight in 1963 as I was on my first one in 1930.

On 5 August, I took one more flight at Hampton Roads, then headed home, still a long way from my gold naval aviator wings. Pensacola ran just two classes a year then, one starting in March, one in October, each with some sixty students. The October 1930 class was already full, so my first chance was the March 1931 group. Meantime, I was sent to the *Richmond,* one of ten *Omaha*-class seven-thousand-ton light cruisers from the 1920s. Actually, she was an overgrown twenty-seven-knot destroyer (the builder, Wm. Cramp & Sons, said up to thirty-five knots) with 6-inch guns. She also had a two-aircraft flying unit, its skipper a lieutenant, his flying team all lieutenant, junior grades.

Our captain was J. V. Babcock. He'd been with Admiral Sims at the Navy War College and in World War I, helping lay down the "mine barrage," a minefield fence to stop German U-boat attacks on our ships in the North Sea. Sims also had sold British Admiral Sir John Jellicoe on using destroyers to escort convoys across the Atlantic. It had cut our supply-ship losses dramatically, a major reason the Allies won. Babcock was a feisty character, leery of ensigns who wanted to fly. He made me main engine officer in the Engineering Maintenance division. The chief engineer was a can-do-it-all guy, Lt. Cdr. Allen Quinn, whose senior assistant, my boss, was a personable Lt. Henry Naff.

We were part of a cruiser scouting force that had turned my division's basic job, making the engines run well, into a contest. It was a crew contest, really, since the ships were a lot alike except in age. Each got a ration of oil. If we met the burn-rate standard, we scored 100 percent. If not, we scored lower. There were two awards, a white Engineering "E" for Best Ship in Class, and a red "E" for Most Improved Ship. Under way, her fouled bottom fighting her engine and the sea, the *Richmond* was a poor actor, a rough ride. So it was the red "E" we were after.

My job was to keep score on a "Form H" of fuel used while at anchor and while under way. At anchor, the game got a little silly with me doing stunts like learning to run the generators off back pressure from the auxiliary exhaust and racing around the ship turning off lights. A few times on midwatch at anchor, we burned only twenty gallons of oil an hour to run the whole ship. I enjoyed the engineering challenge and liked being with the sailors, the grease monkeys, down where rank has no privilege. We did win a red "E," but the competition didn't really reflect how well ship and crew worked together. A lot was just "cheating reasonably," it was said, on Form H. One way was to fudge a report of ship's reserves prior to a refueling. I didn't try that myself—lacked the nerve, maybe. But everyone knew the scheme. Surely some ships cheated. If our admirals thought the scores graded combat readiness, they were kidding themselves.

When not battling Form H, I poked into every part of the ship—engines, hydraulics, fire control—to learn how it all worked. I did that mainly to prepare for tests I was due to take in 1933 and had to pass for promotion to lieutenant (jg). The other result of this cruise would be that by the end of it, I'd be qualified to stand the top engineering watch aboard a ship. But we were a pretty relaxed outfit in those lean years. For instance, my ship, the *Richmond*, was to tour New England until November; go into the Brooklyn Navy Yard for overhaul; then, after Christmas leave, join the Pacific Fleet for its annual Fleet Problem. Until then, our sail would be low on work, high on parties.

The best one was in Newport, Rhode Island, where Muriel Church Vanderbilt invited all us ensigns every Saturday to her Marble House mansion. She'd hire Lester Lanin's orchestra and have in girls to chat up and dance with. We often missed our last boat back to the ship, anchored off Jamestown. But she always served a big breakfast on Sundays to cap off the buckets of champagne drunk the night before. Newport was an open-gambling town then. On only $120-a-month income. I avoided that, but I surely met lots of friendly young gals, and decided Newport was a real "navy town"—provided you were an ensign on Muriel Vanderbilt's guest list.

Next came Bar Harbor, Maine, where I met Louise Brooks, Gen. Douglas MacArthur's stepdaughter. He'd married her mother, a Stotesbury with tons of money. It was awesome to me, seeing how really rich folks lived. I did wheedle out of her grandfather copilot rides in his Sikorsky amphibian. She insisted I meet her at their Palm Beach mansion that winter. Nice, I thought, lovely girl, all that money, a plane to fly. But there were liabilities there I didn't care to face.

From Bar Harbor, we loafed a week off Long Island, then went south for a month of bomb and torpedo runs. My log says "the Admiral" inspected the ship on 19 October and, "Captain commended me to him for my work," what work, I didn't note. On 27 October, Navy Day, begun by the Navy League in 1922 "to call public attention to the Navy's importance to the Nation's security," we were lucky no "Public" was there. Fog shut down our Navy Day "tare" and "baker" drills. Two days later, our plane fired one torpedo, then sprung an oil leak. Repairs at Hampton Roads off Virginia, then foul weather stymied our runs for ten more days. "We're costing Uncle Sam lots of money," I wrote, "with few results. Some 'guardian of democracy.'"

These "tare" and "baker" drills were aircraft torpedo and bombing runs at the ship, which would simulate an antiaircraft defense using its 3-inch guns. The ship's fire control was primitive, the results terrible, with many rounds fired but rarely a "hit" on an airplane. (By 1942, inventions like the proximity fuse would level the playing field some.) Still, a torpedo run was no walk in the park. To put his "fish" on target, a pilot had to attack "on-the-deck" (at an altitude of five hundred feet or less), on a straight-in course at low speed, almost a sitting duck for any fire-control system no matter how crude it was.

That short trip to Virginia was a typical operation back then. Our navy's smaller ships were mostly on the East Coast, our battleships on the West Coast, all rarely far from home port. Our country's policy was "Isolationist, First Class," ignoring omens of war everywhere. Economic depression was making some fragile nations desperate and had others planning aggressive expansion of their empires. We reached the Brooklyn Navy Yard on 15 November, my birthday. It was a soft time, including for me a reunion with Yankee Stadium on 12 December to see an Army-Navy charity football game.

The game reestablished a tie broken the year before in a hassle over player eligibility. Army had been enlisting college graduates for their full four-year program, while Navy obeyed a rule limiting varsity play to three years no matter what schools a player attended. (Army's star, "White-Horse" Harry Wilson, had played three years for Penn State, then four more as a cadet when I was a midshipman. His first game against us was in 1921 when my class's best player, Joe Clifton, was in seventh grade!) The game was "a great battle," of course—which Navy lost 6-0.

On 10 January 1931, a dreary day with icy winds, the *Richmond* sailed past Sandy Hook to join the other scouting force cruisers, picked up stores and our aviation unit in Norfolk, then plowed through a big ground swell, headed for Guantánamo. I remember the ride because they gave us beans for breakfast every day. Our month in Gitmo was all shooting, our inept fire-control system proving to me that the 6-inch gun was poorer than the 8-inch. So I was one of the gang opposed when Admiral Pratt went to the 6-inch on seven new *Brooklyn*-class cruisers we'd been authorized to build under the ten-thousand-ton cap in the 1930 Naval Treaty.

Gitmo was a playground, really, with swimming, ball games, and bus rides up to Caminara, Cuba, to drink outside the reach of our Prohibition. The harbor always conjured up for me a crazy vision of my father's little spit-kit *Yankee* in the Spanish-American war zooming in to fire beanbags at and take Guantánamo fortress. This time, on 15 January 1931, I saw my first naval aviator spin in. Says my diary, "5F7 crashed, today. Poor Mac. How he talked and joked with us last night." And on the 26th, "Another crash, today. Cruiser *Northampton*'s plane, this time."

I jotted down a few other sad notes, too: "Went ashore. Cuba pretty rough. A girl in a pest hole tried to make all the officers who came by, said she wanted to have a nice clean white child. Tragic." "Hotter than Hades. Only fish and damn fools go to sea. . . . Ball game. The *Richmond* lost." In early February, as we headed for Panama, I wrote, "Nearly went to a watery grave. Wasn't the rough sea. I just fainted dead away. Doctor said I'd gotten dehydrated down in the hot fire-rooms. Don't know how they found out I was in trouble, but they did." Three days later, a Saturday, my log says bluntly, "Docked at Colon, crocked at the Stranger's Club. Beer, naked women. Bizarre. Hasn't changed much from four years ago. Still a crossroads of the world. (Sunday, Feb. 8, 1931) No church, again. Mary, Mother of God, help me." (Feb. 9) "Mother's birthday. Done her best for us. Head for Boston soon. I'll tell her so then."

A week later, we went out to do our "problem," me noting, "Under way at 2200 hours. She vibrates and trembles, smoke pours from her stack, full power, racing west." This "problem" was a "scout" drill, cruisers hunting for enemy ships. Back then, most ships' officers, the "black shoe" folks, said aircraft were poor hunters of ships, lacked durability, and had Stone-Age navigation and communications systems. They had a point. Meantime, I'd "sounded off" to the captain and been relieved of duty. The reason was so frivolous I didn't even write it down. Evidently, it was over our chief engineer since I added, "Mr. Quinn is the best chief engineer I know, and Babcock is a typical senior Navy officer, as narrow a mind as I know of. Why do outfits like ours breed that sort of person? If he knew I was writing it down, I'd be in the brig!"

Back in Panama on the 21st, I finagled an airplane and flew all morning, loving it, "Then went to the Union Club that night with Vera," my diary says. I skipped church again the next day, ashamed to confess, but saw Father Brady for the first time in six years. That night, I recorded simply, "He is such a great man."

As a hint to how low-key training was, at the time, we jg's had not only to compete in athletics but also coach a team. My water polo record got me made swimming coach. It was eight days vacation living ashore, swimming, nights at the Union Club. Then came the "Terrible blow," I wrote, "Academy classmate, ship bunkmate Bill Moffett, has orders to Pensacola. I don't. Is it because he's the Admiral's son?" (Rear Adm. William A. Moffett was, and had been since its creation in 1922, head of the Bureau of Aeronautics.) "But I'm not giving up."

On March 1st, a fire, which on a ship can be disaster, burned up our airplane. By the 7th, we were back at "Canal East," in Colon, and I wanted out of this endless, infernal heat. I was very bored and homesick. By the 11th, we were in Gitmo to finish our gunnery work and this time I was "camera party" boss on the target-towing tug *Contocook*. Filming where shots hit is intense during the action but dull otherwise. Not this time. On its first run-in, one ship abruptly aborted for no evident reason. Then, as it ran at us again, when they tested their firing circuits at sixteen hundred yards out, aimed right at us, its guns roared! Dumb mistake. Guns are supposed to be unloaded if a run aborts. Luckily, only one shell hit, slicing off a piece of our deckhouse. But the crusty old bos'n who ran the tug raced aft, cut loose the target, and drove us off at top speed. No more damn firing today, he said. The officer types did not see fit to overrule him.

Half of our six-day sail home was in bitter cold weather and snow, winds whipping the ocean into a rage. It fit my "No Pensacola" mood. At Norfolk, we off-loaded Bill Moffett who advised me, as I later did, to see his father in Washington, D.C. There, bluntly, I asked Rear Admiral Moffett, "Will I make it to Pensacola?" He politely assured me I would, but asked me to "put it in writing." After I had, he wrote confirming what he'd said. His explanation of the delay was, as they say, "not helpful."

In May, the *Richmond* sailed for New York for Fleet Week, but at sea on the 19th, we were told to divert "quickly" to Honduras "to protect Americans." The Washington swamis were afraid, they said, that a Nicaraguan revolt might spread into Honduras. So we picked up our flying unit in Norfolk and went "flank-speed" to Puerto Castillo, a tiny place on a spit of land on the northern coast of Honduras. Across the bay from it is Trujillo, site of a big United Fruit Company operation and, we were told, where the rebellion was.

If it was, we didn't see it. I did meet, in a saloon, the Los Angeles hammer-killer Claire Phillips: "Wickedest woman in the world," L.A. newspapers said. We had no extradition treaty with Honduras at the time, so felons could flee to sanctuary there and she had flown. She didn't seem the type to murder people with a hammer but evidently she was. The heat that June was a thick wool blanket, a "lure of the tropics" I could live without.

And I noted in my log, "These natives are an odd lot. I doubt they're telling us the truth. This revolution is a comic opera."

Still hunting for a war, we went upcoast to La Ceiba, Honduras. The New Orleans–based Standard Fruit Company wanted us to protect their brewery there. "Protecting American lives?" I asked. Back then, a cruiser landing force was drawn from its engineering division. So, stumbling ashore, went fifty bluejackets—mechanics and electricians with rifles—Ensign Hayward in command. That first day's pitch-dark night, my sharp-eared warriors heard movement on our defensive perimeter. First one, then all opened fire, their rifles flashing like Fourth of July fireworks. A fierce battle, the worst of it was my getting them to stop shooting. When they finally did, no sound could be heard except our own breathing. Next day at dawn, we saw the fierce foe we had killed—a pair of cows! The guys aboard ship quickly named our "firefight" Ace's (my) Battle of La Ceiba. But it had nothing to do with the brewery giving us free beer each day at 10, 2, and 4.

A week later, we sailed to Tela, Honduras, to pick up our State Department consul stationed there. We were offshore and I was relieving the watch when a sailor fell overboard. "Man overboard!" rang out and a man on the fantail of the moving ship jumped in after him. It quickly was clear they were not staying afloat, so I dove in to help. Making the one who could swim hold onto my shoulder and towing the other guy, I pulled and leg-kicked us in to the beach a mile or so away. It was a lot like water polo, really, with two guys grabbing at me, and me getting dunked and fighting back up. My happy captain issued a standing order: "Ensign Hayward may swim off the *Richmond* any time he wishes." The sailors, Machinist Mate First Class Robert Doak and Engineman Second Class George Coker, were, they said, instantly my fans, and I was a hero to the rest of the crew. Later, in Pensacola where they caught up to me, I was given the Treasury Department's Congressional Medal for Life-Saving.

I was surprised to see that the consul I was sent to get in Tela was Thomas Stout, who'd entered Foreign Service school after "bilging out" of my class at the academy. Tela was his first post since graduation. After we gathered in Stout, some of us went inland to Tegucigalpa, Honduras' colorful capitol, where I heard a tale of how "Shoot the bull" became part of marine corps jargon. Our assistant naval attaché there, they said, once had been a marine lieutenant colonel who toted a single-shot, 45-caliber rifle disguised as a cane. At a Tegucigalpa bullfight one day, as a bull was about to gore a toreador, the colonel had stood up and "shot the bull." Was that true? As Yankees manager Casey Stengel once said, "If you don't believe it, look it up."

Our next stop was Puerto Cortez, a dirt-poor coastal town. Here Guy Maloney, a one-eyed character out of a Robert Louis Stevenson novel, had started a brewery. Years earlier (as a book, *Les Incredible Yanquis* relates), Lee Christmas, an Illinois Central railroad engineer fired for drunkenness, had recruited six Americans to help him start a revolution to take over Honduras. They tried. Maloney had been their machine-gunner, the team's sole survivor. But he hadn't heard of any recent revolt in Honduras. Not that month. The navy finally let us go home, relieved by the *Sacramento*, a gunboat in our "Banana Fleet" out of Balboa, Panama. On this junket, we were given no specific mission, no precise military objective, no rules of engagement. "Peacekeeping?" Meaning what? Going home, I wrote, "What a fool my country is at times."

It was my first brush with what historians call "Yankee Dollar diplomacy," "gunboat diplomacy." President "Teddy" Roosevelt had used it in 1903, sending warships to Panama to secure a bloodless coup by a Frenchman, Philippe Jean Bunau-Varilla. (Bunau-Varilla had promised to give the United States "sovereignty in perpetuity" over the Panama Canal—which the U.S. Army began building then, and completed in 1912—if he succeeded, as he did, in wresting control of the new "Republic of Panama" away from Colombia.) President William Howard Taft had sent marines to Nicaragua in 1912, to "restore order," as had President Coolidge in 1927.

In the first third of the twentieth century, U.S. investments in Latin America boomed. Most noted, and hated by Latin revolutionaries, was the United Fruit Company of Boston, which dominated the banana industry and whole "Banana Republics" as well, keeping in power, sometimes with U.S. military help, those officials who backed its investments. Ever since, we've had "gunboat diplomacy" to "protect U.S. citizens and vital interests in our own backyard." Latin American nations were then and many still are political eggshells easily broken. During the last forty years, we've sent military forces to Cuba, Haiti, the Dominican Republic, Nicaragua, Grenada, and Panama. I suspect we may again, when Fidel Castro dies or if Panama's dictators give the Chinese Communists the "lock," so to speak, that they seem to want on both ends of the Panama Canal.

Sailing back to Newport, Rhode Island, in late June, we fought rough seas and storms again. Once there, the *Richmond* ship's complement resumed a favored recreation, chasing girls. But the sport was tricky since we lived on our ships, anchored seven hundred to one thousand yards off shore; the girls lived ashore; and the last boat out to the ships left at midnight. One well-remembered time, I didn't reach the dock until 1:00 A.M. I saw only one way to avoid being AWOL. I swam back, my clothes on my head, my shoes laced on my neck. Never ever really warm, Narragansett Bay was cold that night. At the gangway, as I climbed the ladder, the officer of the deck, Lt. (jg) Richard Elliot, bellowed, "Mr. Hayward! Mr. Hayward! What are you doing?"

"Reporting in, sir. As you may recall, sir, some weeks back, the captain said I could swim off the *Richmond* whenever I want." I half-expected a dressing-down for my smart-mouth answer but my spiel worked. Elliot just frowned and a chief next to him said, shaking his head, "The navy must be on hard times when ensigns have to swim to get aboard ship."

3

The Yellow Peril

After my foolish middle-of-the-night swim, I went (with the *Richmond*) back to Boston, where I decided to get a commercial pilot's license before going to Pensacola. Back then, ten solo flying hours were enough for a private license. A commercial license, allowing me to haul paying passengers or freight, required fifty solo hours and passing an aviation-engineering quiz. I had the hours, flying mostly rented planes, so on July Fourth weekend in 1932, I earned one of the first commercial pilot's licenses issued by Boston's Logan Field. I've lost track of it, but do remember its number was very low. (One of my remarkable talents is an ability to forget where I file stuff.)

Then, itching to be home on the Fourth, I rented for $500, with option to buy, a Buhl Bullpup midwing monoplane with a three-cylinder Zekely forty-five-horsepower engine, ten-gallon gas tank, and a top speed of ninety-five miles per hour. The flight, down past New London to Roosevelt Field on only seven gallons of gas, was smooth, the weekend with my family quiet. On Wednesday, 8 July, I cranked up my little bird for the return to Boston. Assa Jordanoff, famed for his instruments-only flying talent, author of two books on the subject, advised me not to go. Bad weather coming in, he said. Looking back, I'm still appalled at how stupid I was to try it.

At takeoff, fog low on the horizon, I followed a railroad track to Greenport at Long Island's east end, then north over the Sound toward New London into a high grey wall of fog. A Bullpup's only navigation aids were a turn indicator, magnetic compass, and pressure-plate airspeed gauge mounted on the wing, no help in that soup. Quickly, I lost a sense of direction. Forget New London, I said, turning back to hunt for the small airfield in Greenport.

Zipping along in fog just above row on row of telephone lines, suddenly I saw a smokestack loom up, dead ahead. I jerked back the stick so sharply the Bullpup went nose-up, dead in the air, fishtailed into a spin, then screamed into a dirt pile in someone's backyard. Hanging upside down in my busted plane, I was amazed to see a man reading a paper by a rear window of his house. He hadn't budged, not even looked up. Quickly, children were swarming around my wreck, helping me out of it. Seeing that commotion, the man jumped up—he was stone deaf, I learned—ran out onto his porch and screamed at me, "Get that airplane out of here!"

There wasn't much to "Get." I sold for $50 what wasn't broken and junked the rest, losing my investment in the process. It had been a searing lesson. I never saw Jordanoff again, but did buy his books right away and any others available on "flying blind." I also vowed never to file a flight plan without a safety valve, a "bailout," in it, and never to fly if the weather is worse than my plane's instruments can handle. Instrument flying was an infant then. Most pilots didn't study it, and many were killed doing something as stupid as I had done.

In Newport back aboard ship, life was navy routine: boiler work, hydrostatic tests, inspections, target practice (including the "camera party" to film the practice), night patrols looking for the "drunk and disorderly" on Easton's Beach, dances in Jamestown, drinking in the Newport Casino, and rarely dating girls—a recreation beyond my pay grade. A kind, middle-aged guy named Pettee did tell me I could fly his Lockheed whenever I liked, which I did often. But my first flight after my crash was in an NB-3. This was a test bed to improve "spin" behavior on the Boeing NB-2 biplane, a single-float seaplane trainer convertible to land use for pilots and backseat gunners who fired a 30-caliber machine gun on a swivel. That NB-3 crashed two days later, killing one man. Once more I'd dodged a bullet.

From Newport, we sailed to Bar Harbor where I virtually lived with the Stotesburys, who got me invited to the wealthy Atwater Kent's posh wedding. And Louise Brooks was closing in. One night, she hosted a dinner for 136 people with me as honored guest. She loved me, she said, but wanted me to resign from the navy and quit flying. Vincent Bendix would give me a fine job in his company, she said. That cinched it for me. I didn't say so, but knew I would not be at her Palm Beach mansion that winter. Or any other time.

Then we sailed, as usual, to Fort Pond Bay where I treated myself to a round of golf at tough Shinnecock Hills, the site on Long Island of the second annual U.S. Amateur and U.S. Open golf tournaments in 1896. (The first of each of those was played at the Norfolk, Rhode Island, Golf Club in 1895.) Then my orders came, posting me to Pensacola. To celebrate, my mother, dad, and sister Margie drove over to the ship for dinner. Next day, I rented an airplane to take Margie up for a look at Long Island from the air, adding in the thrill of one roller-coaster loop-the-loop.

Flying and partying weren't all I did before reporting for flight training. I also finished Naval War College courses in strategy and tactics and in international law. These courses led to a lifelong hobby of reading books on history, religion, philosophy, and international politics; and on when to and not to use military force in the international arena. As I left town, my father said he was proud of me, which pleased me no end, and he urged me to keep studying. My mother, unhappy as mothers were in an era when aviators couldn't buy life insurance, wished me good luck and God's grace, anyway.

Mike Sanchez, a fiery, fun-loving Latin type, a water polo teammate, also was headed for Pensacola. He asked me to go there with him in his new Model A Ford, sharing expenses, on what was my first road trip through Dixie. As we left, a *Richmond* shipmate, Lt. (jg) Herb Riley, asked me to chaperon his girl, Leila Hyer, in Pensacola. I said I would, not sure what he meant by "chaperon." On the way, we stopped in Annapolis to see Navy beat William & Mary. Next day, at lunch with Louise Brooks, she said she still wanted me but on her terms, no navy, no flying. Then no marriage, I told myself as Mike and I left.

Our last overnight before Pensacola was Atlanta, the South's New York City, some said, but the people were a whole different culture. The Civil War, "War of Northern Aggression," they called it (some still do), had ended sixty-six years earlier. Yet the wounds still festered, much like today's Greek-Turkish conflicts, some Arabs' hatred of Jews, and the tribal wars in Africa and Southeast Asia—all centuries-old conflicts. Our diplomats and politicians ignore our own history when they push a quick-fix peace process, "democracy-building," they call it, on someone else. This one-size-fits-all ignores the world's wildly differing cultures, and the fact that many tribes out there are run by mean people. Most of this century's rebellions were not, as mythologists claim, "destitute people rising up in anger." Most were a well-educated cabal trying to seize for themselves another dictator's power. My father, for instance, believed we fought the Spanish-American War "to free poor Cubans from the Spanish yoke." What would he say today about Fidel Castro, the "liberator of Cuba?"

I've lived in an exciting era of hypersonic aircraft, travel in space, satellite communication, and TV reporting world events as they happen. The next century will see even greater advances if we preserve the doctrines of individual freedom and free enterprise that nurtured this century. My parents understood that part. They backed "Teddy" Roosevelt and his pal, Dr. Leonard Wood (army chief of staff from 1910 to 1914), who helped organize the "Rough Riders." They agreed with both when the two publicly criticized Wilson's "pussy-foot" mobilizing of an army for World War I. My parents knew Europe, were appalled at the murder of the czar's family, and knew that two quick, successive revolts had left Lenin's Communists in charge of a new creation, the Soviet Union. Yet, not once did my parents discuss what that might mean for the United States.

Of course, in the 1930s, few Americans saw us defending democracy against a Communist dictatorship. When the Cold War's existence was obvious in 1948 (at least to some people), it was recalled that Alexis de Tocqueville wrote in 1835 that someday the world would be caught up in a "final war" between us and Russia. Some say the Bible predicts it too. But noisier prophets now say, "The Cold War is over." That's too flippant, too glib. When I enlisted, our leaders agreed that victory in World War I had made the world safe for democracy, but few here and none in Europe believed that a League of Nations could assure it. Nor can a UN (United Nations). Though the new owners evidently have trouble keeping track of the warheads, the huge Soviet nuclear arsenal is still there. Other hostile dictators—in Iran and North Korea, for instance—are increasingly active. Teddy Roosevelt's "Speak softly but carry a big stick," seems as sound today as when he said it.

In the 1930s, aviation's war-and-peace value also was a contentious subject in spite of Congress creating the NACA (National Advisory Committee on Aeronautics) in 1915 to fund R&D (research and development) that was too costly for a single private company to risk. Just two examples of what the NACA funded, widely used by the early 1930s, were the Wright "Whirlwind" and the Pratt and Whitney "Wasp" engines. Nor was 1931 the first year ever for naval aviation. In 1918, the Marine Corps Aeronautical Company was the first American aviation unit sent into World War I. In 1919, three navy Curtiss flying boats were the first aircraft to cross the Atlantic nonstop, though only one reached Europe's mainland. In 1920, the navy had been first to successfully test fly a radio compass. In 1922, our first aircraft carrier was commissioned. In a trimotor aircraft, navy pilots had been the first to fly to the North Pole in 1926 (though the truth of that would be challenged inconclusively in 1996) and the South Pole in 1929.

But these were exceptions on a gloomy stage. A 1930 London Naval Treaty, inspired by President Herbert Hoover, would, in its six-year life, leave our navy very inferior to Japan's in the western Pacific, making American interests there dependent on treaties Japan routinely ignored. Our ships did only four to six weeks a year of sea duty because of lack of funds. Our nation was mired in a policy of "no foreign entanglements," and was in the midst of an economic depression. Closer to my life, the battleship admirals had a big edge over naval airpower. One reason was that naval aviation, itself, was young, most of its ablest people junior officers. William Moffett, William "Bull" Halsey, Ernest J. King, and a few others had tried to lure senior officers into flight training, but most of those firmly believed that the way to their admiral's flag was on a battleship.

That didn't worry me in October 1931, however. I just wanted to fly. On the 9th, we reached Pensacola, a small, sleepy town with the loveliest beaches anywhere. It had no direct rail service, the most-used long-distance travel those days, except a once-a-day freight-train run to Jacksonville. Its main street, Palafox, ran down to the docks; its only hotel was the midtown San Carlos. Pensacola's "old money," earned mainly in lumber and selling stores to the navy, lived on North Hill. Hunting and fishing, once just a favorite sport, was by 1931 a way to avoid starving. The town's auto dealers did pretty well, I learned, when I bought an open Ford touring car for $300. It got me around nicely, but girls didn't go for it.

Pensacola had lived under five flags, giving it a rich heritage. Fort Pickens guarded its beautiful, pristine harbor, one the town fathers hadn't exploited yet, like the folks in Mobile, Alabama, had theirs. The navy had been in Pensacola since before the Civil War; but the flight school, the navy's first, had been there only since 1914. It was the local economy's wellspring by 1931. Many local girls had become navy wives, and many retired navy people lived there. It was, in short, a navy-friendly town.

My flight class, like Bill Moffett's March class, was only sixty-seven students. (His elder brother, George, already was a navy pilot.) I spent my first day hunting up academy classmates: Joe Ruddy, Emmett O'Beirne, and old pal Joe Clifton, who promptly took me out to the golf course. Next day, a Friday, we took a stern physical with no on-the-margin waivers, unlike ones I'd had before. I passed, including the depth-perception test, a happy result since it had given me problems before. They also gave us a "Snyder" circulatory-index test. In it, they took the subject's pulse after he'd lain down for five minutes, again when he sat up, again after he'd exercised, and once more after he'd rested. A perfect score on this drill was 18. The Snyder can trip up high-strung types. But I was a low-blood-pressure guy. My Snyder was easy.

We had one other exam new in my experience, a psychiatric test given by a Dr. DeFoney. He asked a ton of goofy questions to help him, I learned, write his opinion of how well he thought each student would handle the rigors of flight training. I also learned, if I survived the one-year flight school, I'd have to go back to hear him read what he'd said at the outset about my potential. Much later, I also learned, if a pilot got a down-check during training and had to go before a Board of Inquiry, what DeFoney had said could decide if the student was allowed to stay in the program. Anyway, after that farce, I drew my flight gear, then played another round of golf.

To start the next day, Station Commandant Rufus Zogbaum gave us a long "fight on!" speech. It was wasted ammunition. We were pumped up already. After the oratory, we were split into two wings. Mine was to go to ground school in the morning, fly in the afternoon. The other wing would do the reverse. Ground school was not just textbook studies in all phases of aeronautics: aerodynamics, engines, ordnance, communications. We also had to tear down, overhaul, and reassemble an engine. We had to do the same to an airplane: take it apart, refinish it, put in a new engine, and fly it. We had to learn how to pack a parachute. And we had to take a "radio" course, learning to send and receive Morse code, which all our aircraft radios broadcast back then. Once a guy shows he can receive twenty-six words of press a minute and send out fifteen words a minute of five-letter code groups without error for three weeks in a row, he can skip that 8 to 9 A.M. session. To do that well takes an ear for music, one of many talents I don't possess. It took more than eight months before I finally hacked the radio problem. Just barely.

Our first flying class, called Squadron 1, started "on the beach." There, we did some ninety-five flying hours in the same canary yellow, two-seat, single-float NY-1 seaplane we'd flown in Norfolk. It was dubbed "The Yellow Peril," since a lot of guys washed out in it. We were taught all the standard maneuvers: spins, wingovers, figure eights around pylons, precision landings. The toughest task was to fly up to six thousand feet, cut the engine, then spiral down to land within one hundred feet of an anchored boat. Designed to develop carrier-landing skill, of course, doing it was vital to passing the Squadron 1 course.

In Squadron 2, we went to Corry Field for thirty-four hours in land-based planes, mostly the NY birds without the pontoon. Squadron 3 was in the Chance Vought Corsair O2U-2 scout plane. Like most navy aircraft back then, it could be a floatplane catapulted off a ship, or outfitted with a hook for carrier landings. Here, under one of the navy's finest aviators, Marcel Gouin, a future admiral, we learned nine-plane formation flying. We also learned dive bombing, a tactic just beginning to inch its way into naval aviation and one none of our planes were designed for. (The navy was having a true dive bomber built, but slowly because of a lack of funds.)

Also in Squadron 3, students were paired, one as pilot, the other to operate the radio and maintain contact with ground stations. Basically, we were learning what doctrine back then said was our role, to be the "eyes of the fleet," to widen the light cruiser's scouting range and spot where the battleships' shots landed. Partly, this straitjacket was the result of a so-called Morrow Air Board ruling in 1925, during the angry Mitchell flap over the promotion of airpower. In effect, the board limited naval aviation to mostly seaplanes and banned army air corps' flying more than one hundred miles offshore, edicts that the start of World War II would prove had been singularly inane.

Squadron 4 planes were big for that era. On twin-engine Martin flying boats, we hauled various-sized loads, did single-engine landings, and flew on instruments to learn how to handle big aircraft in bad weather. Here, we also flew T4M torpedo planes, mainly to learn what dropping the weapon did to the plane's flight behavior. I wasn't eager to be a torpedo-plane pilot, given the sitting duck aspects of it. The Martin boat, on the other hand, was a fine ship, though I admit I was a bit biased. I like big aircraft. We also put in some time in the then-brand-new Link trainer. Squadron 5 was a fighter syllabus at Corry Field in Curtiss Hawk and Boeing F2B-2s and F3Bs. It was stunt flying as much as air combat. I looped, rolled, did spins, had dogfights with everybody, and loved it. But, unlike today, we not only had to fly nearly every aircraft the navy owned, but also had to tear down and reassemble them.

As in Norfolk, in Squadron 1 the first solo came when the student's instructor said so. For me, that came after seven hours. Then I flew five hours solo, practicing what I'd learned. Then he took me up for more advanced work; then I had five more solo hours. In each of the squadrons thereafter, at the fifteen-flying-hours mark, an instructor other than the student's own in that particular squadron gave the student a "check." An up-check was a pass to the next squadron. A down-check meant going up twice more, each time with a different instructor. If either one of them also gave a down-check, it meant facing a squadron board that could allow a student up to three more flying hours to erase the down-check. But if they sent the poor guy to the "Big Board," it was gloom-and-doom since about all they ever did was flunk people out of naval aviation.

We had other problems, too. For one, all the navy's aircraft carriers were home-ported in San Diego, so we had to simulate a carrier landing by coming in over a "fence," as we called it, a rope slung tightly between two poles, to give us a sense of what an approach looked like. But aircraft didn't have brakes in those days, making it easy to lose control on a landing and flip into a ground loop, especially coming down crosswind (a big part of the instruction). And each squadron had its own peculiar set of tough flight tests, for many a final test as well, one that would end their careers in naval aviation.

For instance, in Squadron 2, one exacting test was emergency landings. As a student improved his flying skill, his instructor would spring that on him without warning. He'd take a guy up, often over the worst-looking terrain in sight, and suddenly idle the engine. When that happened, the student had to land in any small field he could find. We dubbed some the "Clay Pits," a clue to what they were like. Some were just grass fields the navy was renting—like crosswind landings, a tricky surface to stop on for an airplane without brakes. Usually, just before touchdown in those contrived "emergencies," the instructor would gun the engine to full throttle, both of us praying it would catch and not flood out.

In Squadron 2, we also had to make full-stall landings, another drill to teach us how to set an airplane down in a very small, designated area. We had to drop from two thousand feet in a dead-stick landing with all three wheels touching down inside a one-hundred-foot-diameter circle—and we had to do that at least four times out of every six tries to avoid a down-check. Most of the guys busted out were flunked for failing this one test. My only crisis in Squadron 2 was on taking off from a "Clay Pits" field, when a small jack-bush snagged in my elevator and jammed it. I notified the tower, and the squadron commander, John Cassady, flew up to see if I should bail out. He radioed me to try to land, "even if you have to burn it on in," which I did, aircraft undamaged. Just another day at the office, we agreed.

In Squadron 1, Warrant Officer Zemp Cornwell was my instructor. I don't know why he picked me. Maybe Admiral Moffett had suggested it. An old-timer who knew all the other old-time navy aviators, he'd been an aviation machinist's mate on the 1919 navy-Curtiss transatlantic trip. He was now a designated aviator, probably our most experienced instructor who was not also a commissioned officer. He was an excellent, tough teacher in a business where those two adjectives are synonymous. Over my forty-some years of flying, his instruction stayed with me, even saved my life a couple times. No student of his ever had flunked out of school, they told me when I arrived. I did not intend to be his first.

At the time, the Bureau of Navigation ("BuNav" in navy argot) was after him. Always anti-anything since year one, their beef was partly "this noncommissioned officer" earning hazardous-duty pay. He received it as long as he flew at least four hours a month. (For me, that extra pay meant a 50 percent bonus to ensign base pay of $125.) I'd heard how much nonflying officers resented that, some bitterly. The bureau bureaucrats also just did not want noncoms wearing wings, a badge "reserved for officers." Zemp asked for my help. I said I would, wondering what help a just-hatched ensign could be. Finally, the "elitist" BuNav minds just backed off. (Zemp would die of heart failure at age forty-eight, leaving a wonderful record for making fine aviators out of so many youngsters.)

My weeks with Zemp were mixed work and play. First time up, he gave me hell for not landing the way he wanted. He jumped me again days later when a sloppy landing busted my pontoon's forward strut. Then a guy named Hutchins spun in from fifty feet, our first crash "on the beach." Instantly, Zemp took me up to show me how to avoid that. If I flew poorly, he'd snarl like a marine drill sergeant. If so-so, he'd growl. Only "perfect" was acceptable. But weekends and some Wednesdays, often with Bill Moffett, he invited me to play golf. He rarely won but, when he did, he'd say he wasn't sure which was worse, my flying or my golf swing.

When I moved up to Squadron 2's land planes, I got lucky again. Twice. First, I had "Bos'n" Baker, another first-rate noncom instructor, then marine Capt. J. S. E. Young. He'd won a Distinguished Flying Cross with my brother's marines, fighting "Sandistas" in Nicaragua and, like Father Brady, was one tough cookie. He too, demanded "perfect" in every maneuver. I don't recall how many barrel rolls I had to do before he was satisfied, but it was a lot. Squadron 2 is the only one where I received a down-check. It ticked me off, drove me right back up to get the two up-checks needed to erase that smear from my record.

In early July 1932, I passed the final check in Squadron 4, a solo run in a T4M. My final check in Pensacola, in Squadron 5, came in September. Instructor John Crommelin graded it from the ground. At six thousand feet, flying over Corry Field, we had to do two slow rolls to the right, go on out to reverse course, and do two slow rolls to the left coming back. Then we went into a roll on top of a loop, then a pull-up into a precision spin followed by two double snap rolls and a downwind "falling leaf" maneuver—all within six minutes. Later I wrote, "Like having wings on my back. They sure do burn up the blue heavens . . . and is it fun!"

By graduation time, the seventeen of us who received our golden wings—out of sixty-seven who started school with us—each had amassed two to three hundred flying hours. I believed then, and today, that the best training school for aviators is the military, and navy training the most thorough in the world. None of us would die for lack of flying skills. Those lost either were killed like Lance Massey leading torpedo bombers in the decisive Battle of Midway in 1942 or died riding in someone else's plane. Truth is, hard as it was, flight school was one of the most fun years of my life.

Or maybe my evolving from grim tenderfoot to happy warrior was because I'd finally met Leila Marion Hyer, pronounced, "Leelee," and written, "Lili," the girlfriend Lieutenant (jg) Riley had asked me to "chaperon." Before her, I was not exactly a hermit. Near our bachelors' quarters was the well-patronized Halfway House run by the Carpenter brothers. Or on nights strolling Palafox, there was the B&B deli for a fancy chicken-liver omelet. And Prohibition didn't bother the natives any, nor us. The local bootlegger was Sam Clipper, supplier of a brew called "Shinny." Some locals made their own too. A distilled mix of rye and corn, and a good imagination could make it taste like bourbon. When we could afford to buy a charcoal keg of it, we'd put it in the back of Joe Ruddy's "Bluebird," a beat-up old Ford, and let it age by riding it around for a few days. It was the preferred nectar for most parties.

We'd gotten focused on that early. It was a tradition then at Pensacola that when anyone soloed, we threw a "Solo Party" that weekend. All hands were invited with, if they had one, girlfriends or wives. We held most of these and similar hours-long bashes at a private club called The Barn. Catering to naval officers, it was run by "Aunt Jenny" and Bill Turtle in a house owned by Lili's grandmother. It offered music on Saturday nights. Predictably, the riots we convened there we dubbed "Barn Burners."

Not that we partied a whole lot. For one thing, on my $187.50 monthly pay, I had to buy food, uniforms, and such. For another, each day before we flew, we had to sign a "Bevo" sheet, swearing we had not drunk any alcoholic beverages in the previous twenty-four hours. If you couldn't sign, no one fussed at you. You just didn't fly that day. The real sin was saying you hadn't had a drink when you had, a lie easily exposed in that small community, one that got you instant dismissal from flight training.

The first time I saw "Lili" was at The Barn. She was, as they say, the "Belle of the Ball." All the bachelor instructors were after her to dance. I told Moffett, no way a fresh-caught ensign can beat out those lieutenants. But in early 1932 at a party hosted by a classmate and his wife, I was paired with Lili—and hooked, deeply and forever. We talked at length of many things that night. Then she had me meet her "Daddy Pete" who owned Hyer Harbor Towing Company, and her mother, "Mommie Floss," a doll, too—as was her grandmother, except that "Granny," a Georgia girl, had doubts about letting a "Yankee" in the house. Lili's sister was a bit snippy with little use for me. But Lili's grandfather, "Waddy," who'd sold his bank just before the crash of 1914, was fascinating, a devout Christian and avid historian. He and I re-fought the Civil War many times that year, and I was invited back regularly. Daddy Pete made home brew, the chicken was delicious, the family delightful; but Lili, not the chicken, was why I came to dinner.

By then, the Depression had sunk into the pits. Banks were going broke. My father sounded near bankrupt. In June, he drove my mother and Margie down to stay the summer. I rented a $50-a-month furnished cottage for them, and Lili was a constant visitor. But Mother, out of the blue, advised me to hold off marriage until I'd been with the fleet a couple years and "can afford it." So I didn't say, as they left in September, that Lili and I had decided not to wait. My version of why is that Lili was afraid if I left unmarried, I'd not come back. Her version is that when I said we'd have to wait, she began to cry and I caved in. That banter aside, we picked Saturday, 15 October, to be married by Father John Royer at St. John's Church in Warrington, Florida. Three days earlier, I was up doing a vertical roll when I quit paying attention and went straight up into a tail slide. My instructor, John Crommelin, forgave me only because he said I had a lot on my mind: orders to join the *Langley* in San Diego and getting married the next Saturday.

My parents were irked at having to make a second trip to Pensacola so soon and at me for not waiting until I "could afford it." But by wedding-bells time, they were all smiles when, wearing my new golden naval aviator's wings, I got married. It rained incessant sheets all that day. The organ played and some gal sang, "I Love You Truly," over and over. I was ready to skip, thinking Lili had stood me up, when she and "Daddy Pete" finally arrived. Later, she admitted they'd been "lollygagging" along the way. When the reception had about run its course, we headed for Great Neck, New York, in a new Buick convertible bought for $1,300 ($53 a month for eighteen months, courtesy the Federal Finance Corporation). It was worlds better, to Lili, than my old Ford touring car, but a drain on my income. Still, I had a job, an asset a fourth of the country's working-age population didn't have. In New York, my parents and sister Margie introduced Lili to the rest of my family, friends, and neighbors. My grandmother became especially fond of Lili. "You take good care of her," she warned, "or you'll answer to me!"

Next, we drove to Akron, Ohio, to see sister Eleanor. She'd married Ray Kitchingman, a war-time army aviator, now a barnstormer who owned an aerial photography company. Through him, I met Tex Settle, a balloonist who'd won the Gordon-Bennett balloon race a couple times and would become a navy vice admiral. Fortunately for his survival, in my opinion, he never acquired an interest in dirigibles. While in Akron, I also saw Jimmy Doolittle again. Over the years, I would work with both men on a few aeronautical projects and get to know them quite well. But in 1932, Jimmy, working for an oil company, had an airplane I dearly wanted to fly. He wouldn't let me. Too dangerous, he said.

I heard that a lot back then. (Later, after I got some navy purchasing clout, everyone wanted me to fly his airplane. I'd as soon have passed on a few like one I called an airborne foxhole, the F-104. And the Saab Aviation one. Saab wanted me to solo it but, with its manuals and cockpit gauges all in Swedish, I insisted on a copilot. Reluctantly, they gave me one said to be fluent in English. Maybe he was. Aloft, he said, "You take it." Coming down, he said, "I'll take it." And that's all he said during the entire flight. Sweden was careening into socialism at the time, and lots of Swedes thought it their duty to trash America. Maybe that explains it.)

Old Route 66 took Lili and me to Coronado, California, south of San Diego. There, for $50 a month, we rented a house, fully furnished excepting linens and silver, near the Naval Air Station. Our landlord, a long-time San Diego resident, was an old ogre, but the house was cozy. On $206 a month income, less the rent, car payment, a $30 monthly food bill, utilities, and such, we were always just a step ahead of bankruptcy. We also added a dog to our family, the first of many over the years, a frisky Bull Terrier we named Butch. And, other than a weak bank account, life was fun. I was married to Lili and flying with the fleet.

Next door in a twin to our house were Lt. (jg) "Chick" Renard, class of 1928, who'd been in my academy company, and his wife, Kit, who, unlike Lili, didn't mind his nickname at all. We soon set up a major once-a-week project, deciding what flavors to add to a gallon of alcohol bought that week from a local bootlegger, George Pandell. It was a chemistry test, really. Over time, we learned to make a good gin, passable bourbon, even a scotch. For variety, like before payday when a dollar was wealth, we drove to Agua Caliente, Mexico, to lively Caesar's bar where, if we bought a beer, they served a free lunch called "Coronado Quail." A hot dog wrapped in bacon, actually, but it played better as, "We went to Caesar's for cocktails and quail."

That year, Herbert Hoover was being blamed for all our ailments. Worse, he acted as if the economy could fix itself. That may be sound John Locke philosophy but it's lousy politics. Franklin Delano Roosevelt was elected president easily in November. But he'd not be sworn in until 4 March 1933—five months! A law in thermodynamics says the longer a disorder goes on, the worse it will get. Surely, for our country's sake, FDR and Hoover, in concert for five months, could have hatched an economic plan or two. Rumor said FDR was afraid that if he did, Hoover would get the credit. Was FDR so small? I don't know. I do know, just as many praise Roosevelt's presidency, time has redeemed the good opinion people once had of Hoover as it has that of Harry Truman, who also left the presidency under a cloud.

In January 1933, Lili announced that I was to become a father in August. I was elated—other than fretting over how to pay for it. Then we found an obstetrician who would handle it all except hospital costs for $100. The only family reaction we got to our news flash was Lili's sister, Mary, saying she'd come out to help housekeep after the baby was born. Then on 4 March FDR became our thirty-second president and on 10 March, I fell into my first earthquake. For the only time while I was aboard, our ship had docked in Long Beach, California—"Wrong Beach," we called it. Not having been there since my summer cruise on the *Nevada*, I went ashore to sightsee.

At exactly 5:55 P.M. as I was strolling Ocean Boulevard, the ground shivered. I thought it was me having a dizzy spell until I heard people screaming and saw pieces of buildings falling down. We went on rescue duty immediately and spent all night helping families with children board our ships. The next threat, the experts said, might be a tidal wave, a "tsunami," they called it, rolling in off the Pacific. That didn't happen but some 120 people died in the quake. A few days later, FDR hit us with another shock.

We had guessed his election might be good for us. Named a navy assistant secretary in 1913, he had badgered President Wilson to boost the navy budget. But in 1933 he did a "Full Astern." He declared a bank holiday, meaning for us no bank loans. Then he froze federal workers in place, no promotions, and cut by 8.33 percent the ensign's pay grade. In June, I was to take a battery of tests in electrical and general engineering, law, strategy, tactics, navigation, and the duties of a deck officer for promotion to lieutenant (jg). I had a huge incentive to pass. Back then, navy policy said ensigns are not married, even when they are, but marriage is recognized for a lieutenant (jg). It meant that, on top of a base-pay raise to $175 a month plus half that again in flight pay, I also would receive a "family" allowance of $40 a month. For Lili and me, it was to be a huge jump in income, enough for us maybe, finally, to pay off our debts, even start a little savings account. In effect, however, Roosevelt had ruled that, even if I passed the exams, I'd not receive 40 percent of the income I had anticipated. Life in the navy had begun to look very grim.

By June, most of us 1931–1932 Pensacola graduates had passed the exams and were lieutenant (jg) in title but drawing less than our former ensign's pay. Our protest, one a bit risky in an era of strict discipline, was to sew on the sleeve above our old ensign's stripe a second one instead of a lieutenant (jg)'s two-stripe emblem. At inspections, the jg's not drawing jg pay stood out like warts on a nose. A few higher-ups groused at us but none ordered us to buy the real thing. We also dreamed up a protest for dive-bomber drills. We didn't yet have a plane for the best angle of attack, straight down. If one of ours tried to pull out of a dive steeper than 60 degrees, *g* forces would have torn it apart. That summer, we did many dives at only 40 degrees, and, if someone bitched, we'd say, "Well, that was a pay-cut dive. If you want me to go steeper, I'll need more pay."

Meanwhile, wars were brewing. After grabbing China's Manchuria province in 1931, renaming it Manchukuo, Japan was massing an army there.

Italy's Fascist dictator, Benito Mussolini, was rattling his saber. Adolf Hitler's Nazi, née Socialist, Party had taken political control of Germany, vowing to rearm and stop the reparations "gouged" from them after World War I, a fee so high it may have helped Hitler's rise to power. Yet Washington had us treading water, basically.

4

Life on the Covered Wagon

Pay cut or no, we stuck to our training program as best we could. The carrier *Langley* was my duty station in Coronado. She'd been a collier, sister to the *Orion* (which had vanished eerily, going from Brazil to New York during World War I). She was an electric-drive ship, the navy's first, so her two engines ran as one. Taking her out was like moving an 11,050-ton, 542-foot-long motorboat in a crowded yacht basin. She usually carried twelve fighters, twelve two-seat scout planes, and ten torpedo planes; but on my tour she housed eighteen fighters and twelve scouts, with a lone elevator to haul them all up from the hangar deck.

At 7,000 horsepower, she could do only maybe fifteen knots. Her bare wooden top deck, 610 feet long by 64 wide at its widest, had inspired the "flattop" nickname for aircraft carriers. Her speed and shape were why we called her, affectionately, "The Covered Wagon." (Today's nuclear carriers run on 280,000 horsepower, cruise at up to thirty-five knots, over forty for short periods, have a flight deck some 1,000 feet long by 252 wide, and can carry ninety-plus aircraft and launch them three at a time.) To qualify for carrier work, we had to do seven good carrier landings and takeoffs in a row. Sounds easy today. Not so on the *Langley*.

We'd been catapulted off a seawall in Florida but the *Langley* didn't have a "slingshot," as it was called. Nor did the *Lexington* or *Saratoga*, our other two carriers. (Their sixty-foot-thrust catapults wouldn't be installed until 1938 to 1939.) We had to take off on our own, and liked at least twenty-five knots head wind to do it. The *Lex* and *Sara*, we called them, with their cruiser-size engines, could do that easily for their up to seventy-two aircraft, but the *Langley* needed Mother Nature's help. Sea swells moving her tiny deck up and down like a roller coaster made catching her arresting cable in landing a sporty trick, and made a poorly timed takeoff, as when the bow was on its downstroke, a dunk into the sea. And that was on clear days with a calm sea.

One break for us *Langley* folks was that she tied up at a North Island dock so we could drive to work over a causeway from Coronado. But San Diego Harbor wasn't deep enough then for the *Lex* and *Sara*. They had to anchor out in Coronado Roads and use powerboats to ferry people in and out. Most of Pacific Aircraft Battle Force (AIRBATFOR) nested in San Diego, as did all our Pacific Fleet destroyers, and most of the big ships in Long Beach or San Pedro. In "isolationist" America, dispersing a fleet to avoid a single enemy blow sinking it was an idea unimagined. For lack of funds, our ships didn't take long trips either, inspiring a joke popular then among seagoing folks. It went like this: an Australian sailing vessel meets a navy ship off San Clemente island in a fog. As the Australians close in, they yell, "Ahoy, ahoy, what ship are you?" and our ship answers, "Ahoy, ahoy, what ship are you?"

Says the Aussie, "This is the *Gladys B. Simmons*, 240 days out of Australia."

Answers our ship, "This is the United States Ship *West Virginia*—out all night!"

Communications was a problem more serious to us than sailing time. The radios we had were my Pensacola bugbear: CW (continuous wave) dots and dashes, no voice. When voice sets did arrive later, only section chiefs got them at first. Still, it was an improvement. Before radio, a navy pilot took a carrier pigeon with him, to be released when he reached his destination. When it arrived back at its home loft, that told Operations the pilot had survived. The pilot had to feed the bird but, I was told, it could ward off starvation in a crisis. Occasionally back then, I envied the pigeon's navigation skills. Eventually the navy did decide the radio was a useful invention. So they tore out the *Langley*'s pigeon loft, located just below the flight deck, and put in posh (by our standards) quarters for junior officers.

We thought the Covered Wagon a happier home than the *Lex* or *Sara*, whose crew lived life, they said, "in the fast lane." We had handball courts in the hold where we were supposed to store the spare airplanes we never had anyway. And in an upper-deck wardroom, at least one poker game was going almost all the time. Most aviators gambled back then, certainly on our ship except for us married folk who, most days, had near-empty wallets. They drank a lot, too, even before Prohibition was repealed in 1933. A pilot avoided the sauce, however, for twenty-four hours before a flight. The "Bevo" list we'd signed in Pensacola was here too. To me, it made eminent sense. Until it was abolished in late 1934, no one I know of got killed flying because he was hungover.

Back then, naval aviators all knew each other, many of them tough survivors of the rough early years flying what amounted to bamboo crates. In my scouting squadron, VS1-B ("V" for "heavier than air," "S" for "scouting"), what impressed me were the veterans. They had a penchant for nicknames, a hint to their high regard for each other. Our commander was R. S. D. "Ross" Lyon, a longtime flyer but not an academy graduate. His XO (executive officer) was Don Smith, academy class of 1922. Operations officer was "Hap" Hazard, 1923. The rest included "Dusty" Rhodes, 1925; "Sock" De-Wolfe, 1926; "Rags" Parish and "Bill" White from the class of 1929; and "Gillie" Carpenter, "Sunshine" Howerton, "Matty" Mathews, and "Beppo" Garcia, all class of 1930. Beppo was my roommate aboard ship—where I exercised my right under academy class-standing to commandeer the lower bunk.

With us in our hangar was VS-2 off the *Sara*. Like all squadron leaders, its skipper, Felix Stump, was a lieutenant commander. Lee Webb, and later Donald Duncan, had the *Sara*'s VS-3, Ralph Ofstie its VF 6 ("F" for "fighter") squadron. Allan Flag, and later Forrest Sherman, had VF-1. "Jocko" Clark led VF-2, the all-enlisted-pilots squadron and one of our best. Our two torpedo squadrons were led by Herman Hallen and Hugh Sease. The *Langley*'s VF-3 squadron was led by Mel Pride, one of our greatest naval aviators, history says, a perfectionist pilot, expert in aviation technology, who never asked any of his people to do with an airplane something that he hadn't tried first. In fact, most of the guys I met then would be key architects in moving naval aviation into World War II and beyond.

When I joined VS-1B, Capt. A. C. Read was the *Langley*'s skipper. As a lieutenant commander, he had led the 1919 nonstop flight across the Atlantic. During my tour in Coronado, he was relieved by Capt. Kenneth Whiting, also an early aviator and an exceptional man. Whiting's XO was Virgil Griffin. As a lieutenant, Griffin had made the first takeoff from the *Langley* in 1922. A Mobile, Alabama, boy, Griffin was predictably called "Y'all," or "Squash" for a reason never explained. But he was a tyrant about protocol, proper uniform, the proprieties, proper everything. And part of one of my more memorable moments on the *Langley*.

Whiting's wife had stayed home this tour, so he belted the booze a lot. One day, heading for a bar, Whiting made a courtesy call on Admiral Lanning, AIRBATFOR commander. A junior officer goes first, says protocol, then a senior officer repays the call. How I fell into the middle of this one starts with my academy class being the first allowed to go directly into aviation. So now we had to learn by doing what two years' sea duty would have qualified us for, standing watch as officer of the deck (OOD). And the morning after Whiting's courtesy call, I went to the ship to be OOD for the 0800–1200 watch. The guy I relieved reported the captain on board. His cabin was way aft, a palatial room crammed with nineteenth-century furnishings including a four-poster bed with thick, velvetlike drapes hanging from the top of its frame to the floor.

At promptly 0900, flag waving from a fender post, the admiral's car zipped onto the pier. I rang the required bells, then summoned the side boys (a sort of honor guard) and XO Griffin to the quarterdeck. I didn't call Whiting. He knew that a ship's captain must welcome an admiral aboard personally on such visits, and he'd heard the bells, hadn't he? The admiral came up the gangplank. No captain. His eyes like ice, "Squash" Griffin barked, "Mr. Hayward, where's the captain?"

Shaking a bit, I said, "He's aboard, sir. In his cabin," which was pure guesswork.

So all of us—the admiral, XO Griffin, the side boys, and I—went aft. We looked in the captain's cabin, in its head. We looked outside. No captain. The admiral stomped away, madder than a drenched hen. Griffin, enraged, wanted to arrest somebody, anybody. And OOD Hayward was a likely target. As "Squash" flailed away, Whiting's marine orderly sidled up to tell me, "Sir, I found the captain. He's under his bunk, sound asleep!" His four-poster's drapes had hidden him from view. Later I learned he'd come aboard ship at 4 A.M. I'm sure, if the admiral had seen him under his four-poster, Whiting's tour as skipper of the *Langley* would have ended instantly.

Still, most of the time the *Langley* just sat at her North Island pier. It was an island in those days—filling in the Spanish Bight came much later—our field one big asphalt mat, no runways, on what we called West Beach. Beside it, off what they called Rockwell Field, the army air corps was flying Martin bombers. My squadron had Corsairs and, because of the magic of class standing, I finally got a Corsair two-seater of my own, and a good radioman to boot. My other delight was Hap Hazard, chief of my three-plane section. He excelled in flying skills, was deadly with his machine gun, and hard on ensigns. I was number three on his lead and he told me, and meant it, if he had to look around to see where I was, he'd punch me out when we landed. So, aloft, I stuck my plane right up by his. Not once did he have to turn his head. (Sadly, though, he had a demon in him somewhere. He ended up a suicide.)

Ours was a good team. It had to be. Typically navy, we had fierce competitions among the squadrons, VSs against VSs, VFs versus VFs. We had five sets of contests: communications; both fixed and free gunnery; dive bombing on a sled towed by a destroyer; and camera "gunnery" by the rear-seat man simulating a machine gun. The best teams received a Battle Efficiency award, allowing them to paint a red pennant with a black ball on it on each aircraft.

We also had individual contests where a pilot got to paint an *E* up by his cockpit if he did certain things. One was to obtain a top score in the bombing runs. Another was making twenty hits out of one hundred shots at an aircraft-towed target sleeve. That wasn't as simple as it may sound. The pilot had to make five runs at the airborne target: one from high overhead; one on an opposite course; one from the high side on the right; one from the high side on the left; and one from underneath, with twenty shots allowed for each run. These were umpired by another squadron, firing up lots of arguments over whether a pilot had obeyed the constrictions or not. There were few arguments in the bombing contest, however, since where the bombs fell was recorded on film. All these exercises occurred during AIRBATFOR's "concentration period" in the summer.

Near the end of my first summer there, on 28 August 1933, our first child, Mary Shelley, was born in San Diego. Shortly thereafter, a call came down from on high. They wanted a junior officer off the *Langley* to join an aerial-photography team on a four-month survey of the Aleutians, hunting usable sites for naval bases. The *Argonne* was to be the support ship with Lt. Cdr. Arthur Radford, class of 1916, commanding the team. He, too, had wanted to fly since he was a boy, and had earned his navy wings in 1921. I might have enjoyed chatting with him, but I didn't want the job. I wanted to fuss over Shelley. Anyway, the Loening Amphibians they'd be using were dogs. Every third time we'd tried landing one on the *Langley,* it crashed. Using a fine old navy tradition, our skipper had "Sunshine" Howerton, "Gilly" Carpenter, and me roll dice to see who would go. "Sunshine" lost, in my opinion, because he won and went to Alaska.

Other than that dodge of a bullet, I ended my two-and-a-half-years' tour in Coronado with more pilot hours, I think, than anyone else there. The reason was that I asked for nearly every extra flying opportunity they had. One was a daily "Aerological" scan, a flight that took off at 7:30 A.M. and lasted an hour. It meant flying to altitude, usually through fog, with a barograph on the plane, then landing to give what the device had recorded to the weatherman. (The radiosonde and newer, better ways to chart the atmosphere didn't exist back then.)

The other way I piled up hours was volunteering to ferry aircraft from San Diego to Norfolk for overhaul and, in Norfolk, pick up one to deliver back to San Diego. The navy paid $6 a day for these trips, peanuts now, but the price of a week's groceries for my family back then. A bonus was the fun of cross-country flying. And, since we were not allowed to fly at night nor in bad weather, the other bonus was the chance to stop in towns I'd not seen before, mainly in west Texas and New Mexico, about as far as a flight from Norfolk could reach, in those days, before nightfall. One intriguing aspect of those trips was to see how curious the local people always were about the tail hooks on our aircraft for carrier landings.

I remember well one return flight from Norfolk in a new Berliner Joyce scout plane. I ran into a blizzard west of Albuquerque, New Mexico, forcing me to land at the Zuni Indian reservation with no way to tell home base I was okay. The Zuni didn't have telephones, and the aircraft voice radios just being installed back then were lousy. On land, their range was very short; and aloft, even on a soft day, just a plane's wind stream made it nearly impossible for either a pilot or ground station to decipher what each other was saying. So I moved in with the Zunis. At night, a Zuni dwelling is, to put it tactfully, overpowering. Between the snoring and the smell, I could not sleep. I spent the night with a deck of cards, teaching the Zuni how to play hearts. During the many ferries I made, that was the one closest to a life-threatening crisis.

Another well-remembered event from those years was the "Fleet Problem," what the army calls maneuvers. Run each fall, after the summer flight competitions, it used all three carriers, usually, and other ships, all sorted into two teams. Each team's skipper was given a strategic objective and a brief on his tactical situation. Each had to make his own plans and decisions on how to use his assets to gain his objective. Because of the shortage of funds for fuel and such, the maneuvers were held inside a constricted area but were very realistic—assuming naval warfare would continue to be, as in the past, a head-to-head shoot-out by the battleships. It meant our aircraft spent all our time hunting "enemy" ships, not in mock attack on each other's carriers and aircraft.

The *Langley* being slow, we usually were in the defender force, the larger, faster *Sara* and *Lex* being with the "enemy." In one 1933 episode, my unit, from a ship position 190 miles at sea, was to "attack" Point Arguello, California, on 15 November. The technology and tactics for night carrier landings were as mature, back then, as my newborn daughter. So we usually took off just before dawn in order to return to the ship in daylight. Since I'd be at sea on my birthday, Lili gave me on the 14th my gift from Mommie Floss, a pair of shiny red silk pajamas. Mission briefing was to be at 3:30 A.M., so I slept on the *Langley* but skipped breakfast since, well, I hate to get up early. When the siren wailed to start the mission, I jumped up, pulled my fur-lined flight suit on over my new pajamas, put on boots and goggles, and ran with my charts for the ready room.

Outside in driving rain and high winds, the ship was bobbing like a cork but we were told, "Weather certain to improve. You will launch at 4:30 as scheduled." I was to fly number three spot and "Rags" Parish number two off Lt. Cdr. William Updegraf, who was not much of an aviator. I knew I was a far better navigator than he was. Since my near-fatal fiasco on Long Island, my study of aerial navigation had been constant. We hadn't much to navigate with, however—just a plotting and a chart board, stowed under the instrument panel, and a drift sight for the back-seat man to use. If the pilot was alone, he estimated that by the sea state and its drift.

We took off just about on time, any delay caused by each of us waiting until the ship pitched up to be sure we didn't fly smack into the crest of a wave. As we headed for the target in tight formation, the weather got worse, the sea waves mountainous. We hit the target at dawn under a low ceiling, which made it more like a torpedo run than dive-bombing. That pass done, we rejoined, checked the plotter board to see where the *Langley* would be at the course and speed she was on when we left, and headed that way. When we arrived there, she wasn't. Nor was any other ship. In a screaming wind and rain, we began to circle and send out messages.

In those days, aircraft didn't have automatic direction finders, only a loop antenna on the wing. If a station is sending MOs (Morse code "beeps"), a pilot maneuvers until the signal fades, then flies on for fifteen minutes to repeat the process, which takes the 180-degree ambiguity out of a sender's location. Or if the pilot sends MOs on a ship's frequency, the ship can get a bearing on him and radio his course home. That day, our MOs weren't being answered and the ship wasn't sending at all. We'd been in the air four hours, and were well out to sea and low on fuel. Suddenly, Updegraf hand-signaled that he was lost and turning command of the flight over to me.

I promptly did a sharp turn, heading east. However far we were from shore, I wanted it as near as possible when we ditched. A few minutes later, 8:33 A.M. on my clock, I waved to the others that I was going down. My fuel tank was nearly dry, and we'd been told, "Never crash-land at sea on a dead engine. A dead-stick landing into a wave is like hitting a brick wall." I powered in but the high storm-driven seas had our aircraft awash quickly. We managed to slide off before it sank, and my radioman freed our life raft. Unfortunately, as he inflated it, the approximately forty-knot wind tore it away from him and it went scooting downwind, out of sight. Left with only our life jackets, we began swimming east. My fur-lined flight-suit quickly became waterlogged, so I shed it, leaving just my pajamas. Time passed, the wind screamed, the rain came down in sheets, and our hope of being found was near zero.

Then, after hours that seemed like days, a lovely light cruiser, the *Cincinnati*, plowed into sight. Nearly ran us down, in fact. It sent out a whaleboat to pick us up. All the talk aboard ship was about my red silk pajamas. "Only a damned aviator," her captain growled, "would be caught out in the middle of the Pacific, swimming around in his pajamas." Maybe the caste-conscious "black shoe" sailors did not have the élan, the get-up-and-get of the "brown shoe" aviators I knew. Whatever the reason, that sour animosity persisted right up into World War II. Our rescue was a big story in the newspapers. That lured Lili down to meet the *Langley* when it returned to North Island. Even when she saw my airplane wasn't there, she didn't worry, she said later. "I knew you'd never get lost." Silly girl.

In an ensuing Board of Inquiry, we learned that as we were "bombing" Arguello, the *Langley* had changed course, hunting for calmer seas. And she had not been sending a position MO because the jarhead with the key to the radio compass had locked it up and left his post to pick up some small stores. Since it wasn't manned all the time anyway, I guess he thought his need for toothpaste must be as important as our need for MOs. Updegraf said loss of the three aircraft was all his fault, not that of his young wingmen, but the Board ruled that *Langley*'s Captain Whiting was equally at fault. And Whiting issued a blanket order that the carrier's radio-compass station was to be manned at all times.

The Fleet Problem for the fall of 1934 was to be an attack on the Panama Canal, what military folks call a "choke point" heavily used by commercial shipping between Europe, Asia's Pacific rim, and America's ocean coasts. The "battle" was like one held when I was on the *Richmond* in 1932. That year, battleships, destroyers, and the *Lex* had been the attackers, our light cruisers and army planes in Panama the defense. Even Coco Solo, Panama, flying boats had been on our side. It was a large war game by that era's standards. We biased aviators on the *Richmond* had believed, and she proved, that our worst enemy was the *Lexington*. Using her fighters to fend off enemy (army) aircraft that never showed up, the *Lex* surprised the defense, "wrecked" the Canal, and "sunk" my ship.

In 1934, the fleet mix was different, all three carriers being in the attack force, but the result was the same. We even used the same tactic, a feint toward South America, then a swing north, close inshore, to hit Panama. The defenders had no chance. Clearly, the carriers had shown that their aircraft could surprise and crush a target even when defenders knew about what to expect. There is no record of what happened to the critiques written on those two Fleet Problems. They just vanished.

At the end of that war game came leave in Panama. Lots of tourists were in Panama City by then, Mamie Kelley's bar one unending party, the Union Club still the Saturday-night place to compete in rum and beer drinking. A week of that sent us off to Gitmo, where oil began to seep into the blower section of our new SU-2 aircraft's engines. No one knew why or how to fix it. Grounded, we were told we'd be off-loaded from the *Langley* in Norfolk— and miss all the free-of-charge parties at New York's annual Fleet Review and Navy Day parade. A letter from Lili, waiting for me in Norfolk, said she was going to New York to surprise me. I was a bit peeved until the lovely, caring girl joined me in Norfolk. We rented a room from a gracious southern lady in a neat boarding house by the Ghent River. The time until Lili left flew by like a second honeymoon. Though it left me broke.

At Norfolk, the *Langley* was to become a seaplane tender, and my squadron was to board the *Ranger*. Named for John Paul Jones' ship, she was the first U.S. ship designed from the keel up to be an aircraft carrier. FDR was using Works Progress Administration (WPA) funds to buy her, a finagle not authorized by Congress. But Congress didn't squawk. Still, she'd not be commissioned until 1935. So plans to convert the *Langley* were shelved, and we were ordered back to Coronado. My flying unit rode the *Sara*, now commanded by Captain Whiting, as far as Gitmo. While we waited there for the *Langley* to catch up, he invited us to go to a bar "with mucho rum" in Caminera, Cuba. When he'd had enough of that, he rode us back to the *Sara* in his gig, he and some others pretty well rummed up though us junior-types were fairly sober.

His XO (executive officer), Cdr. Marc Mitscher, was waiting on a gangway leading to the ship's hangar deck. As senior officer, Whiting went first. Suddenly, at the gangway, he dove overboard and began swimming around the ship. Mitscher, then all of us, yelled at him to come back, which he eventually did. He must have been a very good swimmer. Most fully clothed people can't swim well sober, let alone drunk. He was a fine pilot, an able leader, but he never got his flag. Maybe stunts like that one cost him. A weak or conniving fish can slip past a General Board now and then, but usually for promotion to rear admiral, they sift the evidence very finely.

Sailing to Coronado on the *Sara*, I forgot Whiting's frolic when I lost another A-one classmate and friend, Bob Patten, killed over Nicaragua when he collided with "Hitch" Saunders' Grumman SF in a mock dogfight. For *Ranger* duty, our squadron went from twelve up to eighteen planes in January, and we became extra starters on the *Lex* and *Sara* so we'd know all the carriers. Then in 1935, Ernest J. King ordered an emphasis on night- and bad-weather flying, which told us a jump in our casualty rate was likely. (Full admiral by 1941, he was both commander in chief of the U.S. Fleet and chief of naval operations by 1942, an unprecedented two-hat assignment. His boosting carriers at the expense of battleships in the 1930s was a major reason, some say, for our Pacific victory in World War II.) Having to night-landing qualify on the *Langley* surprised us. If daylight flying off her was an adventure, using her at night had to be a bitch.

For starters, we were assembled by the landing signal officer's platform on the flight deck to watch the LSO, Lt. Walter Holt, and a night-landing veteran, Lt. Giles E. Short, in a Boeing F4-B, show us how to do that. As Short flew in, Holt, with lighted batons, was to signal him, "Too high ... low ... level off," whatever, until two lights on the F4-B were lined up. When they were, that told the LSO the F4-B was "in the groove," and he'd wave it on in. Finally, he'd signal the pilot to cut his engine when that would drop the plane down into the arresting cable. With fifteen knots' head wind, Short looked perfect as he came in on final. But when Holt waved, "Cut the engine," the F4-B dropped like a rock into the water. Only its two outer panels "landed," one on deck, one in a safety net under the deck. A "Plane-Guard" destroyer rescued Short, unhurt but literally "mad as a wet hen." It did not stir in us any enthusiasm for night landings on the *Langley*.

We practiced the night landings over on Otay Mesa until our bosses reluctantly admitted it was just too dangerously unfair to have us night-landing qualify on the Covered Wagon. They ordered us to do that on the *Lex* and *Sara*, whose 880- by 85- to 90-foot flight decks made the *Langley*'s seem a postage stamp. To go from thirty-three-thousand-ton cruisers to aircraft carriers, those two ships had each lost some eighty-five hundred tons, mainly by removing all their heavy guns, but they still had the cruiser's turbines and boilers, most powerful in a warship at the time. As a result, they could cruise at thirty knots, even hold for an hour or so at thirty-four-and-a-half knots. With a crew of 1,788 (169 officers) including aviators, each could carry up to 120 aircraft but, to avoid crowding, usually had about 80 at any one time. Still, for all their advantages over the *Langley,* in their first night-takeoff landing drills, done off both at once, they used minimal lighting. Consequently, as aircraft rendezvoused aloft, I lost another classmate and pal, "Jocko" Kelly. A major risk in night takeoffs back then was the pilot not having a visual horizon to help orient him. He had to fly instruments-only and, most importantly, trust what his gauges told him. If he didn't, if he started looking around, the confusing flash of twinkling lights could cause vertigo that could lead to panic, then to driving his plane into the sea as Jocko had.

At his memorial service, I reviewed the roster of my flight class. We'd lost Frank Higley first, at Pensacola tangled in his 'chute, bailing out. "Mac" Vorhees had crashed on Otay Mesa by not changing his fuel mix coming down to low altitude. Like Patten, "Pumpkin" Hart died in a mock-dogfight collision with "Swede" Eckstrom. Luke Felton, an usher at my wedding, had died sitting in an SU-2's back seat when its propeller spun off, knocking him out and hurling it into a violent pinwheel crash. Vic Gaulin, a 1925 shipmate, usher at my wedding, died when his plane's pilot flew it into a mountain. A pal and excellent pilot, Johnny Burgess, died in the back seat of Oscar Pate's fighter when Pate crashed trying to fly under bad weather, a mistake often made back then. He should have flown up into the soup and gone on the gauges. At the service, I wondered—will we all crash eventually?

My first four landings on the *Sara* went smoothly, but not my fifth. I was on final approach, goggles on my forehead, when my gas line broke, spraying me with high-test fuel. My goggles not over my eyes, where they should have been, was a stupid, painful mistake, I quickly learned. Every second until I landed was raw fear that spewing gas would ignite, turning my plane into a fireball. It did sear my left eye. When that ulcerated, I was locked in a pitch-dark San Diego Naval Hospital room, my head one big bandage, doctors trying to save my eye, me afraid I'd never fly again—while my scouting squadron 1-B played with our new, first-of-its kind aircraft carrier *Ranger*. It all made me an irascible bear for the two months I took to convalesce.

By May, the eye had improved to 20/200, giving me back my wings, and Mommie Floss arrived to check on me. She loved to bet the ponies, so we took her to festive Agua Caliente where she dissected the entries and handed me a ten-spot. Bet it, she ordered, on "Morcharm" to win. The tote board said that was as likely as sunrise at midnight, so at the ticket window I bet it to show. And "Morcharm" ran like lightening while the rest of the field moved like snails. As that nag stretched its lead, Mommie and Lili hollered and screamed and I sank in my seat. I finally had to confess what I'd done. The drive back to Coronado was long, silent, and sullen.

Then, ending three years in Coronado meant transfer to a new duty station. The usual move was to flight instructor at Pensacola or to either the Hawaii or the Panama flying boat squadron. Lili and I wanted Pensacola, of course, and hoped my extra flying hours had earned me an edge in that department. Also, for my eye's sake, I'd been asked to stay off carriers for a while, and Pensacola didn't have one. So on one of my ferry flights east, I went to the Bureau of Aeronautics (BuAer). When the detail officer, Lt. Cdr. Arthur Radford,[1] asked me where I'd like to go, I said, "Pensacola. Anywhere but Panama. I've had my fill of Panama's heat." A bit later, the Bureau of Personnel (BuPers), whose compassion was legendary, ordered me to "Report 1 July 1935 to patrol squadron VP-2 in Coco Solo, Panama!" From now on, I swore, when asked, say, "Any place, any ship."

But first came a thirty-day leave in Pensacola. Mommie Floss groused over my new billet, especially after Lili revealed that she was pregnant again. But the visit helped me come back all the way from my eye injury. The end of June, Daddy Pete drove us to New Orleans, a tiring five-hour ride in a Chevrolet not built to handle comfortably the three of us plus Lili's folks. We rested overnight in the Monteleon, one of the city's best hotels. Next day Lili, Shelley, and I left, to Mommie Floss's dismay, on a Panama-bound United Fruit liner. The ship's Cuban captain doted on us, I guess because Daddy Pete's company had once rescued a fishing schooner of his, blown aground on a Florida beach in a storm.

Judy and "Rags" Parish, our VS-1 personnel officer in Coronado, also were aboard. Class of 1929, seniority-sensitive, he had a persistent urge to pile advice on me as if I were a child. But on this trip, little Shelley avenged me. A playful tyke usually, this time she whined constantly, as little imps can do, and each day flung her breakfast prunes at Rags, an assault he recounted happily for decades. The rest of the trip was a joy, we poor folks acting leisurely like the rich with a stop in Havana to watch the drinking and sinning at "Sloppy Joe's," gambling everywhere.

At the Cristobal base on the Canal's east end, seniority got Rags and Judy low-cost base housing. For us more needy folks they had none. And it was raining. (Panama has two seasons, rain for six months, dry northeast trade winds blowing the other six.) So, alone in the rain, we hunted for a home in grimy nearby Colon, maybe the worst town in Panama for raising a family. All it had at my pay scale was a tiny, dark, hot (home air-conditioning not having been invented yet), two-bedroom, ground-floor apartment downtown where a gaggle of street noises beat on us all night. It cost $75 a month, unfurnished. Literally. Even lightbulbs we had to buy ourselves.

But Panama wasn't all grim. Following tropical custom, our duty hours were 7:00 A.M. to 1:00 P.M., leaving hours for swimming, watching ships ply the Canal, riding the train along the Isthmus, playing golf. Saturdays, Lili and I often ended a night of dancing at The Strangers Club in Colon's harbor with an omelet at the Atlantic Bar and Grill, and a ride home in a horse-drawn *carametta*. And my commercial pilot's license got me work, flying tourists to Panama City in Isthmusian Airways' twin-float plane, low pay but with a second baby due it paid for Clemence, the maid hired to help Lili. Big, coal-black, from Martinique, she spoke French or her native lingo, mostly, amusing us. And I paid $85 for Lt. Jim Baker's 1929 Model A Ford to drive to Colon past Gatun Locks to work, eight miles of the only paved road back then. Its floor was rotted out—"Easier to see the ground," Baker said. Fading memory can gild the good, soften the bad. Yet, all told, Lili and I enjoyed that tour even thought we had little money.

Coco Solo had three patrol squadrons. Mine, flying PM-2s, was VP-2; skipper D. P. "Dippy" Johnson; Lt. "Carp" Doan his XO; Baker on operations; and me as squadron navigator. Rags Parish was in our unit, too. Squadron VP-3 was flying old, slow Douglas P2D-2 twin-float torpedo bombers. Our third squadron, VP-5, had the less-old Consolidated P2Y-2 flying boats, way underpowered for their weight. Their follow-on, the P2Y-3 with bigger engines, made the first full-squadron flight to Hawaii, highly publicized but to me just proof of how idiotic the 1920s Morrow Board was in ruling that the navy must fly only over water, the land belonging to the army. It was petty politics defying common sense when dramatic gains were being made in aviation technology.

Another nutty notion promoted even by some naval aviators was to use seaplanes and flying boats for high-level bombing of enemy ships. That's why our PM-2s and some torpedo planes carried the so-called "secret" Norden bombsight, which the whole world knew about. We young pilots didn't buy that stuff. Flying a fixed-course bombing run, an antiaircraft sitting duck, with its two .30-caliber machine guns a puny defense against fighters swarming in like bees? Nor did we think a fixed flight at ten thousand or fifteen thousand feet over a ship on a zigzag evasion course was likely to score hits even if the bombs we had were big enough, which they weren't.

Dive-bombing, we said, was the way to sink a ship. So, though told we'd be just scouts in a war, our scout planes practiced low-grade, 60-degree-dive-angle imitations of dive-bombing, with a periscope-type sight to steer by. Yet we knew, as German JU-87 *Stukas* had shown in the Spanish War, that the best attack was straight down at terminal velocity. But the *g* forces generated trying to pull out of that dive with only landing flaps would rip the plane apart. The plane needed oversize "dive" flaps. In late 1934, BM-1s, our first true dive-bomber, was deployed on the *Lex*. But it was a poor bird, so few were built. Then from 1938 to 1939, we began to receive the Douglas SBD scout/dive-bomber, which became our dive-bombing workhorse in World War II. By then, we "Brown-Shoe" flyers were lepers to many "Black-Shoe" ship types, quoting all the time Admiral Sims who'd said that in the next war, whoever controls the air over it will control the sea. The "Jutland" virus hurt our submariners too. In Panama, they had only ancient "S" boats designed for coastal patrol. The first ocean-going fleet submarines were only then being built, and submariners didn't begin to get hazardous-duty pay, as we were, until World War II began.

By the mid-1930s, world events argued that our claim might soon be tested. We read that in letters Frank Tinker, academy class of 1933, sent us from Spain. He was one of the Americans who volunteered to fight for the Loyalists in the Spanish Civil War—in his case, flying a Russian *Chato* airplane.[2] In that war, Hitler was testing his new combat aircraft. And in May 1936, Benito Mussolini's troops had taken Addis Ababa, adding Ethiopia to Eritrea and Italy's portion of Somaliland, and renaming it all Italian East Africa. Ironically, France had endorsed his invasion seventeen months earlier, hoping to lure him into an alliance against Hitler. Politics aside, the pertinent fact to us was that he'd beaten with airpower a maybe-better army that had no air force.

In Asia, Japan was beating Chinese Communists and Chiang Kai-shek's Nationalists alike, Chiang's one solid counterpunch being a tiny band of American fighter pilots, the "Flying Tigers." Japan also now had a huge navy, including many aircraft carriers. And we saw Japanese and Germans in Panama mapping the Canal's defenses. But our politicians back home were railing instead, at "Merchants of Death," American bankers and companies who'd "started and prolonged" World War I "to get rich." True or not, to us navy folks in Panama, it was chewing on old news. We wanted them to mobilize for what would be, unlike 1918, a truly worldwide war. The *Yorktown* and *Enterprise* keels had been laid in 1934, and those two carriers were due to be launched in April and October 1936.[3] But in no year since I enlisted had we built as many ships as allowed by treaty, and the navy's 1936 roster listed only 106,000 people. Our 100,000-man army was poor too, troops on maneuvers carrying "rifles" carved from tree limbs. We were a weak military power. Unfortunately, most Americans believed we were a first-class one.

5

The Mid-1930s: Getting on the Step

But in Coco Solo, I hadn't time to fuss over policy. My nail-biter question every day was, "Will my PM-2 fly today?" It was marginally underpowered for its twenty-five-ton full-loaded weight. To get "on the step," up to takeoff speed, we needed at least ten knots head wind, rare in the rainy season when I arrived. Often, I just churned up Colon harbor, trying and failing to get airborne. And if the sea was high, pushing swells past our breakwater into the harbor, it tossed the PM-2's nose up and down like a ping-pong ball. A pilot had to know a lot of seamanship to fly a PM-2.

In October 1935, as the rainy season waned, making it easier to get "on the step," I received a jarring letter from my mother. My father had fallen off a ladder and broken his neck. He was recovering, she said, but could I possibly come home? For six years American companies had been strapped for capital, and had to compete with FDR's "economic recovery" program for what capital there was. Calls for engineers had been rare. Now, he was bedridden. I could only guess at the grief and financial threats they were facing.

To get there, Jim Baker and I offered to fly two old PM-2s to Norfolk, where we were to pick up two new P2Y-3s in return. Then, in Norfolk, I borrowed an open-cockpit F4B-2 biplane from Ziggy Sprague in base operations, and flew home. The reunion raised my father's morale a lot, I think. And to my parents' delight, I was able to visit them a few more times because Norfolk didn't release the P2Ys to us until almost a month after we'd been told to come get them. Worry over Lili made me fly nonstop, eighteen hours, back to Panama. The P2Y could cruise at least twenty knots faster than the PM-2, and had power to take off easily, full-loaded, even on a still day. We did fly in some rough weather, adding to my instrument-flying time. In Coco Solo, I found Lili a "happy camper," as they say. She and Shelley had stayed with and were cared for by the Parishes while I was gone. That was the start of a lifelong friendship.

Immediately, I was pulled back into our flying program, a busy one in spite of the navy's skimpy budget. A long flight in the fall was to the east, in the Caribbean, and in the spring was to the west, in the Pacific, both times charting the Central and South American coasts and offshore islands. As we did, we wrote terrain and shore-facility intelligence reports, for "future national security needs," it was said. The surveys were an "advanced base exercise," so-called since our operating base was a seaplane tender anchored in some remote harbor. On the fall swing the *Swan* was our base and the flight was to Trinidad, down to South America's coast, up to St. Thomas, Barbados, Puerto Rico, Gitmo, and over the Mayan ruins of Tulum on cliffs at Cozumel, Mexico, which itself was a few huts in 1935, not the big resort it is now.

Back in Panama, I was out on a Caribbean patrol on 17 December 1935 when I got a radio message from Rags. Lili had been rushed to Colon hospital. I reversed course at top speed and, after a three-hour flight, met our second daughter, Leila Marion. Cute as a doll can be, she was a citizen of both the United States and Panama, according to their laws. Except as a topic to talk about at a tea party, what good that might do her we could not imagine.

As 1936 arrived in Panama, so did the annual Fleet Problem. My squadron was in the Canal defense force, told to find the enemy before he found us, but the result was the same as before. "Enemy" carrier aircraft, striking at dawn, surprised the army-navy defenders and clobbered us. I stayed shut-up about our not being well deployed for early warning work.

After that, our westward "advanced base exercise" with the *Lapwing* as our seaplane tender went first to Nicaragua's Gulf of Fonseca, then the tiny fishing village of Acapulco, then to the Galapagos with its array of "flora and fauna," reportedly the inspiration for Charles Darwin's theory of evolution. There, the *Lapwing* anchored in Post Office Bay, so named because a barrel on its beach was where tuna fishermen, mostly from San Diego, put their homebound mail and from which the next guy heading home, now and then, was expected to take the stack with him. During our visit, only two people lived on the islands, Dore Strach from Germany and a dentist, a Doctor Ritter, who wore a set of self-made stainless steel false teeth. There'd been a second man, we'd heard, who used to fight the good doctor, literally, for the right to share Dore's bed. Dore fell for the Doc and, it was said, the two of them broke the triangle by murdering the other guy. As we left, Ecuadorian authorities were coming to investigate. We were to fly next to Guayaquil, but were told a black plague was ravaging the city. So we scooted back to Coco Solo.

In addition to regular duties on that 1935–1937 tour, I always volunteered to check out all our land planes, transports to fighters—made a pest of myself over it, I guess. I also was the first one asked to fly mail runs, maybe since I'd acquired an image as a good navigator. As a result, I saw many places fascinating maybe only to me. Near Coco Solo, for example, was Portobelo, the "great emporium of South American trade" in the seventeenth and eighteenth centuries, where famed explorer and privateer Sir Francis Drake died and was committed to the deep in 1596.[1] (In those years, a pirate at sea was a "privateer" back home if his monarch endorsed his raids.) It's also where the legend of the "black Christ" inspired a moving religious feast held every year during Lent.

Farther down, below Cape Nombre de Dios, is a cluster of 365 islands, the San Blas, some so tiny they can hold only one palm tree, many crammed eave-to-eave with thatched huts. Cuna Indians were, and still are, the rulers of this autonomous province. Short in stature, the men worked in Panama but did not let their women off the San Blas. If one did leave, she was banished forever. The Cuna claimed it made them the only "pure-blooded" tribe around. Over centuries, they had amassed a treasure trove—coins from the 1364–1380 reign of French king Charles V, for example—but trying to buy any of it at a bargain price was futile. They knew very well the value of what they had.

Those visits sent me to the library to research Panama's history. There I found in a very old journal the tale of a team a Spanish king had sent to Ecuador to measure a mile's length at the equator so he would know how big the Earth was. They set up a triangle on a plain near Quito, then took sightings of the sun at its noontime altitude to find the length of a minute of latitude. Their result was remarkably close to today's nautical mile. In the end, the British seized the expedition, hauled it to England, forced it to write up its findings, then tried to take credit for its work. I wrote an article about it all, titled, predictably, "How Long Is a Mile?"

Especially when Lili and the kids were in Pensacola, those flights and mail runs sent me to Bogotá, Caracas, Quito, Lima, La Paz, Costa Rica, Mexico, and down El Camino Real, loosely translated, "The King's Highway." Obscured by urban sprawl today, this once ran with few breaks from the northernmost Spanish settlements in California to their southernmost in Chili. In 1937, at the end of it all, a Panamanian travel-agency boss said I was qualified to be an official tour guide. I chilled that idea when told I'd have to mix with the customers and let someone else fly the airplane.

My favorite tropical city, I decided, was San José, Costa Rica. It had a short air strip but its people were friendly, delighting us and themselves with a promenade in the town square at five every afternoon, the "loveliest girls in Latin America" circling one way, men the opposite way.

For panorama, my best flights were up the Amazon, to the top of the Andes mountains and the 12,500-foot-high Lake Titicaca on the Peru-Bolivia border; and once, with our commercial attaché to Venezuela on board, up the Orinoco river where he pointed out its spectacular ten-thousand-foot waterfall as we hunted the back country for but didn't find a lost commercial aircraft. These were lonely flights with rest stops on the rivers and lots of floating junk to avoid during tense takeoffs. But a seaplane was the only practical and relatively safe way to go in those days. In desolate jungles, if a land-based plane went down or its pilot bailed out, no one would ever find him. For most places I flew over, a seaplane was the best, sometimes the only, way to get there.

On many trips, I had to fly instruments-only and land on strange short runways tucked up by a mountain, boosting confidence in my ability to do both. In the States, we had a crude guide, the Adcock Range, a continental four-sector grid of towers. Its radio signal, steady if the pilot was on course, became a broken Morse code "N" if he strayed right, an "A" if to the left. Over a tower, he was in the "cone of silence," no signal at all. South America had a Range of sorts, inaccurate, with only a two-hundred-mile range. On the sight-seeing and mail runs, I mostly was on my own.

That changed my reading habits, too. I'd been studying mostly physics, books like Root's tome, *Mathematics of Engineering*. As physics folks said, "If you can't measure it, you don't know what it is." I began to read up on celestial navigation, back then certainly the best way to fly over oceans. Yet few pilots studied it. Most flew on dead reckoning, dropping float lights to gauge wind drift, a poor way to try finding a tiny island far to sea. The wise way was to go to where the island should be and, if it wasn't there, begin a gradually enlarged square search until it was found. To just fly on if the island was missed was an almost dead-sure course to oblivion.[2]

Ships used a sextant that, to work, needed a stable platform. Stable, airplanes aren't, so aloft we used an octant. It measures a 360-degree circle in $1/8$th increments and, to imitate a stable horizon, has on it a bubble floating in a tube like that in a carpenter's level. Daytime sightings are done at the sun's prime vertical. At night, the best sighting in the northern hemisphere is Polaris,[3] the North Star; and the Southern Cross in the southern hemisphere; then sighting on a star at a 90-degree angle from it. Aircraft vibration shakes the octant, so one reading likely will be off. But by holding on target ten seconds and taking the mean of several readings with an averaging device, a skilled navigator can fix a position within five miles of true. A pilot-navigator also had to learn astronomy, celestial mathematics, and the best stars to use. I became good enough to get published some technical papers I wrote on the subject. More importantly, knowing how to use the Southern Cross saved my life a few times during World War II in the South Pacific.

Meantime, the Navy PG (Postgraduate School) asked for applicants during my Panama tour. Sources like the Institute of Aeronautical Sciences, which I'd joined when it was created in 1932, said technology was expanding rapidly on a broad spectrum. To grab hold of some of it, I wanted a masters-level education in aeronautical engineering. With my petition to the school went excellent endorsements from all my previous commanding officers. In January 1936, I learned I'd been rejected but people below me in class rank had been accepted. It was a bitter, bitter blow. I was having vastly more trouble getting my own career "on the step" than I had with that old PM-2 in Panama in the rainy season.

The turndown did carry one reward. With Rags accepted at the PG, the base now had a cozy beachfront home for us, soft trade winds fanning it all day long, worlds better than dirty, raucous downtown Colon. While painters and such refurbished the place, Lili and the girls sailed for Florida to vacation in a cottage her parents owned on Santa Rosa Sound, east of Pensacola, and I moved into base bachelors quarters. Our idyllic new home didn't ease my anger over the PG turndown, however. A PG degree, I was told, was a ticket to flag rank. I didn't care about that. I did care about being an aeronautical engineer, a title rarely earned without a master's.

I was told my lack of a high school diploma probably hurt me. I snarled back, "Standing fifty-first in my academy class meant nothing?" Lt. Cdr. Abie Rule, a friend of Lili's family, said it was a blessing in disguise. His assessment turned out right. Long off-hours study, to "show them," was to help me often in my career. And as history would record, I was selected for admiral long before any of those who went to PG that year. In fact, my friend, Rags Parish, never did make admiral, a bitter disappointment that I thought, at the time, was to be mine.

I didn't learn all that until much later, of course. But curiosity sometimes leads me to wonder what I'd have done if I'd been turned down in today's era of "political correctness" when the knee-jerk reaction to any imagined slight seems to be a lawsuit? If that had been the aura in 1936, I'd have handled the legal work myself. Back then, all junior officers had to take a course known as Navy Courts and Boards. It covered rules of evidence, qualifying witnesses, contract law, legal precedents, how to conduct Courts Martial, and much more. Material from that course plus international law, the rules of the road for ships at sea, made up one of the longer tests given for promotion to lieutenant.

After taking the course, I got add-on duty as a Fifteenth Naval District judge advocate. We were ruled back then by Articles for the Regulation of the Navy that we called "Rocks and Shoals." Once a month, I had to read one of those to an assembly of officers and crews, but my main job was prosecution, mostly of officers or men charged with stealing or embezzling at the PX (Post Exchange). Like law work, PX bookkeeping also was add-on duty, but we were held accountable. (When Congress approved a Uniform Code of Military Justice for all services in 1953, the navy set up a full-time legal corps and, of course, specialists now run the exchange stores.)

As my six-month judge-advocate duty ended, a law specialist, a Lieutenant Wilkins, took over. Then suddenly I became a defense lawyer after a *Panama American* newspaper headline yelled, "Murder Committed!" Agua de Carbonella had been killed in one of the small open-air cells where Panama's prostitutes lived and did their business. Her head had been bashed in, body covered by a piece of canvas, her room ransacked. A witness of sorts said she'd seen a tall black sailor crawl out the cell's rear window and run away. Answering a Panamanian demand, a Captain McWhorter ordered a lineup of our mess attendants and all the black sailors on our fleet supply ships (called collectively "the train"), in port at the time.

In the second such lineup, the witness accused our William Eaton, so it became a navy case and I was told to defend him. His insistent answer to every one of my many questions was, "I'm innocent." His repeated alibi was that she'd demanded seventy-five cents to have sex with her and when he said he had only fifty, she'd grabbed him and begun yelling for the police. So he hit her with her flatiron and ran. Next day, I learned he later had asked to see a marine officer, then gave him a signed confession to the murder.

The seven-officer court convened, Wilkins presiding, with full-dress uniforms, swords, and white gloves. To first-degree murder, we pleaded not guilty. My problem was the confession. Navy law at the time said an accused must be tried for a lesser crime—in this case second-degree murder or manslaughter—if a confession to a senior officer is admitted in evidence. The court allowed it in. I did hem and haw on "lack of corroborating evidence" but that was just Perry Mason talking. Eaton got three years in navy prison for manslaughter. As he was shipped off, he told me that the whore had given him gonorrhea and, in a rage, he had gone to her cell to get even. He had not robbed the crib nor covered the body with a canvas. Did the witness against him do that? Probably.

My final case was a lieutenant charged with embezzling a ships-service account. Doing these accounts was another add-on job for officers, few of whom knew accounting in a loose setup where fraud could go undetected for months but, once found, was easy to prove. Off my client went to prison and with him, my law career. (Eventually such part-time duty plus law courses taken did result in my receiving a Doctor of Laws degree in 1967 from Providence College in Rhode Island.)

During my legal calisthenics, Lili had given me endless excuses for staying in Florida. I decided to go fetch her and the girls in person. I had leave coming anyway, but was told I had to use military transportation. Luckily, the destroyer *Schenck,* a plane guard for us in my *Langley* days, was anchored in Gatun Lake on a midshipmen's cruise, most of her crew Reserve Officers' Training Corp (ROTC) students, mainly from Georgia Tech. Quickly packed, I got a motor launch to haul me out to her. As I came aboard, her skipper, Lt. George Bahm, ordered me to take the watch and set sail for Charleston, South Carolina. I was stunned. I'd never conned a destroyer. The snap in his voice said he expected this young "flyboy" to say I didn't know how to do that, but the job seemed to me easier than a carrier landing. "Very well," I said, which jarred him. "When do we sail?"

"Right now!" he said. So I got her under way, took her out to sea, and set a course for the Windward Passage, then was relieved and went below to sleep. A few hours later, we began to pitch and roll, bobbing like a cork, pounded by increasingly high seas, wind over sixty knots, nearly the sea state of a small hurricane. Urgently, Bahm summoned me topside. The entire crew is seasick, he said, all his ROTC kids, even the Filipino mess attendants, groaning and retching. Help, he said. So, until the weather eased off two days later, Bahm and I traded off steering the ship and making cold sandwiches. One thing it taught me: avoid duty on a destroyer. The rest of the sail went smoothly. As we neared Charleston, a pool was organized to see who came closest to guessing when we'd pass under the Cooper River bridge. I had the watch then. As a result, to everyone's surprise but mine, I won the pool. Bahm gave me a fine fitness report, one I've loved reading. The Hyers's place on the Sound was clean soft beaches, fine fishing, the children were fun-filled imps, and my leave was a month-long picnic.

Back in Panama in mid-1936, Adm. Ernest J. King and Capt. J. S. McCain arrived. When Moffett died in 1933, King had taken over BuAer. Now he was our boss and would command all AIRBATFOR by 1940. Abrasive as sandpaper, "He shaves with a blowtorch," FDR would say. McCain was to be base commander. Each had done the Pensacola "short course," giving them maybe two dozen pilot hours. (After War II, it was a practice often done for non-aviator admirals. I know of one, Arleigh Burke, who refused, saying gold wings earned that way would be sailing under false colors.)

How each handled that said a lot about them. His first Sunday there, King told me to fly him to the army's Albrook airfield where his son-in-law, Second Lt. (eventually General) Freddie Smith, was stationed.[4] King had a Douglas DC-2 transport with the new retractable landing gear and I was reading the book on it when he showed up. When I told him I hadn't flown a DC-2 and wanted a minute more to study, he said, "Son, are you a naval aviator?"

"Yes, sir."

"Then get in the airplane. We're going."

So off we went—King; his aide, Lt. Sam Dunlap (class of 1925); my crew chief; and myself. Once I had us airborne, King took over. At Albrook, he had us coming in too high and fast, suddenly telling me, "You land it!" So I pulled us up and around to do a proper final while he yelled, "Damn it! When I bring the plane in and tell you to land it, you land it! Right now!"

But later, Dunlap whispered to me, "If you ever let him land that thing, you're crazy."

McCain was another kind of guy altogether. Like Mommie Floss, he loved to watch the ponies. So Sundays he had me fly him to the Juan Franco racetrack in Panama City. But he told me, if he got in my way, I was to say so, "And under no circumstances," he said, "are you to let me fly this DC-2." I grew very fond of him, one of the nicest bosses I had in my entire career.[5]

As 1936 ended, I was itching to do the lieutenant's tests. More pay was one reason, of course. (Having lived it, I, like most veterans, think it an outrage that thousands of married sailors qualify for welfare while Congress, the cause of that, votes itself six-figure salaries, five-figure pensions. Congress was sneaky-sly in 1933 too, when it set the Social Security retirement age one year older than average life expectancy was.) A second reason was that the navy had ninety-nine hundred officers in 1936, only five hundred more than in 1933. Even as a lieutenant, I'd have a bunch of people ten to fifteen years ahead of me in seniority who had to move up or out before I could. The idea of having to excel just to stay in grade until age forty-five was not attractive.

A better way, of course, would be to grow a bigger navy, as our likely enemy was doing. When the 1935 Naval Conference collapsed, FDR had inched up our shipbuilding rate, but Japan was on overdrive, making more submarines and more combat, troop, and supply ships than we were. Hitler had put troops in the Rhineland in 1936 and was raising a giant army equipped with more and better hardware than ours had. And as we, Britain, and France diddled in diplomacy, he signed an Axis Pact with Italy and an Anti-Comintern Pact with Japan.

Hitler's raging oratory, "spellbinding," his fans said; and his audience, one hundred thousand at a time, Hitler Youth and the rest, shouting "Sieg Heil! Sieg Heil!"—their equivalent to "Hail Caesar!"—in perfect military cadence, gave me the shivers. By early 1937, their political weapons in place, all three dictatorships were clearly after an aggressive expansion of empire. We were headed for trouble and neither our nation nor our navy was ready for it. The navy was mired in "battleship" dogma, our weapons inferior to our potential enemies' in nearly all categories.

As the war clouds rose, I got new orders. First, I was to go to New York for thirty days leave. My girls and I left Panama at navy expense on a Grace Lines ship, a soft voyage until our last day out, 6 May 1937. We were off New Jersey when, like a gray-white ghost, the huge *Hindenburg* dirigible swept silently overhead. Then at dusk, we saw a huge, billowing black cloud rise up in the west. Later in New York, we learned that a sudden explosion of fire had destroyed Germany's "Luxury Ocean Liner of the Air" as it neared its Lakehurst mooring tower. We nearly had seen firsthand the disaster that killed all public interest in using "balloons" to travel. It didn't stir us up much though. We were just too delighted to be back in the States.

My orders also said, "Report 1 June 1937 to the 4th Naval District Commandant" in Philadelphia for my lieutenant-promotion exams, then join the *Philadelphia*. She was a new ninety-seven-hundred-ton, square-tailed *Brooklyn*-class cruiser then being built by the navy, using WPA money, the same emergency program that in 1933 had paid for the *Ranger*. She had a 14,500-mile range at 15 knots cruise speed, a 32.5-knot top speed. Her 6-inch guns, three in each of five turrets, used the new "case," or "semi-fixed" as it also was called, ammunition. I was to be second in command of her flying unit to Lt. Ed Ecklemeyer, class of 1927. He had just gotten a Navy Post Graduate School master's degree in ordnance but said he hadn't flown much since earning his wings. The lead in all our flying work would be up to me, he said.

The aircraft setup was new, the best I'd seen, two catapults instead of one and hangar space for six, not just three or four, aircraft. Those were new too, single-float, single-engine seaplanes convertible to land use. Clones of the Curtiss SOC-1, they were designated SON-1s since our Philadelphia Naval Aircraft Factory (NAF) had built them. (For several reasons, Congress had enacted a law requiring that the NAF build at least some, up to 10 percent, of any new aircraft the navy bought.) We also were to use a new aircraft-recovery tactic. Before, we'd taxied up to the ship, dead in the water, to be lifted aboard by a crane. Now, to make the ship more difficult for submarines to hit, we'd taxi into a net cast astern and be lifted up in that with the ship under way at twenty knots. It meant landing close-in. Otherwise, at taxi speed, we'd never catch up to the ship. With ship and tactics both new, it would be a very busy shakedown cruise.

Lili and I rented a house in the suburbs and I joined a car pool, "driving squad," he called it, run by AEDO (Aeronautical Engineering Duty Only) Cdr. Lloyd Harrison, head of NAF production, then took my promotion exams. As I'd had Navy War College courses in strategy and tactics and in international law, those tests were easy. The others were in ordnance and gunnery, engineering, communications, and navigation. We had to do the navigation for a submarine on a twenty-four-hour cruise, crossing the equator at the international date line where losing track of the date or Greenwich mean time makes a mess; answer questions on turbines, storms and weather, and ship handling; sketch and describe in detail how a power amplifier works; explain how to light off a boiler; and much more. In all, the tests—tough and thorough—took a week.

Helped by eleven years of test-taking, I was sure I'd passed but bureaucracy must be served. The papers had to be sent to BuPers (Bureau of Personnel) in Washington, D.C. Finally on 30 June 1937, I was a lieutenant, jumping my pay to $660 a month plus living allowances. We got our SON-1 seaplanes then, too. Our ship was to go into service quickly for a reason not explained, so her shakedown cruise was just to Cuba with a stop in Bermuda. Our skipper was Capt. Jules James, another guy who thought aviators were invented by Satan. Maybe that's why, until we reached Gitmo, all we did was exercise our catapult gear and fantail-to-hangar deck elevator.

Juan Batista, the army sergeant who had just seized dictatorial control of Cuba, visited the ship in Havana, and I chatted briefly with him. A nicer, more hospitable guy we agreed we'd never met, though history says our glowing assessment wasn't very perceptive. FDR joined us briefly, too, and told James to sail to a specific spot where he wanted to fish near Havana. Aviators being idle crew on that ship, I was told to take FDR out in a whaleboat—as James ran the ship aground on a coral reef. Ordinarily, he'd have received the kind of nasty reprimand that can ruin chances of becoming a rear admiral. This time, nothing at all was said because it was the commander in chief who had ordered the *Philadelphia* in there.

Finally, off Gitmo after FDR had left, Ecklemeyer was told to do our first aircraft launch. Once set in the catapult cart, he started his engine, all routine at that point. But as the engine idled, we were shocked. His plane began to move slowly down the track. Weirdly, as it toppled out of the cart into the sea, the crew fired the catapult. That broke the link holding the cart in place and it went over the side, too. Quickly ordered to "Report!" I as quickly discovered the farce was caused by a dumb human error. The crew hadn't set the holdback lever designed to keep the plane in the cart as its engine runs up to speed. James erupted like a volcano, hurling flak at the catapult crew but mostly at Ecklemeyer. I'm sure that got Ecklemeyer a nasty fitness report, which may be why he didn't fly in World War II and never made admiral.

The rest of the cruise was humdrum for us, no aircraft net-recovery runs while under way at twenty knots. Everyone seemed loathe to try it. I felt we'd better practice it since we'd have to do it in war, but I was odd man out. And junior officers do not tell captains how to run the ship. Lili, who was "due" again, had gone home when our cruise started. The ship had just reached Philadelphia, in time for Christmas leave, when I was jolted by a wire from Pensacola. Our first son, Charles Brian, named after my father, had died only three days after birth. A Dr. Webb mumbled vaguely about his being born with breathing problems and dying of pneumonia. I thought then, still think, he'd have survived in a big (for that era) modern hospital in Philadelphia or Boston. Lili said he'd been "so cute." He should not have died that young, the first bitter tragedy to hit our family. We spent Christmas in Pensacola, Lili's folks helping to soften our grief.

Mainly from hauling lots of fuel and asphalt for Standard Oil people in Kentucky, Daddy Pete was prospering; but the Haywards were not. We lived paycheck to paycheck, and I sensed my father was near broke. He'd helped the Wright Brothers win a lawsuit over their patent for lateral-control ailerons, but had not figured in the payoff. He had some real estate and some IOUs from a company. But in 1928—as 1929 revealed, maybe the worst time to do it—he built a second home in Great Neck, New York. In 1937, he told me he'd sell except that my mother and unmarried twenty-year-old sister Margie, living at home, were dead-set opposed. I couldn't blame them. It was a lovely home with Manhasset Bay its backyard. His message told me the Depression had a choke hold on him. (Unemployment was fifteen million strong that year, and history now says the Depression would last another year.) It was a gloomy trip Lili and I took that year back to Philadelphia.

In January 1938 the *Philadelphia*'s crew began fixing what her shakedown cruise had said she needed to have fixed. If there's a more dismal job than being on a ship tied up in a navy yard, I'm not aware of it. Then, the end of the third week in February, at 2:00 A.M., the officer of the deck woke me. I had a phone call, he said, from "a Mr. Browning," my parent's neighbor and family friend. His shocking message was that after an evening playing chess with him, my father had collapsed, walking home, and died of a heart attack. As fast as I could dial the phone, I reported the emergency to the proper people, then raced for Great Neck to be with my mother.

When I arrived, she told me that as they drove home a month ago from the city, my father had hit one of the pillars holding the local elevated train's tracks up over the street. She was sure he'd been having a stroke at the time. Today, if that happens, it's called a warning, tests are done, and medicines, operations, and diets prescribed. In 1938, a survival with no apparent aftermath meant simply going back to work. His death was a sad, hard loss. In ways not present when I was growing up, we'd become quite close, this bright man and excellent father of mine. Now the mutual warmth we had begun to build was ended. Losing both father and my son, his namesake, had turned 1937–1938 into the rottenest time in my life. His burial was in the family plot at Woodlawn where his father and most of the rest of our family lies.

And a look at her finances told me mother had to sell the Grace Avenue house in Great Neck, the Station Road house on Long Island with its big (for 1938) $23,000 mortgage, and all her other property, too. She and Margie agreed to rent an apartment in Great Neck and I set up for her a $50-a-month allotment from my pay. It meant a tough go, but Lili's smiling willingness to carry that load is why taking care of her, my first priority, shoved me into another hard look at leaving the navy to shift over to "big-bucks" private industry.

Then BuPers gave us a break. In August 1938, they said, I'd be named senior aviator on another *Brooklyn*-class ship, the *Phoenix*, being built by the New York (now Cramp) Shipbuilding Yard in Camden, New Jersey. She'd anchor in Philadelphia at first, they said, but when she joined the fleet, I'd stay at the Naval Aircraft Factory as assistant chief engineer for instruments. In effect, BuPers was giving me most of four years to help my mother, tinker with all kinds of airplanes, do my graduate studies, and be with my family every night. It was unique: BuPers anchoring me in one place for two tours in a row, and telling me well in advance. I'm sure it was because they knew our problems at home. It was Lt. (jg) "Chick" Hayward (wasn't it?) who first said, in Panama, "The Bureau of Personnel is all heart."

Our navy yard had left junk all over the *Philadelphia* when we got her. Not so the *Phoenix*. When the Camden folks delivered her in mid-1938, she was as clean as a just-waxed new Cadillac right from the dealer's showroom. At ten thousand tons, she was a bit heavier than the *Philadelphia* but had the same 32.5-knot top speed. She could carry six aircraft with all the necessary spares on her hangar deck but what they gave us was the usual four. While she was readied for her shakedown cruise, we used the NAF field. And once in those thumb-twiddling weeks, I got to paint another "E" on my aircraft after a shooting contest at a towed target off Forked River, New Jersey. It surprised me, truth be told, since I'd not fired my guns for weeks.

The *Phoenix* shakedown was to be long, from December to late January. Since I'd not be home for Christmas, Lili was going to Pensacola. Her sister drove up to get them, arriving just before the Army-Navy football game. Winter came early that year, dropping a heavy snow on the stadium. But Navy won, inspiring loud, liquid celebrations at the Bellevue Stratford hotel, a traditional post-game watering hole back then. Next day, Lili and company left, planning a stop at Parris Island, South Carolina, to see my brother, Dick, a junior-level officer at the marine corps recruiting base there. A day later, Monday, the *Phoenix* left for Port of Spain, Trinidad's capitol. It was a fast trip down, aided by fairly calm seas. On the way, my unit did some net recoveries, not perfectly but acceptable. Trinidad was more play than work, soft blue tropical skies, scenery a rainbow of color, a unique native culture, dozens of grog shops. We did do a little serious scout-flying, one day watching a German training ship, an old cruiser, fire its guns. Our Washington folks were startled by that report. I was amazed that they were surprised.

In 1936, I'd learned that Germans ran most South American airlines with mostly German pilots. Now we saw that Germany had political and economic power in much of the rest of South America, too, especially in Argentina. But their ship off Venezuela was small compared to the fires Hitler had lit. In March 1938, his army had taken Austria. In September in Munich, he and Mussolini had met with Prime Minister Neville Chamberlain of Britain and Premier Edouard Daladier of France. With Czechoslovakia not invited, Hitler demanded and the "peacemakers" gave him its Sudeten region, nullifying a French mutual-defense pact with the Czechs. They also promised Poland and Hungary a piece of Czech territory at some vaguely defined future date.

To justify carving up Czechoslovakia, the only democratically run nation in central Europe, Chamberlain claimed it would assure "Peace in our time." Munich had been his third trip to Germany, trying to rein in Hitler's greed. That and not "talking tough," historians now say, made him appear weak and Hitler strong. Mussolini followed Hitler's success with a demand, in early 1939, that France return "Italian lands" seized by Napoleon 150 years earlier. Except for a small army mobilization, Daladier's reply was just to visit Corsica to assure it that would not happen. And, as a warning, the Brits and French signed with Poland a pact saying an attack on it meant war with them. Hitler was not impressed. He knew he had them all outmanned and outgunned.

While that began to evolve, we on the *Phoenix* sailed to Santos, a coffee-shipping center and port city for São Paulo, Brazil's "New York City." By then, I'd been told part of our job on this cruise was to "Show the Flag," to offset in South America "the German and Italian predominance," whatever that meant. This was a ploy first used when President Teddy Roosevelt (1901–1908) sent a navy "White Fleet" on a world cruise. To be on time for those diplomatic dates, we didn't fly much while under way, but I did find out that our skipper, Capt. J. W. Rankine (pronounced "Rank-in"), six-feet-one or so, brown hair and eyes, was a fine ship's captain and even finer gentleman.

He also was unique among 1930s ship's captains. He thought aviators had better judgment than the average line officer, especially in an emergency. For that reason, he said, he often made me senior watch officer, usually when we faced a difficult harbor entry. Doing that told me, my pilots, and our aircraft mechanics that we were an integral part of his team. At least, that's what we thought. It boosted our morale sky-high, a major reason, I think, why we won the Moffett trophy in 1940 as the "best and safest" ship-based flying unit in the navy.

The Japanese having shot up his first B-24D Liberator patrol-bomber, nicknamed "Fats," during the first half of his 1943–1944 combat tour in the South Pacific, VB-106 squadron commander Chick Hayward inspects the artwork on its replacement, *Chick's Chick*. U.S. Naval Historical Center, Washington, D.C.

Naval Historical Society photo files say, without noting the date, that this is the first flight of a P2V, the only aircraft the navy had at the time big enough to carry an atomic bomb, yet small enough to operate off an aircraft carrier—here, the *Midway*. (Inset: Hayward at the controls.) But Hayward's log says he did that first flight off the *Midway*-class *Coral Sea* on 7 March 1949. U.S. Naval Historical Center, Washington, D.C.

In November 1949, Commander Hayward, far left, and navy escort officer William Rowberger, bracket "the arch enemies of naval aviation," left to right, Secretary of the Air Force Stuart Symington, Secretary of Defense Louis Johnson, and General of the Army Omar Bradley, chairman of the Joint Chiefs of Staff. After the photo was taken, Hayward flew them in a P2V from the *Midway* carrier to Washington, D.C. What Hayward said to Johnson just before takeoff prompted Chief of Naval Operations Forrest Sherman (not pictured) to snap at Chick, "Don't I have enough trouble with the secretary of defense without you getting into the act?" Naval War College Historical Collection, Newport, R.I.

This is the first "family" photo of Hayward, skipper of the escort carrier *Point Cruz* in 1953, with "George Cruz Ascom," the name given to a part-Caucasian infant found in a trash can at an Army Storage Command (ASCOM) sick-bay tent near Inchon, South Korea, on 28 July 1953. CBS-TV made a Christmas-season movie of the event. First shown in December 1997, it has been rerun at least twice since. U.S. Naval Historical Center, Washington, D.C.

After his retirement in 1968, Hayward took his wife, Lili, on a month-long trip to Europe to visit long-time friends. While there, he also arranged for the two of them an audience with Pope Paul VI, shown here greeting Lili. Naval War College Historical Collection, Newport, R.I.

On the bridge of the *Enterprise,* the navy's first nuclear-powered carrier, with Secretary of Defense Robert S. McNamara in 1963, Hayward's face reflects his views that "McNamara affected a knowledge of the navy he didn't have," and "He treated us as if he was management and we were labor." U.S. Naval Historical Center, Washington, D.C.

Hayward hosted Princess Grace Kelly of Monaco in July 1962 aboard the carrier *Independence* off Cannes, France. Of the visit—she brought along the Monaco symphony orchestra to entertain his crew—he said succinctly, "She was a doll." Naval War College Historical Collection, Newport, R.I.

Hayward, rear admiral commanding Carrier Division Two in the Mediterranean at the time, is escorted to the bridge by Capt. Walter E. Clarke, commander of the carrier *Franklin Roosevelt,* on 22 March 1963. Once there, Hayward received a certificate attesting to its being his three-hundredth landing on an aircraft carrier. It was noteworthy then, but supersonic jet aircraft have made three hundred landings on a carrier something a pilot can now achieve relatively early in a career. But jets also cut the standard flight time for missions to just two hours, so it is not likely that any future flag officer will ever attain the thirteen-thousand-plus pilot hours Hayward logged in his career. U.S. Naval Historical Center, Washington, D.C.

As senior naval officer on duty at Pearl Harbor on Christmas Day, 1965, Vice Admiral Hayward, commander of U.S. Naval ASW Forces, Pacific, at the time, hosted Cardinal Francis Spellman, who performed a wreath-laying memorial rite at the USS *Arizona* Memorial there. Naval War College Historical Collection, Newport, R.I.

When he retired from the navy in 1968, Naval War College President Hayward, wearing his French Legion of Honor medal, addressed the college's faculty and class at change-of-command ceremonies in front of Luce Hall. Naval War College Historical Collection, Newport, R.I.

Commander Hayward being interviewed on 22 June 1944 in Hawaii by NBC radio war correspondent Morgan Beatty, just after Hayward returned from nine months of combat operations in the South Pacific. U.S. Naval Historical Center, Washington, D.C.

In December 1962, after the *Enterprise* had been his flagship for some time, Rear Admiral Hayward invited his friend and submarine nuclear engine expert, Rear Adm. Hyman Rickover, aboard to inspect her. Naval War College Historical Collection, Newport, R.I.

6

Peace in Our Time?

One of our "best-and-safest" scores was earned when we were three hundred miles out from Rio de Janeiro and Rankine asked me to fly in there to pick up mail and State Department orders our embassy was holding for him. We'd not have a disaster like the one off the *Langley* in 1933 because this captain gave me his course and promised to move at a steady twenty knots until I returned. With time to refuel, I figured a round trip of five hours. I decided to use two aircraft with Lt. Groome Marcus flying the other one. We catapulted off easily enough and joined up, flying west at 160 knots. Two hours later, we saw Corcovado, a 2,310-foot mountain south of Rio, circled, landed, and tied up at the Pan American Airways Yankee Clipper flying-boat pier.

An hour later, longer than they should have taken to pump gas and deliver mail, we taxied out for the long chase after our ship. She was some sixty-five miles from where we'd left her, by then a speck in a trackless ocean. With our navigation gear merely a small plotting board and grease pencil, finding a ship three hundred miles out, moving at twenty knots, would not be easy. And, because of Rio's slow-motion service, it would be sundown when we arrived—if we did. That meant a night recovery, which we'd never before done. The sun was setting when I sighted her, hull down in a fairly heavy sea and high winds. I called in, telling them I'd arrive within thirty minutes.

As we circled, Rankine, bless him, turned his ship 45 degrees, into the wind, so we could land in her wake. If he hadn't, to catch her in a long taxi in heavy seas and at twenty knots might have been impossible. Landing in tight, I was brought aboard quickly on her net's first cast. Her next one reeled in Marcus. This proved to me that the SON-1 was the best rough-water seaplane of its type we had. Rankine praised our work, but he certainly deserved credit, too. He'd swung the ship to help us land close in and let the recovery be done in rough weather. I knew many skippers who'd not have allowed our six-hundred-mile flight in the first place.

Next day, we docked in Santos, São Paulo's port. Both cities were bright and busy. Their official and civilian "elites" treated us to protocols, salutes, and receptions, entertaining us as if we were royalty. This was the first time I'd seen São Paulo, Brazil's financial center, its biggest city and one of South America's most modern. The going-away party our hosts gave was dinner at 2300 hours, lots of booze, laughing, and dancing the samba. At sunup back aboard ship, I flopped into the sack. Shaken awake minutes later, I was led to Rankine's cabin. There, he said he'd offered "a high-ranking Brazilian" a ride in my airplane and the man had accepted. Would I do it?

"Sure," I said, sounding nonchalant, but inside I was awash in worry. Was I up to this after an all-night party? Will this guy do what a backseat man does, hook us onto the hoist as we taxi into the net? At noon, as we set up my plane, we learned we'd launch at 1330 hours, when our guest, Brazil's President Getúlio Dornelles Vargas, we were told, finished lunch with the skipper. (Vargas, who had been elected in 1930, would serve until 1945 and again from 1951 to 1954.) Evidently, the news had spread. Thousands more folks than usually show up to stare at our ship were crowded at pierside by 1330. Had Rankine's English been poorly translated? Maybe. Anyway, Vargas now saw what a catapult launch was to be and it was a taut, trembling man we helped into my plane's back seat. What kept him from backing down was all those Brazilian voters watching. Checklist run, I waved to my launch crew and they fired us off starboard with a roar.

Once aloft, I gave him a long look down at São Paulo, then flew him over the countryside, him grinning wide-eyed at the scenery below, me worried about how to get us back aboard ship. Later, as we swung into the Santos River basin to end the flight, a *Phoenix* signal said she was ready. But Vargas wasn't. My gestures, saying I'd need him to grab the hook, received back a blank stare. I'd have to do it solo. We landed smoothly, heading upstream against the fast-running river's current. On the ship, Bosun's Mate First Class Lightfoot, a smart sailor, was at the crane. To help me, he let its hook down softly like bobbing for fish in still water. I taxied up past it, cut the engine, and let the river push us back under the hook. As it did, I jumped up and snagged the hook, luckily on my first try. We quickly were hoisted aboard, Vargas thanking me, congratulating me, and all that, while the pierside mob cheered, clapped, and yelled.

When I tactfully could, I went to bed, relieved this out-all-night aviator had not botched the drill. Next day, after I'd been summoned again and all parties duly assembled, President Vargas awarded me Brazil's Southern Cross medal. Given the anguish he'd suffered, hanging in there when he really wanted to go hide, he should have pinned the medal on himself, I thought. Still, that "Show the Flag" exercise earned our image a gold star in Santos. Then we went south to the mouth of Argentina's Rio de la Plata and, from there, some 120 miles upriver to Buenos Aires.

I was senior watch officer again for that run and, as we docked, across from us were two new Italian cruisers, the *Duke D'Aosta* and *Duke De Savoia*. Their sleek lines were typical of that era's Italian-designed ships. We saluted Argentina first, then the Italians, the latter duly returned. At our courtesy call, the Italians dropped on us endless brags about "Il Duce" who, they said, would do what the Romans had done two thousand years ago, make the Mediterranean *Mare Nostrum*, "Our Sea." (To do that, Rome had put troops on their galleys and crushed the Carthaginian fleet, ruler of the "Med" since the Phoenician's days.) Other than the baloney, they were a friendly lot who looked as if they might give a good account of themselves in battle.

On the other hand, if their cruisers were crazy enough to start a fight, we had them severely outgunned. They each had six 6-inch guns mounted in twin turrets, but the *Phoenix* had fifteen 6-inch guns whose rate of fire was at least twice theirs. Still, they were a popular bunch. We also heard lots of toasts of "Viva Il Duce." Most folks in town seemed to admire Mussolini as much as they praised Hitler, while we generally got the cool, aloof treatment. But it wasn't all grim. Fact is, we were guests at dinner and dancing often, in places like the Hotel Alvear roof with its spectacular starlit view. Our "Show the Flag" was paying off, said our Argentine emissary, Ambassador Douglas. He was unknown to me but he had clout in the White House. He'd sold FDR on our trip, I learned, and decided when we'd visit where, including in Buenos Aires when we did because he'd known the Italian ships would be there. A widower, he and his pretty daughter, Catherine, treated us to a splendid dinner party with captivating senoritas—armies of them, easing the unhappiness at not being home for Christmas.

My own holiday highlight was a visit specifically to see me by Capt. Marcus Zar, the top man in Argentine naval aviation. Six feet tall, handsome, dark hair and eyes, "a Latin lover" in movie jargon, his heritage was a mix of Spanish and English common in the Argentine. In his midforties, he had been to our flight school at the end of World War I and married a Pensacola girl. Instantly, we were friends, as happens sometimes. He'd come to offer me the use of his Lockheed JO transport. He especially wanted me to see southern Argentina, their wheat and cattle country, he said, and by air was the best way. I snapped up his offer, checked out in his plane the next day, and a day later, escorted by another Argentine naval officer, flew south along the coast. Well down over the pampas, my passenger pointed to an airstrip on an "estancia," their name for their Texas-size ranches, and suggested we land.

I did, happily, because he said he'd arranged an "Asado," the locals' big feast. Its centerpiece was a butchered cow put upside down in a pit with the fire built under its rib cage. As it cooked, we watched, drank champagne, and chatted up the senoritas. They were descended from Spanish, English, and Italian settlers, proving what marvels that can produce. The buffet was roast beef, steaks, several side dishes, more champagne, and more time with the girls. I told my hosts I could learn to love the gaucho's life. I meant it. Well, at least a little bit.

In early January, we ended that memorable visit to Buenos Aires for another, 150 miles to the north in Montevideo, Uruguay. As we neared our pier, which was a striking piece of architecture, thousands of people there roared, "Roosevelt, Si! Hitler, No!" Evidently, Adolf didn't hold all the high cards. It was midsummer and Punta Del Este, Uruguay's tourist mecca, was full. Again, it was parties 'til dawn and friendly folks eager to entertain us. This time, I hauled out clubs always with me on a ship and played golf a few times on a long, neatly manicured course—even shot a two-over-par 74, once.

On the last leg of our cruise, the long nonstop run up to Guantánamo, we tested all our systems every day: aircraft launch and recovery at twenty knots, gunnery drills with my planes aloft using direct radio contact for the first time to give the ship instant reports on where the shells hit. We also were equipped by then to spot at night as well as in daylight. At Gitmo, the heavy firepower Rankine ordered amazed me. I was one of those who'd claimed bag-powder 8-inch guns were better than the case-fired 6-inch guns Admiral Pratt had championed for the *Brooklyn*-class cruisers. (For a way-over-simplified example, "bag powder" is like loading a Civil War musket. The "bullet" is put in the butt end of the barrel, then the powder "bag" is shoved in behind it and ignited to launch the shell. "Case-fired," on the other hand, is like the bullet in a revolver. Powder fires the "slug" at a target but the "cartridge" case remains in the gun.)

The 8-inch, bag-powder gun shoots a bigger warhead farther, but case-type guns can fire faster. At Gitmo, in its first drill, the well-trained crew on the main battery fired fifteen rounds a minute from its 6-inch guns. That said, if all the ship's guns fired at once, they'd rain 225 shells on a target every sixty seconds. (In World War II, our sister ship, the *Boise*, would do exactly that to destroy a Japanese battleship off Guadalcanal.) At that rate, a major problem would be keeping the guns supplied with ammunition. Admiral Pratt had been right and I had been wrong.

An order to come home cut short our Gitmo show. They wanted our post-shakedown work done fast, the orders said, so we could join the *Boise* and *Honolulu* in California, as a new Cruiser Division Nine, Adm. Husband Kimmel commanding. I was not inspired. He was another guy who thought, and I quote him, "Airplanes sitting on a ship's quarterdeck dirty the fantail."

The order meant homeport in Long Beach, not Philadelphia, making us all scramble, changing plans. I suggested to Lili that she ask her sister to help drive her and the girls out there. "Plan to arrive when I do," I said. "We'll rent a house and see what the future brings."

While the ship was in dry dock in Philadelphia, I flew to Pensacola weekends to be with Lili. Weekdays we used Lakehurst field for gunnery work, both "fixed" where a routine is laid out and we just fly through it, and "free" where Blue and Red teams play offense and defense and each team skipper decides how he'll do that. And I earned another big white "E." Maybe rumbles of war explain why this one did not stir up the excitement earlier awards in quieter times had.

After the *Phoenix* put to sea, we showed Rankine that a cross-deck launch off our catapults was a quick, safe way to deploy aircraft, a tiny gain in combat readiness that only he noticed. A week later, the usual layover party in Panama was passed so we'd be on time in Long Beach for Battle Fleet exercises. There, in a night-battle drill, firing at targets in the twenty-thousand-yard range, I was smugly pleased, I admit, to see our guys do exceptionally better than the other ships. By then, we knew Kimmel preached the old "Battle Line" sermon. For aviators, that meant no war games, just a cruise to Portland, up the Columbia River, around Puget Sound. Only once were we told to do a launch-and-recover. Of course, it fired up a race to see who finished first. On Kimmel's flagship, the *Honolulu,* Cdr. Bruce van Voorhis was its flying unit boss but had his guy, Herschel Smith, fly the launch-and-recover. The seas were rough that day and, as Smith landed, a swell flipped him on his back. Kimmel, Van Voorhis, all of them howled in angry dismay. We didn't improve their mood any by winning the race, recovering the *Phoenix* planes in record time with no trouble.

Back in Long Beach, Lili and I rented a house in Belmont Shores, so close to the ocean that big waves rolling in shook the place. We liked it. But in midyear, Hitler and Joe Stalin signed a Non-aggression Pact, "giving" Adolf most of Poland, Joe the rest plus the Baltic states. Then on 1 September 1939, Hitler invaded Poland. Two days later on our radio, we heard King George declare war on Germany, as did France that same day. Within two months, Hitler's blitzkrieg had crushed Poland, Stalin had attacked tiny Finland, and Congress amended our Neutrality Acts so FDR could sell war material to Britain and France. Later that year, Hitler awarded an Iron Cross to Gunter Prien for sneaking a U-boat into a British anchorage to sink her battleship *Royal Oak*. The material loss was small, his audacity a shock to British morale.

Events in Asia were no less ominous. In 1937 from Manchuria, the staging area it had seized years earlier, Japan had invaded China, held most of its coastline by 1938, and was moving inland, fighting Chinese Nationalists and a rebel Chinese Communist army, as well. (Japan's Gen. Hideki Tojo had a messianic hatred of communists—no surprise, really, since history says one thing a dictator can't stand is competition from another dictatorship.) Japan called it "creating a Co-prosperity Sphere" of influence but the only prosperity it cared about was its own.

Still, we kept selling Japan oil and scrap iron, our politician leaders vowing we'd not get involved. Cruiser Division Nine was not that confident. Our January 1940 Fleet Problem was to be in Hawaii. Since I was to transfer to the NAF (Naval Aircraft Factory) after that, Lili and the girls left for Pensacola as the *Phoenix* left for Hawaii. Pearl Harbor was the focus of one Problem as it had been in 1938. In that one, Admiral King's carrier aircraft had surprised and "wiped out" the Harbor. The result was the same this time—including Pearl being caught asleep again. (Japan would use the same tactic in 1941, even attacking when King had, on a Sunday.)

Then we went up LaHaina Roads to Maui to make the *Phoenix* combat-ready. It meant off-loading all unneeded gear including the Moffett trophy, a big ornate silver plaque. We put it in a warehouse and it disappeared (I suspect it was stolen), making my unit its last winner ever. The Hawaii Territory was a paradise in those days—neighborly, few tourists, the Royal Hawaiian and Outrigger Canoe Club lively, the navy popular. But while I was there, FDR jarred us by ordering three of our battleships to the Atlantic and telling the other eight, most of our Pacific fleet, to stay in Hawaii since Japan's navy was threatening French Indochina and the Philippines.

CINCPAC (Commander in Chief, Pacific) J. O. Richardson defied FDR's order, insisting that our main defense line must be the West Coast. Besides, "My ships aren't combat-ready," he said, and, "my small West Coast supply train can't support a fleet twenty-three hundred miles away," in Hawaii. FDR fired him, and made Kimmel CINCPAC. Our ships docked in one place, Pearl Harbor, seemed a plumb too juicy for any self-respecting dive-bomber to ignore. But what president or barnacled old saltwater admiral cares what a thirty-one-year-old lieutenant aviator thinks?

Anyway, I was told to report to the NAF on 1 June 1940 to be assistant chief engineer for (aircraft) instruments. Being away from family a lot is the worst bugbear in a military career. Flying to Florida for my girls, then to the factory, I fussed over how soon war might separate us again. At the NAF, I renewed studies begun in 1937 at Temple University and at the University of Pennsylvania's Moore Graduate School in Science and Engineering. Initially, the reason was to prepare for a "big-bucks" industry job. The navy had taught me the sort of mechanical engineering used in industry, but in 1937 I'd asked the Moore School dean what he thought would be most in demand. Nucleonics, he said. So I studied nuclear physics, Einstein, and the rest, and realized how little physics I knew in this new era soon to be named "The Atomic Age."[1] But now I just wanted the knowledge. World events had erased any thought of leaving the navy.

On 9 April 1940, after a calm named the "Phony War," Hitler swept into Western Europe. Denmark gave up that day. By 26 May, the British army had retreated to Dunkirk. During the next ten days, with luck and a lot of heroics, the bulk of it escaped to England. Then Italy's "brave" Il Duce declared war on Britain and France. By June, Hitler had taken Norway, Belgium, Holland, Luxembourg, and France, letting aged French Marshal Henri Philippe Petain occupy southern France with a puppet government in Vichy under Hitler's watchdogs.

In September, after Britain snubbed Hitler's demeaning peace offer, he swore publicly to invade England, and began near-daily bombing of British defenses. Then, as the Royal Air Force cut down his Luftwaffe, he shifted to bombing London, evidently just to wreck British morale. By the spring of 1941, RAF fighter victories had forced him to "postpone indefinitely" his invasion plan. Among the reasons our ally won this "Battle of Britain" were the grit of RAF pilots, the plucky British people electing Winston Churchill as prime minister, the breaking of the German military code (without the Nazis aware they had), and a string of coastal radars telling the RAF in advance how many German bombers from where were headed their way.

America now was alarmed enough that FDR could begin to act openly. Stressing "hemispheric defense," he sold Congress on having "The Arsenal of Democracy" build fifty thousand military aircraft annually and two hundred warships. He also ordered a "Neutrality Patrol" deployed in the western Atlantic, citing increased German U-boat traffic from Maine to Mexico. And he signed a "Destroyers-for-Bases" pact in September. A predecessor to the Lend-Lease Act of March 1941, it gave the British fifty of our old destroyers, more escorts for their Atlantic convoys. In return, we got ninety-nine-year leases to put air and naval bases in Britain's Bermuda, Newfoundland, the Bahamas, Antigua, St. Lucia, Trinidad, Jamaica, and British Guiana—a nice move, I thought. A U.S. Naval Air Station at Bermuda's airport on St. David's island meant defending a bit of my family's heritage.

Meantime, I was up to my eyebrows at work. Under the direction of the Bureau of Aeronautics, the NAF did more than build airplanes. It had an Aeronautical Materials Lab that also tested engines; my Instrument Development unit; and a Ships Experimental unit for mostly catapult and carrier arresting-gear work. With reports to an Acceptance Board, the NAF used them to test army aircraft for carrier suitability and navy aircraft to see if a new plane met navy contract specifications. (With Patuxent Naval Air Station not yet built, our only other flight-test field was at Anacostia Naval Air Station in Maryland.)

One of my jobs was issuing multimillion-dollar contracts for aircraft instruments, then making sure they reached a dozen assembly lines in as many states—heady stuff for a thirty-two-year-old lieutenant. I also had to sign up second-source suppliers to give us a broader industrial base, and team with the army in buying piece-parts or aircraft. For example, except for the tail hook on ours, our North American two-seat SNJ and their AT-6 (the Allies' trainer of choice in World War II) were the same. And the navy-designated PB4Y-1 was really an army B-24D Liberator. Our Douglas SBD Dauntless dive-bomber the army began to buy in 1941 as their A-24. And, because the air corps purchasing system was much more streamlined than the navy's, I gratefully hitched a ride several times on air force contracts.

I also was sent to Washington my first week at the NAF to "coordinate," officialese calls it, with the British RAF. There, the BuAer folks introduced my "colleagues": RAF Group Captain, later Air Vice Marshall, Tony Ragg, who became a pal; and U.S. Army Air Corps Col. Tommy Thurlow, navigator on Howard Hughes' 1935 round-the-world flight, who was at Wright Field (Wright-Patterson Air Force Base today) what I was at the NAF—my "teammate," our hosts said. (Thurlow was a warm, easy-to-like man. I lost him in 1942, killed in a plane crash in Texas.)

Aiding the RAF was like playing hooky with teacher's permission. Running for election to an unprecedented third term in "Isolationist" America, Roosevelt had to be very careful what he said and did publicly about gearing up for war. A much-quoted parody of his speeches in 1940 was "Ah hate wahr; Eleanor hates wahr; we *all* hate wahr!" On the other hand, FDR clearly wanted us to help the British every way possible short of going to war ourselves.

The RAF's very first raid, 18 December 1939, had sent twenty-four Vickers Wellington twin-engine bombers, the RAF's best, on an armed reconnaissance of Schilling Roads and Wilhelmshaven. They went alone since

RAF fighters couldn't provide air cover at that range. German fighters shot down ten; two ditched at sea, returning home; three crashed in England. Out of the shot-down-in-flames business came a push, both there and in the States, to put self-sealing tanks in combat aircraft, a relatively long-term proposition. A more immediate result was that the RAF switched to making its bombing runs only at night.

Their major night-flying problem was few pilots knowing how to do celestial navigation. Since we had the same weakness, anything we or they did would help us both. One help was the RAF's "Astrograph." Europe had a galaxy of stars to navigate by. RAF experts picked the easy ones to spot, then on the ground drew their altitudes as star-curves on a sheet of Plexiglas, periodically redoing them as star sites changed with the seasons. By laying the Plexiglas on a map, all the navigator need do to stay on course was take a star-sight. The mathematics had been done for him. The idea, of course, was to simplify the job and reduce navigator-training time.

With England under fire, the RAF moved its navigation school to Port Albert on Lake Huron in early 1940, flying all its students there from Britain. Earlier, we'd said we'd support the school if they did that. Consequently, from 1940 to 1942, I went there often to relate celestial navigation techniques I'd learned, such as the square-search rule and how to use our Bendix-built averaging octant. My main job, however, was to oversee equipping our aircraft bought by the British purchasing office in New York or later under Lend-Lease. They seemed happy with the whole deal, which, I guess, is why, at the end of my Naval Aircraft Factory tour, King George awarded me the Order of the British Empire.

We had less success in our own navy. To determine where one's ship or aircraft is, a navigator must know both what time it is and the position of the sun or, at night, the stars relative to one's own position. Since 1830 as the Depot of Charts and Instruments, the U.S. Naval Observatory (NO) has been our official source for both. As the sun and stars move or, more accurately, seem to move with the seasons, it issues a *Nautical Almanac* of tables listing their celestial coordinates in outer space. To compute a line of position using the original almanac format of tables took a while. They finally did devise a faster way to compute but it still was geared to thirty-knot ships. We said their almanac was as useless to us as a sextant. We wanted one made for two-hundred-mile-an-hour aircraft. The NO (an apt acronym, I discovered) bureaucrats didn't want to do it. It took a war to convince them.

Meanwhile, I had to find us a good watch. Ships each had one, an elaborate, shock-mounted chronometer carefully set on the time broadcast by the NO folks at noon each day, no good for airplane operations. Yet, just as an octant sighting may be off, a watch just one minute wrong can put the pilot at least a nautical mile[2] off his true latitude-longitude site. Also, if its "tick" rate is not the same as the Observatory's clock, that misleads. And, I learned, as the chronometer went from full-wound to run down, it could lose thirty seconds or more. So I set a strict isochronic requirement. Longine Wittnauer and Hamilton met it best and I ended up buying hundreds of watches from them. The flaw in these for lazy or forgetful folks was that these watches had only a twelve-hour spring and needed to be wound twice a day. Of course, with today's atomic clock at the Observatory, any ordinary watch can keep perfect time, once set accurately.[3]

Beyond that, our carrier-based aircraft had really inadequate flight-control instruments (a reason I preferred big airplanes.) Few had voice radio. Those we had were short-range, voices easily garbled by static and aircraft noise. The answer, of course, was to install radios designed for an aircraft environment. That we did, though saying it was a great deal easier than doing it. But it was not as tough a technical problem as some other instruments. The basic ones were the altimeter, airspeed gauge, compass, rate-of-climb indicator, and turn-and-bank indicator. My shop's orders were to improve them all, to create a whole new package. Initially, none lent themselves well to night and bad-weather flight. That era's trend was to all electric-driven instruments, more reliable than air- or vacuum-drive. But if the generators failed or, in single-engine aircraft, if the engine failed, the pilot was in trouble—though some early-1940s planes did have a device to drop into the aircraft's slipstream to run a generator on wind power.

My Bullpup crash in 1932 had made me do instruments-only celestial navigation work whenever I could. But I was an exception. In general, only guys with a tour in flying boat patrol squadrons were skilled at it because that's what they did: long flights in bad weather spiced by occasional hurricanes and hits by hail in a thunderstorm. Like the RAFs, most of our pilots hadn't done that. They proved it in 1936 after we began to emphasize night and bad-weather flight training. The increase in accidents was dramatic, most of them caused by a pilot not knowing the limitations of his instruments. At one point, BuAer even banned such flights except by the few of us flight-testing new gauges and ideas to solve—we hoped—the problem.

At the time, the Civil Aeronautics Administration was devising a Jepperson instruments-only plan, using the old Adcock Ranges, runway marker beacons, and airport operating minimums. The navy sent me; the army, Thurlow; the marines, Col. John Wehle, to be copilots for three months in commercial DC-3 aircraft. I rode American, New York to Chicago; Eastern,

Washington to Atlanta; United, cross-country. We just wanted to see firsthand how they handled night and bad-weather flying.

But back at the NAF, our first priority was to build a better compass. A magnetized needle on a round-card base, balanced on a post in our big aircrafts' instrument panels, it was set by the Bureau of Standards at the latitude of Washington, D.C., relative to the Earth's magnetic field. So, as magnetic attraction faded on east-west flights, the card tilted toward vertical until finally it just spun slowly on its post, pointing nowhere. The RAF's compass, bigger than ours, set on London's latitude, was stabilized by setting it on the floor between the pilot's feet.

Vlad Reichel, an affable Russian immigrant as of 1919 or so, devised the answer we used. I'd met Vlad, vice president at the Pioneer Instruments division of Bendix, in 1937 as I boarded the *Philadelphia*. When I joined the NAF, I used him a lot. He was a brilliant engineer who knew how much harder it is to invent a simple gadget than a complex one. This time, he made us a "gyro flux-gate," he called it. He put a coil around our compass ring, stabilized by a gyro that ran on four-hundred-cycle electricity. In the coil, he put tap-offs, "flux-gates," at 120-degree intervals to make their exhaust relative to the cosine of the flux-gate's angle to the Earth's magnetic field. Thus, off a tiny piece of the magnetic field, a pilot could compute direction very well.

Its weight and its need for its own power source were too much for small aircraft, however, and anyway, Sperry had most of the patents and market for gyro-type gauges. But by 1941, with the NAF running as if already in a war in order to jump-start contract competition and production, we simply ignored the patents. Our lawyer-types got mad at us, and who won the lawsuit if there was one, I don't know. I was not part of that. I did witness the haggle over where to put the device in the cockpit, just proving to me that not all stubborn mules have four legs.

Meantime, I did a compass-accuracy test. Flying north, up past Churchill, Canada, to the magnetic North Pole, we circled five times at specific latitudes to photograph all at once a row of compasses and a clock I'd arrayed on my plane's cockpit floor. On the way up, my declination gauge indicated, far south of the geographic Pole, that there were near each other three magnetic poles, not one as we had assumed. I didn't have time to research it, and still don't know if there really are three. Back home, the pictures told us the RAF's P-1 Compass was far better than ours. All this may seem quaint in this era of radio-guided autopilots and orbiting satellites instantly giving a ship or aircraft its exact location, but the 1940s automatic pilot was way too big for most aircraft, and electronic computers and satellites were just a dream.

We used big planes to lead across the ocean the smaller ones Britain was buying, and wore RAF uniforms to avoid being shot as spies if we had to ditch and a U-boat picked us up. The run was Montreal-Labrador-Greenland-Iceland to ground control in Scotland by CW "Morse code" radio, a dicey test after flying all night in formation. Our loss rate was high until all those airports put up radio beacons in 1942. And again, as with octants and gyro horizons, Vlad pioneered here, too, this time with Stark Draper and Lt. Cdr. "Deak" Parsons in promoting use of Schuller's Pendulum Theory, postulated in 1923, to develop a gyro-type electronic device to track distance and direction to a point in space—the basis for today's inertial navigation and inertial guidance of missiles.

Next, we tackled the altimeter. The early ones used barometric pressure, normally 29.92 inches of mercury at 59 degrees Fahrenheit. At takeoff, the gauge had to be set on the destination's pressure. If not, on a blind approach, a pilot could plow into the field thinking he still had altitude. Of course, in flight, he never was sure how high he was. We tried radio waves, first. The Bell Labs one I tested put two antenna on the wing. One pulsed downward at four hundred megacycles. The time its echo took to reach the other one was converted electrically to feet-height shown on a cockpit's cathode ray tube. Parsons, a Bureau of Ordnance (BuOrd) wizard, confirmed my flight-test results. But at two hundred pounds, it was too much for small planes. (We had airborne radars, today's answer, in 1939 but they were X-band search-types, not height-measuring.)

Still, some instruments—the airspeed, rate-of-climb, and turn-and-bank gauges—we couldn't improve much. Hardest to react to correctly was the turn-bank indicator. Vital in landings, during level flight its needle pointed up at the center line in an arc of lines, swinging left on right turns to the 5-, 10- or whatever-the-turn-degree line and, of course, right on left turns. One needle width on the T&B indicated a two-minute turn, the routine instrument-approach way to get lined up with a runway. Below the needle in a crescent-shaped tube was a bubble like that in a carpenter's level. On turns, if the bubble slid off center, the plane was in a skid. A landing can involve several turns. While sashaying back and forth, a pilot must keep the needle on the turn-line he wants, keep the bubble centered, navigate, and talk to air traffic control, all at once. To do all that well required hours of practice in the air and the Link trainer.

From 1940 to 1942, besides instruments and RAF work, visiting companies coast to coast to justify or deny contracts, I also flew lots of planes—

Lockheed Hudson bomber, Beechcraft GB-1, Grumman F5-F, XPBM and XPBY seaplanes, Bell P-39 Airacobra ("P" for "pursuit" in air corps lingo), Republic P-47 Thunderbolt—all for carrier suitability or other navy purposes; and, for Britain, a Curtiss P-40 they named Tomahawk. I flew my first Hudson, a twin-engine, 250-mph bomber fitted with our directional gyro, to Montreal for the RAF. It vanished over the Atlantic so I stayed with the next one all the way to Scotland, an RAF pilot flying while I navigated. And I flew the single-engine, reverse-stagger wing GB-1, the Walter Beech company's first product, first flown in 1934. It was a soft ride for five passengers, had a five-hundred-mile range and two-hundred-mph speed, fast back then for its aircraft type. My 1939 test of it was like dancing a waltz. So, as did the army, the navy bought ten for "transport of senior military officers."

The XF5-F ("X" for "experimental"), Grumman's first twin-engine fighter, was one of my favorites, faster than its 320-mph F4-F, more guns and range. But it failed "carrier-suitable," so we didn't buy it. The P-39 had a liquid-cooled engine set aft of the cockpit, 380-mph speed, 650-mile range, carried a 37-mm gun, two 50-caliber and four 30-caliber machine guns, and one 500-pound bomb. But its tail hook model also failed carrier-suitability tests in May 1940. So did the P-40, also a liquid-cooled-engine type, slower and less well armed than the P-39. Yet more of these so-so P-40s were built during the war than any other fighter, excepting only (at 15,740 each) the far better Republic P-47 and the North American P-51 Mustang.

In the stormy debate in 1940 over liquid- versus air-cooled engines, the navy stuck to the air-cooled plan we had endorsed in the 1920s. It meant putting the engine up front to let the aircraft nose be a big, blunt air scoop; while liquid-cooled engines could be set back, permitting a tapered nose and less wind resistance. But by 1942, for many reasons, air-cooled was the choice for most U.S. military aircraft, the one stellar exception being the liquid-cooled P-51. Its fastest model, the P-51H, had a 487-mph top speed, two-thousand-mile range, forty-two-thousand-foot ceiling, and was rated by many aviation history books as maybe the best single-seat fighter of anybody's air force in World War II.

During all that prewar action, lots of guys hit me with "a wonderful idea," they all claimed, for some new instrument. The most insistent focused on our wanting an airspeed meter that also told us airspeed at ground level. I was bombarded by schemes for an electrical device to cut through the lines in the Earth's magnetic force field in order, the peddlers said, to generate a current reflecting an aircraft's ground speed. I'd studied enough of Einstein's relativity theory by then to know that wouldn't work. But when I said, "You don't understand relativity," they'd call me the "kid lieutenant who's afraid to take a chance on anything new."

So I had Vlad Reichel take me to Princeton to see his friend, Dr. Albert Einstein, a kindly, shy old man, a bit disorganized, part of him politely chatting, the rest of him off thinking about who knows what. But he hadn't missed anything. The result of a contract I gave him was an incisive analysis I had copied at the NAF. Then, when a pushy type came in, I'd dare him, "If you think you're right, disprove what this man says in his paper." None of them ever did.

But we at the NAF also were party to a few crazy ideas. One was that they had us put into a giant PBM-1 Mariner, tagged the XPBM-2, a five-thousand-gallon fuel cell, enough for a nonstop flight to Berlin and back. Its "B" for bomber label was, I suppose, just someone still wanting us to do strategic bombing like the air corps. But the XPBM was to be for long-range patrols. In the PBY, we had only an eight-hundred-mile range out of Hawaii. With Japan controlling the Marshall Islands, and the Pacific a huge ocean, we wanted a great deal more. However, getting "on the step," hard in a PBM, would be impossible with that fuel load. So, "We'll catapult it," the NAF said, and built an eight-hundred-foot-long ocean-going barge to do just that. On test day, its crew put me in the catapult cradle and coaxed the barge into place. I revved up to full power, waved "go," and was snapped aloft by the catapult's 230-foot cable—while its cradle roared on like an angry bull, shattering whatever was in its path. That retired the XPBM-2.

My other 1941 farce was in a PBY. In 1935 when the P2Y flying boat was re-engined, it flew so well that we asked Consolidated to convert it to a patrol-bomber, PBY in navy lingo. Then the NAF hung under one a forty-eight-foot-wide aluminum ring. An electric generator in the PBY would turn the ring into a strong vertical magnet. The idea was to explode a new magnetic mine Germany had begun to deploy by making the mine think it was being approached by a ship, a weakness Germany quickly fixed. Testing this monster, I realized at takeoff that the ring was blocking my ailerons. A sporty hour was spent over South Jersey, juggling tail flaps and airspeed, able to do only square turns and pray, until I was safely back home. The ring also cut the PBY's speed to only 110 knots. At that speed, if it did explode a mine, that probably would have wrecked the plane too. Fortunately, advances in more effective technologies ended that project.

All told, for variety, 1940 to 1942 were my most unique years in the flying business. Back in Philadelphia in June 1940, the third time in four years, we'd rented a nice house on the commuter rail line to downtown. And on

29 November, Victoria, our third girl, was born. Then for a year, my job was endless—flights to Canada and contractors coast to coast; at home, graduate school at night after a day at work. Lili was a good sport about it, dreaming up adventures for her and the girls, like learning to ice skate.

It was a soft oasis in a violent world. Hitler had invaded Russia in June 1941. Japan had seized Indochina in July, provoking us to freeze Japan's assets in America and to embargo trade, while Japan signed a pact with Germany and Italy, creating the "Axis." In September, Japan said it had only a year's worth of oil reserves and began its drive to grab Dutch East Indies oil fields. By December, Hitler was in the front yards of Leningrad (as Saint Petersburg had been renamed) and Moscow. Then on 7 December at 7:55 and 10:00 A.M. Hawaii time, 363 Japanese fighter-bombers and torpedo planes struck Pearl Harbor. By 11:00 A.M., losing only twenty-nine aircraft and five midget submarines, they had killed or wounded 3,581 Americans, demolished or damaged 241 of 349 air corps and navy combat aircraft, and sunk or leveled three light cruisers and three destroyers.

Eight of our nine Pacific Fleet battleships, what "Jutland" dogma still called our centerpiece, were hit worst. The *Oklahoma* was capsized, the *West Virginia* flooded to prevent fires exploding her magazine. The *Nevada*, under way, was hit so badly she was beached to keep open the Harbor channel. The *Maryland*, *Tennessee*, and, in dry dock, the *Pennsylvania* were least damaged, but only two were reparable locally. The old *Utah* was sunk, no loss since she was now a radio-controlled target ship. (Japanese pilots told Tokyo she was a carrier.) And huge explosions lifted the thirty-three-thousand-ton *Arizona* out of the water before she and all 1,477 of her crew sank.

Only our carriers *Enterprise*, *Lexington*, and *Saratoga*, and some cruisers, all at sea, were spared. The Japanese also ignored our fuel- and ship-repair plants, so recovery was quick. But we were stunned. And in spite of Tojo's samurai ethic, "Give a sleeping enemy warning before you attack," he had not. Japanese diplomats, told to "warn" Secretary of State Cordell Hull, hadn't done so until after the attack had begun. But we needn't have been surprised anyway, except that an army radar man who saw the first wave coming was told by his boss to ignore it. But what really mattered was that our strongest ships out there now were aircraft carriers. I promptly requested squadron-commander duty on one. The answer was, not until my war-level NAF contracts all are let and my relief is found, "probably a Navy Reserve guy too old for combat." By early 1942, the Japanese had beaten us everywhere and my British friends were afraid we'd focus there, neglecting Europe. We might have if Hitler had not declared war on us after Pearl. If he'd stayed shut up, "isolationist" pressure, still rife in the States, could have forced FDR to give only Lend-Lease aid to Europe.

While I waited, Ross Allen, the NAF chief engineer, told me BuAer had hatched an idea to put full-span spoilers on the F4-F's wing "to make it more maneuverable." I was leery. If one wing drops coming in, full-span spoilers magnify the tilt, making a spin-in almost certain. I'd seen the OS2-U full-spoiler seaplane do that. If it bounced on touchdown in a rough sea and one wing dropped, the other flipped up, and over and in she went. I told Ross, I'd not fly it even if ordered to. So they had Grumman's top test pilot, navy reserve Lt. Cdr. Jimmy Taylor, fly it. He died when it spun in on his first turn. It should have stopped there but BuAer sent a new F4 to Anacostia, to "show" us, I guess, with orders not to fly it until they'd "investigated" full-span spoilers. Sadly, marine Col. Bill Saunders taxied it too fast, got airborne, tried to circle and died when it threw him into the Potomac. That ended BuAer's full-span-spoiler romance.

In January, my good news was promotion to "lieutenant commander, temporary." Then came a shocking report, the RAF's analysis of 128 raids in 1941. On moonlit nights, only two of five aircraft got within five miles of the target; moonless nights, one in fifteen. In its largest raid, 7 November, of 400 aircraft that were bombing targets from Oslo to the Ruhr, they lost 37. Worst was the 169 sent to Berlin. Only 79 found the city, very few of their bombs hit it, and they lost 21 aircraft. Most weren't shot down. The pilots just got lost, ran out of gas, and had to ditch at sea—and this is two years after their war had begun! It was a sad day, seeing how poorly our help here had translated into capability there.

Bad news from the Pacific had surfaced far earlier, about 1900, when "War Plan Orange" began to evolve. A joint Army-Navy War College study by 1901, it assumed that Japan would one day go to war with us. In August 1914, Japan had declared war on Germany to be, on paper, our World War I ally. The only hard result was that Japanese forces seized the German-held Marshall, Mariana, Caroline, and Palau islands. In 1919, Japan wanted to annex them as Britain, Australia, and New Zealand had done with New Guinea, Samoa, and Nauru. But we would allow only that they oversee those "independent mandates," not fortify them—a pact Japan promptly broke, building navy ports, airfields, and supply depots all over that huge expanse, defended by Japanese occupation troops.

By 1941, our War Plan Orange had become real-world, including plans to reinforce and hold the Philippines. But saying is not doing. By 6 May, when our last Philippine troops surrendered, the Japanese had taken southeast Asia and invaded New Guinea. In the Coral Sea, 4–8 May, the first time two fleets fought without seeing each other, Rear Adm. F. J. Fletcher's Task Force sank one and crippled one of Japan's six first-line carriers to abort a Japanese invasion of Port Moresby. But we'd lost the *Lex*. At Midway, 3–4 June, the *Enterprise, Hornet,* and *Yorktown,* attacking a Japanese task force, lost thirty-six of forty-one TBD torpedo planes, and the *Yorktown* was crippled. On 6–7 June, a Japanese sub sank her but we sank two Japanese cruisers and four carriers, Japan's best, with scores of its top pilots killed. Only later did we realize that had shoved Japan's navy onto a slippery slope to hell.

7

Preparing to Fight while Fighting

Right after the Midway fight, I was ordered west to accept delivery of the first B-24D Liberator bombers being built in San Diego by Convair, the name Consolidated had given itself after it bought Vultee. I was to do that, my orders said, until 1 October when I was to report to Headquarters Squadron, Fleet Air Wing Two, Hawaii. I assumed I was given the B-24 task because of my Aircraft Factory work and hundreds of flying hours in big aircraft. Or maybe just because Admiral King, chief of naval operations since March 1942, knew me. In any case, my job was a result of a deal King had cut with army air force Chief of Staff Henry H. "Hap" Arnold.[1]

The B-24Ds were a giant step away in combat patrol from relying on PBYs, which were inept birds anyway, pilots in the war zone said. Though its twenty-three-hundred-mile range was adequate, the 1941-vintage, two-engine PBY's speed was only one hundred knots and it could carry just two tons of bombs or depth charges. But its five machine guns can repel any attack, its fans said. Japanese Zeroes had popped that bubble in December 1941, prior to Japan's invasion of Luzon. Three times faster than a PBY, armed with two 7.7-mm machine guns, two 20-mm cannon, and a 132-pound bomb, they had cut down PBY Patrol Wing Ten like a "weed-whacker" and half the rest of MacArthur's air force too, most of it still on the ground in spite of being told an aerial attack was imminent.

In 1940, Boeing, maker of Pan Am's Clipper, started to build us the largest twin-engine flying boat ever made, an XPBB-1 Sea Ranger, to replace the PBY. First flown in 1942, it probably would have been easier to shoot down, really. The Arnold-King pact said we'd shelve it so Boeing's new Renton, Washington, plant could build B-17 bombers for the army. In return, we'd get B-24Ds right off the assembly line to put them into combat right away. Navy-type writers then and still insist on using its navy designation, the PB4Y-1, but that's a quibble. It was the B-24D. It had a three-hundred-mph speed, 2,850-mile range, and four turbosupercharged engines to lift its fifteen-ton top payload—crew, fuel, seven tons of bombs—to above thirty thousand feet. (At 18,475 copies, the B-24, in several configurations, was the most-built of all World War II big aircraft.)

So once more, we packed. The navy'd move us for the war to any hometown I picked except my overseas-embarkation city. Lili chose Pensacola, predictably, and we took the three girls there to live with the Hyers. Then, since I'd be in San Diego until October, Lili and two-year-old Vickie came out to be with me that summer. We stayed with my sister Eleanor and her husband, Col. Dan Pollock, a paratrooper like my brother, Dick. (He had introduced them after her first husband died in an air crash. A much-decorated warrior like many marines, Dan would earn a slew of combat medals including the Navy Cross at Iwo Jima.)

But only the nights belonged to Lili. My days belonged to the B-24D. I test-flew it every day everywhere, out over the ocean, the desert, the mountains. I liked it but it was not easy to fly. Its Davis wing, so-called because it was devised by a guy named David Davis, had a high aspect ratio, 110-foot span by 18 feet tapered to much less at the wingtips. It was a unique airfoil, giving good range at high altitude but hard to lift off full-loaded at a top takeoff weight of 71,200 pounds. We had a training unit there for VB squadrons 101 through 110 that the navy was creating. I kind of fell into running that, too. And once more, my pal Bill Moffett, now a commander, had gotten in ahead of me. He was to skipper our first one, VB-101.

I also had the air force B-24D changed a bit for us. Theirs had two 50-caliber machine guns in the tail, top, and sides, one aft of the bomb bay, and a tenth poking out on one side of its nose, behind the flight engineer's seat with just a 50-degree arc of fire. Maybe a Swiss carillon player could see a use for it. I couldn't. It was thus, the bemused army told me, since B-24s go so high and fast that a frontal attack is not likely. Nonsense, I said, and had an ERCO (Engineering Research Company, father of Emerson Electric) turret put on the nose. It mounted twin 50-caliber machine guns under a power-operated Plexiglas dome for an almost 180-degree arc of visibility and fire.

We'd just begun the ERCO fix when news came of our worst defeat to date, I thought. On 7 August 1942 in our first offensive of the war, Maj. Gen. Alexander Vandergrift's First Marines landed on Guadalcanal (code-named "Cactus"), Gavutu, and Tulagi in the Solomon Islands. Their target was an airfield the Japanese were building on Cactus. The enemy fled, but the orderly landing became chaos, tanks stalled inland while supplies piled up on the beach and second marine waves fumbled among the stacks looking for someone to tell them what to do.[2]

Then at midnight on the 9th, Vice Adm. Gunichi Mikawa's task force caught asleep Australian Rear Adm. V. A. C. Crutchley's picket line off Savo Island, twenty miles from our Cactus airfield. They sank all his cruisers and his *Canberra*, and our *Astoria*, *Quincy*, and *Vincennes*, and a destroyer—without taking a single hit! Luckily, they then withdrew. If they'd stayed to shell our marines, we might have been driven off the island. Mikawa had come to Savo down a narrow three-hundred-mile-long channel called The Slot, a sail not doable unseen if our PBYs had been on Cactus, as they wanted, instead of Espíritu Santo six hundred miles away. As bad was learning that Australian Coast Watchers had told Brisbane of Mikawa's course but nobody'd warned Crutchley!

By 31 August, I'd logged 186 pilot hours on the B-24 business in three months. In late September, we left for Hawaii, a teary farewell, me off to a war we were losing at the time, Lili sure she'd be a widow before it ended. The round-the-clock, kill-or-be-killed war raging around Cactus said she might be right. I'd heard of guys who went to war with a kind of fatalistic faith in their own immortality, but I was afraid. Flying west, I decided all I could do was my best and let Lady Luck or the Heavenly Host—or a Japanese fighter—decide if it was good enough.

With a pickup crew, we flew, full-loaded with fuel and spare parts, to Hamilton Field, San Francisco, 180 miles closer than San Diego to Hawaii. There I picked a night departure so we could navigate by star-sights. Back then, all long-distance ocean flights did a "how-goes-it" drill. Once an hour, we fixed our position, then measured our fuel. On the course is a point of no return. Once past it, the pilot's only option is to try to reach his destination. In a storm against head winds, that "point" can come up on the chart pretty fast. This time, at five hundred miles out, I tuned my automatic direction finder to Honolulu radio station KGU to let it guide us in to end our twelve-hour, twenty-three-hundred-mile flight to Barbers Point. After I gave the B-24 to Supply for delivery to VB-102, I rode over to headquarters, Fleet Air Wing Two in Kaneohe, to see my new boss, Rear Adm. Marc A. "Pete" Mitscher. I'd first met "Uncle Pete" in 1934 when he was the *Saratoga*'s XO.

We'd met again when I was at the NAF and Cdr. Mitscher was at BuAer. I'd gone to Anacostia, to test-fly a new SO3C amphibian, a lousy airplane. I went to his office, wet flight suit and all, to say so after it dumped me in the Potomac, nearly killing me. Since then, he'd called it "Chick's airplane." Now, he boosted my morale. It was inconceivable to even think we might lose the war, he said. His only doubt was how long it would take to whip the Japanese.

He was a little guy, maybe 150 pounds, but in courage and intellect, he was a giant. Academy class of 1910, naval aviation pioneer, in 1919 he'd flown one of the three Curtiss seaplanes on the first transatlantic flight. He was skipper of the *Hornet* in April 1942 when Doolittle's sixteen B-25s used her to bomb Tokyo, shocking Hideki Tojo who'd told Japan that could not happen. He was on her at Midway to see the tragedy of aircraft types not teaming up. (Off all the carriers, flying at varied altitudes and speeds, his fighters and dive-bombers lost track of each other and the enemy, leaving all his fifteen torpedo planes, going in alone, to be shot down.)

I asked him for a combat job. No, he said, "You have to stay here and organize this thing so our aircraft know where to find supply-support. Agree to do that and I promise you'll get a combat command afterwards." A clue to the flinty-eyed old man's leadership style was that he didn't order it done, as he could have. He explained it instead.

At our leisurely prewar tempo, patrol groups had taken their repair people and spare parts with them, replenished by supply ship or seaplane tender. Now, with new bases opening up and aircraft landing all over, our whole logistics train had to be revamped. We had to put PATSUS (Patrol-Bomber Support Units) and CASUS (Carrier Support Units) wherever our planes might be, from Midway, thirteen hundred miles northwest of Hawaii; to our southern anchors at Espíritu Santo and Efate, thirty-five hundred miles southwest of Hawaii. We also would need units in forward bases as they were captured or built; fuel, ammunition, spare parts, and so on shipped in; and the people flown in. And, after I praised the ERCO turret, "Uncle Pete" told me put it on all our B-24s. Then, for seven months, I covered a lot of ocean including to Cactus where putting in a PATSUS meant preparing to fight while fighting, no holidays, and morale often at low tide because Europe seemed to have first call on FDR's "Arsenal of Democracy."

That feeling changed in October when King transferred Vice Adm. Robert L. Ghormley to his own Washington, D.C., staff. As the war on Cactus waffled, King and CINCPAC Adm. Chester W. Nimitz had decided that Ghormley, good planner, good diplomat, was too reluctant a warrior. In his place as commander, South Pacific forces, they put Vice Adm. William F. "Bull" Halsey. The nickname was from Halsey's bulldog-like features, but also fit his aggressive combat record. Academy class of 1904, he had earned his wings in 1935, captained the *Sara* and *Enterprise,* and knew what pilots faced in the Pacific. His Task Force Sixteen had escorted the *Hornet* for Doolittle's raid, then hit the Japanese on the Gilbert, Marshall, and Wake islands on his way back. His arrival had an instant, euphoric impact on morale. By coincidence, though it seemed cause-and-effect, that month VB-101 also arrived and our Pacific fleet's size began to rise rapidly.

As B-24s came in, PBYs now flew only rescue missions or where we had air superiority. Before that, PBY Catalinas had done "Black Cat" patrol, so named after pilots camouflage-painted their PBYs black. A daily run up The Slot and east of it to see if a "Tokyo Express," enemy war or troop ships, was headed for Cactus was a lonely hunt as far out as eight hundred miles. If a PBY saw a Japanese carrier, its only defense was to hide in clouds, if there were any, and radio back the sighting, sometimes adding, "PBY to base: Trailing enemy carrier; Please notify next of kin."

Nicknamed "Dumbo" like Walt Disney's big-eared elephant, the PBY did run some sporty rescue missions, aided by Australian Coast Watchers, a group who acted as if heroics were just the ordinary, unremarkable thing to do. Not only did they report Japanese ship movements, they and our submarines saved a lot of our shot-down pilots. In sharp contrast, the Japanese didn't try at all to save theirs. They just sailed away and if we tried to pick up a Japanese pilot, he'd dog-paddle out of reach until he drowned. The Japanese policy was heartless and stupid. By not trying to save them, the Japanese eventually lost, and could not replace, their best, most experienced combat pilots.

In late October, Adm. Isoroku Yamamoto sent four carriers, four battleships, ten cruisers, thirty destroyers, and twelve submarines in to retake Henderson field. A Black Cat alerted Cactus and, on 25 October off Santa Cruz, we sent to meet them the *Enterprise* with Jimmy Flatley's fighters aboard; and the *Hornet*, a battleship, six cruisers, and fourteen destroyers. The Japanese wrecked more of our ships, shot down more of our planes than we did theirs, and finally on 26 October, Japanese "Kate" torpedo bombers so damaged the *Hornet* that we had to sink her ourselves. It was a tactical victory but a failed mission for Japan. They withdrew, deciding they hadn't enough left to shove us off Cactus.

On 30 November 1942, in another try off Florida Island, twenty-five miles from Henderson, our cruiser *Northampton* was sunk and three more cruisers damaged against the loss of only one Japanese destroyer. (At that stage, Japanese ships were a heck of a lot better than ours at night fighting.) It was another Japanese victory but we still owned Henderson. The *Northampton* had been sunk by a very good ship-launched torpedo we called the "Long Lance"—though it wasn't christened that until more than a year later when its capability was confirmed. (Oxygen-powered, weighing more than six thousand pounds, launched from cruisers and destroyers, it carried a 1,000-pound warhead and could be run at any of three speeds. At its forty-nine-knot speed, it could travel nearly eleven miles.)

By December 1942, Charlie Mason, our first commander, Air Guadalcanal, was just plain worn out. Mitscher took his place, leaving me pretty much the top supervisor of the PATSUS business. I did take a brief recess to fly a PBY-5A to Midway to scout for the Seventh Air Force Christmas Eve raid on Wake. The Japanese had seized it just one year earlier and we wanted to show them what was in their future. Wake is twelve hundred miles west of Midway, the weather that time of year a constant flow of severe cold fronts. The raid tore up the Japanese, gave minor scratches to us; but we lost planes coming back, ditched when they couldn't find Midway.

In January, now-Rear Admiral Mitscher took command of Fleet Air/Solomon Islands.[3] In early February 1943, the last thirteen thousand Japanese left Cactus. In the six-month battle, we'd lost fifteen hundred marine and army troops, had forty-eight hundred wounded, and had twenty-four ships sunk as had the enemy. But twenty-five thousand Japanese had been killed, six hundred of their aircraft lost. "Bull" Halsey didn't stop to celebrate. Within the month, he had invaded the Russells. We were taking the fight to the enemy.

That, we would learn, was what Admiral Yamamoto had feared in 1941. From 1918 to 1926, he had roamed America, attended Harvard, been a naval attaché in Washington, and visited our Naval War College. He knew our people and our industrial power far better than did most Japanese. He did not share Tokyo's opinion that Japan's "disciplined one-race culture" was superior to our "querulous mix of races with little care" for America, with "no collective will" to defend it. FDR's antiwar election campaign and a military-draft law passing the Senate by only one vote in September 1941 had reinforced that folktale. As in most dictatorships, Japan's "elites" lived well but the people were impoverished. Tokyo thought only our navy might stop its sweep into south Asia to grab rice, raw materials, and oil Japan didn't have at home. Yet Yamamoto had planned and led the attack on Pearl reluctantly. It must be decisive, followed soon by the offer of an armistice, he said, because, "We can not win a prolonged war with the United States."

The numbers argue that he was a bright guy. In 1939, the navy/marine corps aviation had some twenty-one hundred aircraft and eighteen hundred pilots. At Japan's surrender on 2 September 1945, the navy's seventy thousand planes were part of three hundred thousand military aircraft that U.S. industry built in the war, nearly five times what Japan was able to build; and we, army and navy, had trained some 655,000 pilots, fifteen times what Japan produced. In 1941, we had fifteen battleships, four carriers, one escort carrier, and a handful of other warships. By 1945, we had two *Midway*-class, sixteen *Essex*-class carriers, some fifty escort carriers, and 5,788 other warships. Pearl Harbor, shocking as it was to us, had not done what Yamamoto said Japan must do. Thus, as the rest of Japan glowed over his "great December victory," he wrote, "I fear all we have managed to do is awaken a sleeping giant and fill him with a terrible wrath."

But in 1943, the box score had us behind and, as "Uncle Pete" Mitscher moved upstairs, he kept his promise. I was ordered to be in Camp Kearney (now Miramar) by May to take command of patrol-bomber squadron VB-106. (He also awarded me a Bronze Star, why, I don't know. Just for hanging around, I guess.) My April leave was a debrief in Washington, then a flight to eight-months-pregnant Lili in Florida. That same month, navy intelligence intercepted a message telling exactly where Yamamoto was going on an inspection tour. And on 18 April, two Army P-38s in Mitscher's command area shot down Yamamoto's "Betty" bomber, killing him. Our spy-guys said his loss was to Japan like us sinking an entire Japanese task force.

Our fourth girl, Jennifer, was born in May 1943 as my lieutenant commander rank became permanent, "retroactive to 30 June 1942." A month later, Lili, at my request, came out to San Diego with Jennifer. We lived at the Coronado Hotel, posh then as now but much cheaper, Lili a haven from the storms raging at work. And on 1 June, VB-106 was commissioned: eighteen flight crews and fifteen twin-rudder army B-24Ds. (The navy later built a single-tail model designated PB4Y-1.) My squadron's warts and wheezes were legion.

Our basic eleven-man crew included a PPC (patrol plane commander), copilot, a navigator-bombardier, radar-radio man, machine gunners for the five gun positions, and a noncom crew chief. Each chief was his aircraft's mother hen. Each aircraft also had, counted as part of the crew, a ground-crew boss. He was a maintenance-engineer officer who had to know how to fix whatever needed fixing on any B-24. The squadron also had two intelligence specialists to interpret aerial photographs, write mission reports, and keep our war diary.

In training, we were to follow a Fleet Air West Coast syllabus, approved by Navy South Pacific and Air Pacific, covering night instruments-only flying, navigation, landings, takeoffs, Link trainer work, friend-or-foe recognition, and combat intelligence. It all could be done within a month but only 37 percent was as we left for Hawaii nearly three months later and I got, on 20 August, a "T," for temporary, boost to commander. Part of why we were late is the people I was given. Of my patrol plane commanders (PPCs)—three lieutenant commanders, twelve lieutenants (some jgs), an ensign, and a chief aviation pilot—eight had PBY time in combat areas. Except for Ensign Bill Snead with only six hundred, each had about three thousand pilot hours.

But my copilots were just out of flight school. Six of my navigator-bombardiers were adept at both skills. The other twelve had 250 hours' bombardier training, little in navigation. My radar-radio men had no radar training. I had untrained machine-gunners and we had only 141 enlisted ground-crew folks. Not until the day we left for Hawaii did we reach 167 ground crew members, one shy of our intended complement. Many, rated machinist's mate or whatever, were zero-trained in aircraft maintenance. Several times, I urged BuPers to create an "air crewman" rating and they, ever the arch rival of common sense, spurned the idea every time.

The other block to completing the syllabus was that two-thirds of our fifteen aircraft were hung up in A&R (aircraft overhaul and repair) to put on the ERCO turret. I also had the G-11 solenoids replaced by Mark 5 side-plate ones in all except the belly and tail guns, which hadn't room enough to put them in. The solenoid sparks the firing mechanism. Against an enemy fighter closing at a mile a second, the solenoid needs good magnetic material to make the gun work instantly. The G-11s were lousy. During the three weeks that A&R took, per plane, the crews went to Convair's and our Transition-Land-Plane schools. Only three days before we left for Hawaii did we get the last of our fifteen aircraft. As I reported nastily later, a report co-signed by VB-104's skipper, Harry Sears, for training we should use B-24 training planes, not our own. The worst of it was, not until late August could my crews each begin to train as they would be in combat, a team, each man learning to work with and trust his teammates. (In 1944, when I tried to tell BuAer and BuPers how vital that is to combat effectiveness, they changed the subject.)

In Hawaii, my people did some 140 hours in gunnery, both at moving and fixed targets and in the 3A-2 trainer; skip- and low-altitude bombing; navigation including radar approach to islands in bad weather, standard in the South Pacific; night landings and takeoffs; tactics working with friendly fighter planes; and oxygen-equipment checks. We also ran flight-emergency drills; and training in air-sea rescue, swimming, and lifesaving. And I had Honolulu's Bishop Museum give my boys show-and-tell lectures in how to survive on tropical islands. This training cycle also took too long. One reason was having to put aircraft in A&R again: once, for a new bomb-release because ours wouldn't release; later, to replace the Ceco carburetors burning out our engines. (We put in more expensive Strombergs to stop that, proving anew that low price may not be lowest cost.)

We also had to fly three eight-hundred-mile patrols off Oahu each day, another "glitch" in the training plan. A few times, we flew VB-101 planes, borrowed from Cactus, to give ours a rest. Still, by the end of September, our engines had been run some two hundred hours, 20 percent of their useful combat life as it turned out and one more reason to use training planes, not our own, for training. It would be four months before the high price of this aircraft abuse, of diverting my boys away from concentrated training, would show up. But, in spite of the kind of slapdash VB mobilization plan devised by BuAer, or whoever, VB-106 had begun to shape itself into a team.

Our roster now listed each PPC and his crew, by job title, under each PPC's aircraft tail number. I had 32087 on whose nose I'd had painted a sultry, naked girl and the inscription, "Chick's Chick." My XO (Executive Officer), Lt. Cdr. Ray Bales, had 32078; Lt. Cdr. Rude, 32101; and Lt. Cdr. Ragan, 32088. My other PPCs, all lieutenants or jgs, were Hank Surface in 32082; "Dagwood" Davis, 32095; Allen Seaman, 32092; Tommy Birch, 32144; Gordon Ebbe, 32091; Sam Patella, 32102; Fred Tuck, 32072; "Johnny" Johnson, 089; Ben Mitchell, 085; Frank Holt, 32084; and Barlow in 094. Much as I wanted to hold them in training another month, I had to admit they were as ready for war as possible short of being shot at by people who want you dead. Stalling, whatever the alibi, would have been simply neglect of duty. It was time to go to war.

And when we reached Canton Island in September, we got our first combat missions. One was for six of us to form a two-hundred-mile-wide screen three hundred miles in front of Admiral Radford's carrier task force as it left Midway to plaster Wake Island. We were to stiff-arm the enemy's scouts, scout the three-island atoll for Radford and, after his strike, bomb the place. It was a fourteen-hour, fifty-five-hundred-mile round-trip through storms, the longest combat flight of the war so far. My boys were a marvel. Only three of our bombs fell outside the target area.[4] Later, in our debrief at CINCPAC, Hawaii, I was told that the day we hit Wake Island, the New York Yankees had dedicated the first game of the 1943 World Series to me, then won it, 4 to 2![5]

Our second mission was my other nine B-24s escorting VD-3 (D for photoreconnaissance) flights to the Gilbert and Marshall islands "until relieved on 8 October by VB-108." VDs put cameras in places where we had weapons, so we went along to defend them. Back from the raid on Wake, my own first VD-escort went out to confirm, which it did, that a day-earlier raid had pretty well trashed enemy bases on Tarawa and Makin. Refueled at Baker Island, the two-thousand-mile, fifteen-hour flight had left us dead tired, our belly gun jammed. After parking our plane by a revetment, I told its nineteen-year-old gunner, Boose, so small he fit easily in the turret, to wait 'til morning to fix it. Then, too tired to walk to a hut, I sank down right there to sleep.

An instant later, air-raid sirens blared and I was up, racing for a ditch. Evidently, a few Japanese floatplanes, one small bomb apiece, had followed us home. It was a nuisance raid, really, but painful for me. As I dove for cover, a burst of shrapnel peppered my butt. A corpsman tweezered out the shards and, back where I'd lain down earlier, one of my guys, smiling, asked where I'd been hit. "Where a combat pilot usually gets hit," I said, and went to sleep. At dawn, I was jarred awake, briefly, by 50-caliber gunfire. Quickly back asleep, I soon was shaken awake, this time by a soldier howling, "You're under arrest!" My "crime" was that Boose's guns, as they cleared, had let go a burst, hitting a colonel's hut, bouncing a large stock of now-broken canned pineapple tins. The colonel was mad. So was I, now. "I don't have time for your colonel," I snarled. "We're tired. You make us sleep on the ground and we have work to do. Beat it!"

Then on 18 October we lost Sam Patella's 32102. It just disappeared, flying into a severe cold front north of Funafuti on a run at Tarawa, a mission he'd requested. Two days earlier, flying that plane, I'd noticed his gyro horizon gauge was awry. Storm turbulence had tumbled it, probably, and thrown him into the sea. We did a three-plane sweep off his last known position but saw no trace of the plane. The rules said we had to list them as "Missing in Action" but in the way I phrased my letters to their next of kin, sent care of BuPers, I hinted that Sam and his crew probably were dead. Losing people in battle is hard to take. Losing Sam the way we had hurts more.

Eventually, after a 50-caliber hit PPC Dagwood Davis' hand, leaving his copilot at the helm, and then later my copilot was killed, I asked BuAir to put two gyro-fluxgate gauges, on separate electric circuits, in our B-24s. That way, if one guy is lost, the survivor has right in front of him the gauge he steers by. Also, if one gyro horizon tumbles in a storm, the other probably won't. "I well-remember going through a front at 20,000 feet," I wrote, "when my horizon went out, hurling us into a steep, spinning dive at tremendous speed. To get the turn-bank needle on center and pull us out, I had to use the ailerons. I lost 5,000 feet in less than 20 seconds and became quite a Christian about being up high, going into fronts. I've logged 7000 hours, often on instruments, but in the South Pacific where any 1000-mile flight usually runs into at least one storm, frankly, I was scared to death."

In Fiji on our way to Cactus on 31 October, I totaled my flying hours in VB-106: 286.6 in June through September, an average of 71.6 a month, and 122.2 hours and twenty-five thousand miles in October. My policy was that my XO, Ray Bales, and I had to lead in the flying, each of us going up at least every third day. We'd seen one VB outfit just collapse, its PPCs regularly aborting flights, "bad weather" being the usual alibi. The real reason was that their commander, a deskbound paper pusher, hadn't flown one mission since they'd arrived in the area. In our nine combat months, not once did one of my guys cancel a mission. On the 31st, my guys already on Cactus ran a good raid on Tarawa with no fighter opposition, only antiaircraft fire and my diary says, "The Japs must be having troubles of their own. Let us hope they have lots more."

On 1 November, marines landed at Empress Augusta Bay in Bougainville in the Solomons, behind strong Japanese defenses on that island's southern coast, and our patrols began to fly mostly the New Britain, New Ireland area, four hundred miles northwest of it. We covered eight side-by-side sectors, each one's eight-hundred-mile leg at an 8.5-degree angle to its other leg. To use our radars' best range, a fan-shaped sweep on a thirty-mile radius, we flew at eight thousand feet. All sectors had to be scouted every day, up one leg, down the other. We took off at sixty-five thousand to sixty-six thousand pounds with four 500-pound bombs, twenty-seven hundred gallons of fuel to give us a four-hundred-gallon reserve at the end of the eleven- to twelve-hour trek. Our most efficient cruise speed was 160 knots[6] on thirty-two inches manifold pressure, the propellers at twenty-two hundred revolutions per minute. But we discovered that two B-24s on identical power settings could vary in speed up to twenty miles an hour. We fussed some with the tail and ailerons but that aerodynamic enigma remained a mystery.

We radioed to Cactus the size, site, and heading if we saw a target, then recorded on our K-20 cameras the result of our own attack on it. Sometimes we also hauled people with special cameras to fill a specific request. I thought their best work, made in five sorties, was a set of low-altitude, close-range three-dimension obliques of New Ireland's entire coast from Borpop to Kavieng. But most sorties, notably in sectors three to eight, were just boring trips over empty ocean. So as often as possible, we did refresher training in gunnery and bombing using a 3A-2 trainer we'd stolen from Kaneohe as we left Hawaii, and IFF (Identification, Friend or Foe) recognition tests.

After several days' patrols, I wrote to BuAer, "Until we can get accurate fuel gauges, we need flow-meters in every plane." A far worse flaw was our gas tanks. The good part was that they were self-sealing, absorbing bullets like a sponge. The worst part was that Convair hadn't put in the cells' stiffeners and, in ramming them into tight spaces in the airframe, had wrenched their seams. Every landing hammered them again, of course. All our level of repair could do was Band-Aid work. During 1944, I'd proposed lots of fixes like that in the B-24. The number of replies I got from BuAer would have fit through the eye of a needle.

At year-end 1943, we began to get Very Long Range Search missions. A VLRS, one thousand miles out, two-hundred-mile crossover, one thousand back, took sixteen to eighteen hours. Its point was to support a fleet attack, usually in sector one, up the Solomons' east side; or sector two, out past Rabaul to Kavieng, nine hundred miles northwest of Cactus. But they never told us when or where the raid was due—worried, I guess, that if we were captured, we'd get our fingernails pulled out until we blabbed. In any case, adding VLRS to ongoing aircraft repair meant needing twelve aircraft to cover three days of sector flying.

My 6 November record says, "Had hot sector 1, today; weather stinko. Ben Mitchell shot up a Rufe," a Zero-type seaplane, "and a ship at Kapingamarangi," a big Japanese seaplane base where we'd lost Bruce van Voorhis. Skipper of VB-102, he'd taken on, alone, a seaplane tender; and a swarm of Rufes had shot him down on his fifth pass, killing everyone. (Exhumed, unidentifiable, from a common grave after the war, they were buried as a crew in St. Louis, an aloof navy giving them only Purple Hearts. That riled me into writing a citation Mitscher signed to give Van, posthumously, the Congressional Medal of Honor, his team the Distinguished Flying Cross.)

On the 9th, *Chick's Chick* beat off four Japanese fighters over Kapingi, then in rain, sixty-knot winds, and two-hundred-foot ceiling, barely squeezed home past Indispensable Strait, aptly named that day. Then we did a week of PBY "Black-Cats" off Empress Augusta and five hours on station off Rabaul-Kavieng, with bad weather and lots of Japanese. And Dagwood bagged a Pete, a seaplane submarine hunter for Japanese convoys, and skip-bombed a Truk-bound convoy. (The army said a B-24's best bombing altitude was twenty-five thousand feet but we attacked ships, not fixed targets. So we each used our four 500-pound bombs to skip-bomb a lot, rarely bombing from above eight thousand feet in any case.)

About then, relaxing one day at Henderson Field with Japanese still holding out on the north side of Cactus, marine Maj. Bob Kriendler came by. His guys had been hit hard the night before and needed a drink, he said, adding, "And I figured if anybody on this island had any whiskey, it would be the aviators."

I gave him what I had, a near-full bottle of Three Feathers we called "The Black Death." Smiling, he said he and his brother owned a little place in New York City and, if we got back alive, he'd treat for lunch. He made it sound like a small, family-owned Italian cafe. (When I checked one day twenty-five years later, I learned his cafe was the Club 21, one of Manhattan's most popular restaurants. I was the only guy in town who didn't need a reservation to get a table.)

Then, back in my B-24, I was sent out to spot for a *Saratoga* raid on Rabaul. The *Sara*'s assault already had begun when we arrived, Zeroes and Betties swarming like bees all around. Suddenly they acted as if *Chick's Chick* was the most attractive flower there. Joe Clifton was on the *Sara* in one of the new F4U Corsair fighters.[7] As aggressive a pilot as he had been a star on our football team, he was a bit weak on book work. I'd helped him pass his lieutenant (jg) promotion tests. This day, he repaid me. His much faster, better-armed F4U shot down three or four Zeroes and chased off the rest as we scooted away, thanking him for saving our butts.

Harry Sears, skipper of VB-104, lost a plane on the 14th. We hunted but failed to find it. "They were in an old crate left here by VB-102, forced on him by AIRPAC [Naval Air Forces Pacific], the Scrooge of the South Pacific," my diary notes. To soften up the Gilberts' Nauru island, a prime Japanese source of phosphate, we and three carriers blasted it on the 19th. Next day, the marines landed on Tarawa and Makin instead, while one of my boys, on a damage-assessment run to Nauru, saw a submarine he assumed was ours. Closer in, he saw a Rising Sun on its sail. As he swung around to attack, it submerged. Since antisubmarine warfare wasn't our job, we didn't carry depth charges. My Intelligence guy footnoted the event in his report to CINCPAC, copy to CNO King in Washington, D.C. King promptly shot me a wire: "Comply with my order SPO-220 unless otherwise directed." We had no rulebooks at all, on Cactus. What the blazes was an SPO-220?

On the 25th, I flew up to Munda, 250 miles northwest of Cactus, due to be our next homebase and "still stinking of dead Japs," my diary says. In spite of still-fierce fighting near Empress Augusta, 250 miles farther up, I also cased its Kneely Field even though no one had said we'd base there soon. While I was there, Cdr. Tommy Blackburn, from a Cactus-based F4U fighter squadron (highly decorated by war's end), arrived on a "beer run" from Cactus, his cockpit awash in broken beer bottles. He'd flown too close to Ballalae, a tiny island south of Bougainville, he said, and been nicked by "The Professor," a nickname we'd given Ballalae's very sharp antiaircraft gunner. We needled Tommy about it, a uniquely wartime kind of humor like telling jokes in a graveyard.

Back on Cactus, Hank Surface hit a submarine and Fred Tuck damaged his own plane at Kapingi. He'd been at 150 feet when one of his four-second-delay bombs blew up too soon, "And," I wrote, "our ammo on Cactus exploded. Went on for 24 hours! Steel mats where we must park are ruining our compasses. No mail for days, but no sickness, yet. And we lost escort carrier *Lipscomb Bay* off Makin. Of the skipper's staff, only John Crommelin survived. Whitey Ostrum went down with the ship." (Two dozen freighters were converted to "Baby Flattops" during the war, to haul aircraft to a combat zone and back up the big carriers in battle. They were fine for both missions but as easy to sink as the supply ships they'd been.)

On the 30th, malaria erupted but I flew to Espíritu Santo in the New Hebrides anyway, to ferry to Munda NBC radio reporter H. V. Kaltenborn and a Captain Boak who said "thank you" with four cases of liquor. And our sluggish takeoff made my crew chief admit he'd stolen and hidden aboard fifty more cases. Next day, I wrote, "Sad news. 'Mushmouth' Morton's *Wahoo* boat is missing. He'd sunk 20 ships, our 7th-best Pacific submarine score. And our VB-106 patrol-bomber 'Information Service' was cited today for helping take Tarawa and Makin."

But VD photos were shot at high tide when a reef around Tarawa is under water, a fact overlooked when marines landed on 25 November—at low tide. Half the 984 Marines killed, 2,072 wounded, in the three-day battle were hit then, wading across a five-hundred-yard wide, shoulder-deep lagoon under heavy fire. Still, of 5,000 Japanese troops and Korean laborers defending it on the first day, only 17 wounded troops and 129 laborers were still alive at the end. On 7 December, Pearl Harbor's second anniversary, I noted, "It's monsoon season but we might take Rabaul by April. If we do, we'll control all of New Britain, New Ireland, the Solomons. Our gas tanks have begun to disintegrate, five aircraft now down with faulty cells," and "Fatigue—even our bones ache—is becoming a real threat. We fly more hours than any other units in the area."

On the 10th, my patrol had to fly through a 150-mile-wide tropical front, surface winds sixty to eighty knots and I noted, "Guessed right, going over at 9,000 feet, smooth sailing in torrential rain, return the same. My years of instrument flying are paying off, even in small stuff like being able to tell a cold front from a warm one, knowing storms never go above 10,000 feet at the Poles due to the cold but can be 60,000 feet high in the tropics. Our basic school is fine but it doesn't include actual bad-weather flying. Yet, that's the best instruments-training there is and, though BuAir claims not, it can be done safely if a bad-weather veteran is co-pilot. . . . A raid smashed Kwajalein today but with eight aircraft, all types, shot down by 'ack-ack.' The Japs are getting good at it. And fighting for Bougainville as hard as they had tried to re-take Cactus."

Then on a PBY stop at Red Oliver's seaplane tender *Tangier*, I saw an SPO-220 order. It said that a sighted submarine is to be attacked at once and until and/or if it signals it's ours. The SPO-220 was for the Atlantic Tenth

Fleet with its big U-boat headache but, since most Pacific boats were ours, Nimitz had ordered, "If in doubt, don't fire." And our IFF recognition setup was a bad joke. For the range needed to reach base, we used CW Morse Code on VHF (Very High Frequency) but our subs had only VLF (Very Low Frequency) gear. They'd been told to always track our VHF and could have if properly equipped, which they weren't; and if surfaced, which they weren't all the time. So our VHF ship-contact reports went Cactus to Noumea to Brisbane to Submarine Task Force Seventy-Two at Milne Bay, New Guinea, then two days later, maybe, to a submarine.

One submarine mission was to help the PBY and Coastal Watch people pick up downed pilots. To do that, though they had no submarine "safe" zones in our area, they did in the open sea north of the Solomons. If a submarine at or on its way to its Life Guard station was in the center of a "safe" corridor, it was supposed to have twenty miles of "safe" water on each side of it. One problem was that our submariners navigated by DR (dead reckoning), so they didn't really know if they were in a safe lane or not. The other flaw was that their location went CINPAC to Brisbane to Noumea to us. By the time we got a listing, a boat could be four hundred miles from where the listing said it would be.

I talked to a submarine skipper in Tulagi, then went to see a Commodore Fife, the Task Force Seventy-two boss. We agreed, starting with the *Balao* and *Blackfish*, that we'd begin putting VHF gear from wrecked B-24s in his boats, and he'd have them send IFF signals on our frequency. And Fife served eggs for breakfast, first I'd seen in months, a delightful way to start a project. "Excellent chow," my diary says, "but I would not be a submariner. What goes up must come down but everything that goes down need not come up!"

On the 16th, we lost Al Seaman's 092. Shot up on a raid, her left landing gear was carried away as she touched down at Cactus. Bullet holes had bled pressure out of the cylinder, unlocking the gear. Luckily, months after asking, we finally received a B-24D to replace Sam Bales' airplane. And my diary says, a week later, "A brave man, Butch O'Hare, was killed, today when Al's new B-24 got shot up over Kieta."

Then on the 22nd, we were told to double up on Very Long Range Search missions in sectors one to three, the first eighteen-hour flight to go at 0200, the second at 0800. Our morale hit the floor. Nonstop flying was not why, stress-filled time wasted was. In 1942, if our base had been Cactus, not Espíritu Santo, six hundred miles south, the Japanese force that surprised our picket line at Savo Island would have been spotted at least three hundred miles further up the Slot. Later, when we were based on Cactus, Japanese convoys, Truk to Kavieng, could not have sailed most of the way unseen if we'd been based on Munda, instead. In short, we were spending half our time not in enemy areas but just getting there. Now, these new VLRSs were to watch for a strong Japanese battleforce moving down from Truk. Adm. Mineichi Koga, Japanese Combined Fleet commander since Yamamoto died, had radioed Tokyo—unaware he'd told us, too—that he planned to run a "Tokyo Express" at us, hoping to lure us into a "knockout battle" to "turn the tide of war in Japan's favor." My diary comment: "We shall see."

Then on Christmas Day, Bill Snead crashed on takeoff, killing him and two crewmen. He had feathered a runaway engine, dropping his manifold pressure. A B-24, its engine exhaust driving its turbosuperchargers, needs its maximum fifty-eight inches of pressure to lift off full-loaded, as he was. Lose an engine, lose the airplane. He'd hit the water at 160 mph, no flaps, hinting he just froze. My least experienced PPC, he'd have flown copilot if I'd had a choice. Walker, with two broken legs, pulled Briggs from the bow turret. Six others also survived. Bill was a grand guy, good man. My letter to his dad, Emerson Electric's treasurer, said so. Some "Merry Christmas."

Shortly after that, for the only time in my career, I fired our chaplain. His aloof disinterest in helping us handle Bill's crash was not all of it. He simply was unfit for the job. From day one with Fathers John Brady and Herbert Dumpstrey in Newport, I'd seen how a good chaplain can inspire both officers and men. They'd been very ecumenical, too, since we were a religious mix. (Only long after he retired in 1950 did I learn Dumpstrey was a Reformed Church pastor.) This guy didn't mix. He was a Seventh Day Adventist first, our chaplain second. Aloof most of the time, he spent his Sabbath—Saturday of course—nagging my guys on morals, drinking, and such. They didn't want to hear that stuff. All they were trying to do was stay alive. It was the men themselves, who convinced me, turning away when he approached, their language becoming more foul the closer he got. So I got us a new chaplain, a good one.

Then on 1 January 1944 with 130 flying hours in December, I wrote, "Sector 1 rough, yesterday, bad weather all the way. Still waiting for a Tokyo Express to show. I think Germany will be whipped soon. We may even avenge Pearl Harbor. They say we'll take Kavieng, but I doubt it. Game not worth the candle." And on 4 January, "Flew today through my first 'tropical disturbance,' they call it, 85-knot winds. In Florida, we call that a hurricane. Night landing on return, my radar out, of course. This is a living?"

On the 8th, back on eight-hundred-mile search, we saw a marvel, "Unless seen," I wrote, "wouldn't be believed. Shot up at Borpop, Ray Bales, on Munda approach, 300-foot ceiling, flew into the water, then out again, to land safely!" Then Dvorchak in VB-104 was reported missing. Our search, begun in a storm, heard weak signals on five hundred cycles, saw nobody. We hunted for a week, our most draining task, praying they were just adrift somewhere. Finally, I wrote, "Dvorchak's crew has joined the Missing-in-Action, more victims of the lack of will to stop the Axis monsters six years ago when it would have been easy. What fools we have been."

Heavy rains in January prevented bombing much, but one of my boys flew out of a squall right over a submarine, machine-gunned, and left. It had been the *Skate*, running to reach a Life Guard station prior to a carrier raid on Palau but 340 miles off any position we'd been given for friendly boats that day. Nimitz fired Gen. Douglas MacArthur a message: "That's the seventh attack this month by an American aircraft on one of our submarines. I want it stopped!"

Then two colonels visited me in Nadzeb, growling, "Commander, what is this?" We were only obeying Admiral King's order SPO-220, I said. Taking on the CNO was not in their pay grade. They left, heads shaking, but I told my boys, "From now on, you do not see any submarines. Even if one waves a Jap flag at you, you will not see it."

Late that month, an intercepted Japanese message to Tokyo said, "Rabaul is doomed." Bone-tired by then with no relief in sight, I sent my crew and Bill Snead's survivors to New Zealand for a rest, planning to steer all my boys, a few at a time, into that routine. All but me. "A skipper must stay at the front, in my opinion," I wrote. On 3 February, we hit the Marshalls and Al Seaman shot down two Zeroes and bombed a DD (destroyer). "He surely is a fighter," I wrote in my log. A Patrol Wing Ten survivor in 1941, he'd landed his PBY on an island on his way out of the Philippines, set up his ever-present chess board, and a tribal chief had challenged him to a game. The chief had won the game handily and "invincible Al" was obsessed with getting a rematch. I asked him, grinning, "Was it you who got MacArthur to say, 'I shall return'?" (At the time, Nimitz, wanting to head right at Japan, was jawing with MacArthur, who insisted, "according to War Plan Orange," that the navy must help him retake the Philippines first.)

A foray in early February in sector one proved that even crude submarine–B-24 teaming works. We'd just downed a Jake, listed "probable," when, late in the day, we saw a five-ship convoy leaving Kavieng for Truk. We reported course and speed, not knowing that my academy classmate Bill Nelson's *Peto* was surfaced seventy-five miles off, tuned to our frequency. Next day, we sent a plane out and, after an IFF linkup with him just beyond the convoy's line of sight, Bill submerged. As my guy traded shots with the Japanese, Bill slipped inside their perimeter and blew up a Japanese DD with his first fish. Helped by the panic that caused, he then sank all the rest of the convoy.

Back on VLRS by then, with my boys not happy and awful weather every day increasing the stress, I did a VLRS of Nauru and Ponape for Navy Central Pacific, a nice trip with no Japanese fighters. And my dairy adds, "Lousy break: Lieutenant, j.g. Corbey and his gunner, shot down on Christmas day off Kavieng, spent 27 days in a rubber raft, drifted onto an island to live with natives 14 days until we found them. Today, they died in a SCAT crash [Southern Command Air Transport, our logistics airline]. Lost, rescued, now listed again as missing only because no bodies were recovered. Will their families believe us after all that or keep hoping they will be found, again?"

In mid-February, on VLRS again, my diary snarls: "Flying over 100 hours each month, full-loaded 0200 take-offs from a marginal field, turbo-superchargers screaming, manifold at 58 inches, afraid he won't make it and ought to abort, then flying 18 hours in bad weather, under fire—the strain is starting to show. Yet, our Old Fuds seem oblivious. Hard to face my boys with their dumb decisions. The Army gets rid of its antiques by promoting experienced young aviators into a chance to run the show. We don't. The Army sure has it over us in this arena."

After I burned out two engines on one VLRS, I decided the "Old Fuds" didn't know the B-24D, either. They'd made us take off with all the juice we could carry, thirty-two gallons of oil and thirty-three hundred of gas. For takeoff safety, at my insistence, they did let the 0800 bunch go on only thirty-one-hundred gallons of gas, but weight and weather can burn up to two gallons of oil per hour. Obviously, at that rate, after sixteen hours, the pilot is flying on the margin. And B-24s didn't have an oil gauge.

8

From Air Combat to Atomic Bombs

An old adage says, "If it's going bad, wait a while. It will get worse." That month, as we suffered through the VLRS missions, I got a note from a BuPers junior officer, four months after we lost Sam Patella, asking, "Any change in the status of those people?" That airhead had chosen not to forward our sympathy letters to Sam's and his crew's families. All of us at VB-106 were enraged. First chance I got, I snapped at the BuPers top man, "It's things like that get us sort of rabid when someone mentions the Bureau of Personnel."

Then Ben Mitchell's plane, shot up strafing Nomoi, crash-landed at Cactus. His crew intact, I put them in 32084. We now had only eight operational aircraft. We'd lost four; and 078, 114, and 088, all shot up by Zekes, the Zero's other name, were in for major repairs. Yet my begging for replacements got nowhere. The navy had the planes. Ford Motor Company's Willow Run plant alone was rolling out a new B-24 every fifty-five minutes. The problem was no systematic routine to move them into combat. I snarled in my log, "Is anyone fired for this? No. They get medals, promotions, a desk job back home. When, oh, when, will we dump this deadwood!"

My next editorial added, "The real foul-ups in this war are BuPers, BuAir, Bureau of Supplies & Accounts. Their decisions imply they work in opium dens. Admiral Mahan (1840–1914) said it: since the Bureau system's creation, its people covet a status quo, resent and resist all ideas for improving the Navy. S&A mistakes would fill a library and BuAer's were summed up by Winston Churchill, 'Too little, too late.'" On 15 February, Sears bagged another Betty. It was Japan's main bomber, top speed three hundred miles per hour, three-thousand-mile range, but to do that, its fuel tanks were not self-sealing and it had a thin skin. We called it a "flying cigarette lighter." We had a contest with VB-104 to see who could shoot down more Japanese airplanes. For a Betty, we added "Lighting one with the fewest bullets." We were behind in both races at the time.

Then I heard I'd be COMNAVAIRPAC (Commander Naval Air Forces Pacific) chief of staff when VB-106 went home, but I wanted to fly, not run a war. And, I griped, "Now AIRPAC [Naval Air Forces Pacific] wants to kidnap all my bombardiers in March, for pilot training! Do it after a full bombardier tour. They'll be better pilots for it. And deployment policy is nuts. Crews should go as a team, not be replaced a man at a time. Rotating in an all-new VB is wrong, too. New B-24s should join veteran VBs a few at a time to avoid making a totally green outfit fly a mission." And: "No excellent young officers promoted ahead of anyone, yet. Nor will they be. The FUDS reign supreme."

In mid-February, Harry Sears' VB-104 shot down its seventh Betty, then left for rest in the States. My diary notes, "Hate to see Harry go but VB-104 rates it. His men think he's the Navy's best B-24 pilot. So do I. He earned the Navy Cross for taking on, alone, an armed Jap convoy, sinking its fuel tanker. He got a DFC—after the FUDS gave a Lt. Cdr. the Navy Cross for 'liaison work and sound counsel in the Far East campaign.'" Later in Washington, I growled to King's staff, "Word of awful inconsistencies like that moves down the line like a firestorm. Our people can read as well as you and I can. It's bad, horrible for morale."

On the 20th, Arleigh Burke said he wanted a spotter for a dawn attack on the 24th on Kavieng harbor. I volunteered. Skipper of a nine-ship DD squadron he called "The Little Beavers," the name of an impish comic-strip Indian boy, Burke intended to use five of his DDs and five from another squadron in a two-pronged attack. With Kavieng six-hundred-plus miles away, I had to be off by 0300 hours to be on time. By 1:30 A.M., we were under a howling hurricane, ceiling zero. My boys didn't think I should or would go, but Burke wanted us. So, with Walt Vogelsang as copilot, we cranked up my B-24, scared as usual, shoved her to full power, and went, buffeted by gusting wind, fighting to avoid being blown into high hills just left of the runway.

We were over the target, spotting for his DDs at the harbor entry and facing heavy but inaccurate "ack-ack" fire, but we were barely there when enemy aircraft swarmed in. As we ducked and fought, scared, impatient Hayward radioed Arleigh, "Hurry up and get it done. It's getting hot up here."

Arleigh snapped back, "Shut up, get off the air! The *Fahrenholt* has just been hit by a shore battery and I need this radio!" So for two hours, we dodged Japanese fighters while sinking two AKs (armed cargo ships). The attack was audacious, us bombing and spotting for the DDs while they shelled and fired torpedoes into Kavieng. The first salvo blew up an AO (aviation oil tanker) at a dock. "Beautiful!" my log says. Later, when Nimitz awarded me a Distinguished Flying Cross for that action, I learned Burke was the one who put me in for it.[1]

By mid-March 1944, the ledger for two years showed that we and the Japanese were about even in ships sunk, planes shot down. But enemy replacement rates were much lower than their loss rate, new aircraft just copies of ones lost; while in 1942 and 1943, when we lost the *Yorktown* and *Hornet*, we got eight new *Essex*-class carriers, including a second, bigger *Hornet*. For every plane lost, we got ten new ones, faster with more firepower. We were starved for B-24s, but the army flow was so large they were retiring theirs at one thousand flying hours, to return to the States as training planes.

My 19 March log reports, "Anderson is gone, shot down by an AK at Kapingi. Searles is missing in bad weather. Ray Bales and Al Seaman sank the AK that got Anderson. To my awed amusement, we got three new aircraft this week plus two back-up crews. Took only five months to get them. Rumor is, VB-116 will relieve us in late May. It also says we'll move to Green Island, then. I don't believe either one. The action is moving west, by-passing it, Truk, Rabaul and Kavieng, too. Our PATSUS set-up has gone sour, badly needs a Mitscher-style shake-up."

Next, we spotted for a big task force in a final bombardment of Kavieng. Poor Japanese devils took a murderous hit. I loved watching. And our first replacement crew flew into a Cactus mountain, killing them all. On 25 March, MacArthur took Manus in the Admiralties, Halsey attacked Guam and Saipan, and Japan was retreating everywhere. On the 26th, VB-106 was sent to Nadzab, New Guinea, in the Markham Valley west of Lae with the crocodiles, to join Southwest Pacific Command under MacArthur and ADVON 5 (Advance Headquarters, Fifth Air Force). We now had served and fought in all three Pacific Command theaters.

The Fifth Air Force pilots were young and eager but had only about four hundred flying hours; and flying in tropical storms, over the fifteen-thousand-foot Owen Stanley mountains hidden by clouds, is not a boy's job. In a single raid, bad weather had cost them thirty aircraft lost. Our planes being radar-equipped, we began to take them out, over the range, back by a seacoast route. And life in Nadzab was hell—sleeping on the ground, our belongings in cardboard boxes, dysentery, bad food, worse water, and most of us sick. Morale hit bottom, but operations were distinctly better. The army put us up front, giving us our full range over the combat areas, and our Betty shoot-downs began to catch VB-104's.

I'd totaled 292.6 flying hours for the first three months of 1944. If we had to fly that much, I argued to the brass, either send out more squadrons or relieve them at six months. My PPCs had averaged over one hundred hours a month for seven months. That, I said, just beats down pilots. We were lucky we hadn't lost any because of it; but Harry Sears, his last three weeks in combat, had lost four crews, all because of fatigue. Both Harry and I wanted an air group set-up like the army's, two squadrons assigned to an area. One twelve-plane VB can't cover eight sectors every day, anyway, but an air group could do five hundred thousand square miles each day with far less strain on aircraft and crews. Unless they have twenty-four-hour, long-range air cover, I said, our ships should not weigh anchor. Our plea got nowhere, of course, provoking another editorial: "Navy policy on big aircraft stinks: Carrier aircraft guys get medals while we fight alone out at the end of our range where a hit can mean not getting home. I swear, the only way my boys will get a medal is if they kidnap the Emperor."[2]

A Commodore Combs, our new boss, came in to chat, then I flew him up to Manus in a PBY. The Japanese were still shooting there so I tied up to the *Tangier* again. There, I learned the air force was moving us to Manus, "making us," I noted, "the fastest moving-around outfit of this war. They say we won't go home in June as I thought we might. I will not tell Lili."

But on 30 March, a message-intercept said Admiral Koga had ordered the Japanese fleet north "to defend the homeland." That told us their losses in ships, hundreds of aircraft, had canceled Koga's Tokyo Express "Plan Z" (same as Yamamoto's for Pearl Harbor). Order sent, Koga took off from Rabaul on the 31st in a four-engine "Emily" seaplane for Davao, the largest Japanese enclave in the southern Philippines. Next day, Tokyo said he'd been "lost in a high, thick fog at sea," and died. I knew that sector well. It didn't get that kind of fog. And Ben Mitchell, in there that day, had seen an Emily "just doping off," made a high pass and shot it down in thirty seconds. It probably didn't see him, never had time to send an S.O.S. Admittedly circumstantial evidence says it was Koga but, rather than give Ben a medal, our leaders chose to believe Tokyo.

Two days later, we sent six planes out—Al Seaman, "Dagwood" Davis, Frank Holt, "Johnny" Johnson, Fred Tuck, and Ben Mitchell—on a heavy night-strike of Wadke Island. I'd have gone too, but I'd been grounded by malaria, told by our boss, an army general, not to fly over enemy lines again until we returned to Manus. For months, I'd been lining my boys up every day to hand them atabrine, the only way I could be sure they took that antimalarial medicine. The shepherd obviously hadn't cared for himself as well as he had his flock. "Dagwood," wounded, got his B-24 back to crash-land at Nadzab. At Wadke, his squadron had wrecked more than three hundred aircraft and started many fires. I put Seaman and Davis in for a Distinguished Flying Cross and Davis' crew for Air Medals.

My 7 April log says, "ADVON [the army bomber command headquarters] operations are a marvel, generous with medals, too, even put us in their Presidential Distinguished Unit Citation. Still, this valley is the pits, my boys really beat, efficiency in a nose-dive the past two weeks, morale shot. And six months flying in tropical heat and humidity has rotted our electronics so badly it's no longer cost-effective to repair. These aircraft should have been shelved a month ago. We've lost six planes and 28 crewmen killed since being commissioned." Then, after a six-plane night raid of Wadke, I wrote, "Too tired to go back for a head count."

We went up to Manus then, again living in mud, the airstrip soggy and making a heavy-load takeoff more risky than on Cactus, but our SeaBees were building a big-bomber runway there. And they had food worlds better than we'd had in New Guinea. Maybe that explains why, during the next ten days, Fred Tuck and Morrison each got a Betty; Bales and Davis each sank an AK; "Dagwood" also strafed a sub; Mitchell bagged a Betty, a "Mavis" seaplane, and a "Pete"; Seaman, two AKs; Holt, two ships; and Porter, three. My boys had become a red-hot outfit.[3]

For the army's invasion of Hollandia, we began the tiring VLRS bit on the 20th and saw nothing, but I did burn out an engine once. On the 24th with our fleet off Hollandia, Bales and Johnny Johnson earned the Soldiers Medal, pulling several men from a crashed, burning bomber. By 30 April, with no word on when we'd be relieved, we heard that Secretary of the Navy Frank Knox had died that day, making his under secretary, James Forrestal, the secretary of the navy; and VD-1 had a plane down three hundred miles north of us. Its Dumbo rescue also crashed, so we had two crews to pick up. Then Bales spotted a Japanese carrier with planes on deck, a cruiser, a DD, and two large AKs. We should have gone after them, but hoping we'd soon go home had us all asking, "Can I stay alive until then?"

Next day, Al Seaman flew to Biak Island where he skip-bomb sank a fifteen-thousand-ton Maru with four 500-pound bombs but was jumped by eight Japanese fighters. He got two but the others shot out three of his engines and set fire to his left wing. "One of my best pilots, bravest men, a leader," I wrote, "he's also one of my best friends. Times like this, being skipper weighs a ton. Search starts tomorrow, Davis to leave at 0200 hours, a 'Dumbo' at dawn, Holt and I at 0800, Porter and Trusso at 10, hoping."

Then on 2 May, Lt. Cdr. Dick Somers, under 5-inch gunfire, taxied his PBY Dumbo into Biak harbor to rescue anyone still alive. He found only two. Three others, wounded, their life jackets inflated, had drowned, unable to keep their heads above water. Al, his copilot, and radioman had been pinned in and gone down with the plane since a B-24 has no cockpit escape hatch. (All the pilots in our last four combat crashes had died that way, simply by being trapped.) I insisted Somers be awarded—and he was, eventually—the Navy Cross. On the 3rd, I wrote, "Heartbroken, bitter tears. Al is dead. His son, born while he was out here, will never see him. How Barbara must ache, her husband surviving the 1941 debacle of PatWing Ten and now this. Al will never know if he could have beaten that Filipino Chief at chess."

We mourned for a week. Then Lieutenant Hardy's crew bagged a Betty with only forty-five rounds, a record! Like beating Army by six touchdowns. And Davis, fending off four fighters for an hour, hit three AKs and a Yubari cruiser. "We began this war with lots of inferior weapons," I wrote. "Our edge was in numbers of brave, dedicated people. I've met dozens of them. We're a close-knit outfit. I'm proud to be part of it. People like 'Dagwood' are why we're winning this war."

In May, we moved six planes up to Wadke, putting its sector one coverage over fifteen Japanese airfields. My diary notes, "Our seniors criticize MacArthur but these Army folks understand the power of an up-front long-range search, to bomb enemy bases, shoot down their aircraft." I'd beefed about that before but what gold braid pays attention to fresh-caught commanders?

And on the 20th, *Chick's Chick* finally bagged a Betty because I had on board Roger Schreffler, the best radar operator in the fleet. Japanese bomber pilots, we'd noticed, rarely checked to see if anyone was on their tail, maybe since Japanese fighters never made tail attacks and Betty assumed we didn't, either. So Roger got us in behind it near enough to put a few bullets in it. It promptly went down on the deck, as they usually did under attack. That dumb move was what we needed. Its 300-mph top speed was a tad faster than ours, but put a B-24's nose down in a shallow dive from eight thousand feet or so and she can run at more than 340 mph—which is what I did, on fifty-five inches manifold pressure and my props spinning at twenty-seven hundred rpm. Forty minutes of that got us close enough to strafe the Betty and drop it into the sea.

On my final mission, I got a Jake. Added to Porter's Betty bagged earlier, that made fifteen Betty aircraft VB-106 had shot down, one more than Harry Sears' boys scored. Our boss, Combs, in a note, told me, "Leave the cripples to the PBY's," reminding me that downing a Jake wasn't much of a trick. Then on the 18th, I was told we'd be relieved in early June. The smiles vanished, however, when we had a man killed, one wounded, that day, skip-bombing a Japanese ship. Then from Biak on May 26th, Frank Holt flew a photoreconnaissance of Leyte and Mindanao, the first such mission to go there since 1942. On his return, he'd come out of a cloud above and behind a nine-plane flight of "Rufes." He could have swept in and got them all, he said, been an instant hero, but, he smiled at me, "I knew how badly you wanted these photos, Chick." More likely he decided as I probably would have, "Why risk getting killed a week before we go home?"

On 27 May, VB-115 arrived to join VB-101, they said, under a group commander. I guessed, hoped, that Harry Sears, in his debrief, finally had convinced CINPAC that's how it should be. We were four shy our fifteen-plane complement, five of the rest had over 1,200 hours on them, and 1,120 was our eleven-aircraft average. Commercial airlines let a crew fly no more than 85 hours a month, and that's with good maps, beacons to steer by, well-marked airfields, and nobody shooting at them. I'd averaged 135 hours, most of my PPCs the same, often in storms at night.[4]

We gave our three least-hours B-24s to VB-115, but I still had a problem, one I'd backed into. Knowing the tropics will turn a B-24 into an old man, air force policy, ignored by the navy, was to store them on Cactus after one thousand flight hours, to ferry home to be training planes. So when we went to New Guinea short four planes, I'd sent a crew to get from Cactus the army B-24 in the best shape. "We'll use it as an extra," I said, "for your 'Fat Cat' runs to Brisbane." (A "Fat Cat" was a swap with our Aussie friends, "Shack troops," we called them, of our Japanese souvenirs for "Aussie stuff." Their brandy, I recall, was awful.) Since our combat fatigues and flight suits looked like the army's and our B-24s were like theirs except for tail numbers, no army supply guy on Cactus noticed a B-24 missing when my guys stole it.

Then I was told to give all my B-24s to Training Command. I could explain the three I'd given to VB-115. I could go to jail for having the army's. To get it on our roster, I sent it on a low-risk patrol. My kids groused, "Skipper, just declare it lost in a storm. Then it'll be our own private airplane." Still, they flew the mission but said I had to fly it to the States. Then, on the 31st, VB-106 was relieved.

I left enormously proud of my men. They had set VB-squadron records for ships sunk and damaged (ninety-seven), aircraft shot down (fifteen). The grim side was that 25 percent of the aircrews who went to war with me had been killed, half in planes shot up, unable to return to base, in a business where 10 percent is considered high. In Hawaii, as I taxied up to park at Kaneohe, a trim, beribboned guy in a neatly pressed short-pants uniform was there to greet a VIP, I assumed. I dropped through the open bomb bay doors, a B-24 pilot's exit, and hit the ground, tired, dirty, wearing a rumpled flight suit. The VIP walked over, smiling. It was Admiral Nimitz.

Most of our tour, especially in 1944, we'd lived like the infantry, wearing old fatigues, sleeping on grimy sheets, meals eaten out of cans, our gear stuffed in boxes because we never knew where we'd land from day to day. Nimitz greeting us did wonders for my boys' morale. Mine, too. I later learned that this marvelous gentleman tried to meet all his unit commanders as they came out of a combat zone. One reason was to hear firsthand about events in the war. The other was to size us up, get an idea of where we might best fit next in his command. He'd tracked VB-106's work since our Wake Island raid, he said.

He took me over to CINCPAC headquarters, gave me a "red rug" welcome, then asked me to attend his morning staff meeting to recite my views on "good" and "bad" in my sector, he said. Next day, 5 June, Nimitz; his Pacific Submarine Force guy, Vice Adm. Charley Lockwood; Vice Adm. Forrest Sherman, head of his War Plans section since late 1943; everyone was there. Before the meeting began, I told Sherman, who'd run VF-1 in my *Langley* tour, I wanted to go back out, be a carrier air-group commander, not what rumor said I was to be, chief of staff to COMNAVAIRPAC. (He'd been chief of staff to Commander Naval Air Pacific, himself, before he joined Nimitz.) His answer was a cold stare.

After a run through his routine agenda, Nimitz put me on stage, a "Commander, T" for temporary, facing all those stars. But years ago, I'd decided to do as my favorite boss, "Uncle Pete" Mitscher had advised. "One characteristic of a good leader," he'd told me, "is integrity, which includes always giving frank, honest reports to superior officers."

To start, I praised the air corps for putting its bombers up forward under group commanders, flying mostly over enemy territory while ours spend half their flight time just getting there. Moreover, I said, army policy is to limit flying hours to what planes and crews can handle effectively. Navy seems to run a VB way past a B-24's useful life, take weeks to replace those lost, and run our crews like robots that can't break. In sum, "I know carrier air is the navy's glamour girl, but until bombers are made integral to naval operations, we won't have nearly the powerful weapon as we could and should have."[5]

I beefed about all our problems with BuPers: holding our letters to the families of Sam Patella's crew; sending untrained machine gunners; sending me as navigators pilots who wanted to fly, not navigate; and so on. We had to train ourselves, pretty much, I said, during the crews' off-hours between missions. I summed up the warts we found on the B-24: leaky fuel cells; putting in the ERCO nose turret; lack of a pilot's escape hatch, which cost the lives of at least three of my PPCs. "But," I said, "its rugged durability is legendary. For instance, once, when a Jap fighter came at one of us head-on—they always attacked head-on—it ran right down our gun barrels. When it blew up, its engine sliced off nine feet of one wing. But we returned safely to base."

The next busted-plane tale I told to make two points. "Another time, in a night attack on a Jap base, the PPC dropped most of his thirty-two clusters of six 20-pound bombs at four hundred feet, 195 knots airspeed indicated. By then, antiaircraft fire had cut most of his hydraulic lines, blown big holes in his main gas tank, knocked out one engine. Fires snuffed, they turned for home, hoping they had fuel enough left to make it. On final, they cranked down the gear and dumped the flaps by hand, kicked out the nose wheel, ceiling three hundred feet—which they didn't know. Ground control hadn't answered their radio-emergency request for weather. Its fluid all leaked out, their brakes failed, and they coasted into a ditch at the end of the runway. The plane had four hundred holes in it."

One point to that story, I said, is, "The army gave that entire crew a DFC (Distinguished Flying Cross) within three weeks. There seems no rhyme nor reason to what the navy awards medals for. The air force method may be a bit generous, but it's certain. They award an Air Medal for five combat missions, a DFC and leave after twenty-five missions. And they make the awards right away, not six or nine months later, which seems to be the navy way." (The army put in citations for all my men for one medal or another that were never received. After the war, I tried to find out why. A building in St. Louis with all the records had burned down, the army said. The navy, after the war, did try to give strike-flight awards as the army had. But they had lost records, too. I did get a Navy Cross for Dagwood Davis and for Al Seaman's wife but, overall, some got medals but no citations, some a citation but no medal, some nothing at all. My guys laughed it off. No more war, everyone home, and who cares about medals and such?)

"That story's other point," I said, "is we did a lot of night-bombing for Army ADVON-5; none for the Navy. 'Cactus' wouldn't light up the runway. Adm. King made us train at night in the '30s. The Japs do it. If we had, I think we'd have pushed them much farther north by now than we have. Certainly, the 1942 night disaster off Savo Island, when Jap planes dropped flares to light up our ships for their guns, would not have happened." I almost added, but didn't, that I thought someone deserved a court-martial for that.

With that, Nimitz asked, "What was your worst problem?"

"Submarines," I said, and recited the whole story about the overfly of a Japanese boat; then being ordered to comply with SPO-220; then after learning that we had strafed the *Skate*, getting a visit from MacArthur's colonels that inspired me to tell my guys, "From now on, you don't see a submarine even when you see one."

Nimitz, angry, incredulous, demanded to know who had issued SPO-220. To keep King out of it, I said "someone in BuAer" had. Then I described what we and Lockwood's man at Milne Bay were doing, slowly, to fix IFF communications. His skippers, he said, all navigated by dead reckoning since most boats in the Pacific are ours. DR, I said, can put a boat way off any position we might have been given for it, and a radar blip just tells us we have a target, not who it is. Nimitz asked Lockwood, softly but pointedly, "Can't you do something about this?" Whatever he did, our planes never again, to my knowledge, attacked any of our own boats.

Then Sherman asked, "How about Jap aircraft? Any problems there?"

"No. Why, on our way here, we shot down a Jake easily. 'Course, he may have been just some staff guy up getting his time." Sherman froze. An ongoing pilots' beef was desk jockeys flying four hours a month of easy time for their flight-pay bonus, and aviator Sherman—who had married a Pensacola girl and knew I had, too—was now one of those staff guys. "Commander," he snarled, "I think you're overdue for transfer home."

Next day was D-Day on Normandy. The Chris Holmes estate hosted a wild bash for the officers and crews of VB-106. I took off for San Diego twenty-four hours later, to report in to Fleet Air West Coast. John Crommelin was there, in charge of training. For all my gripes, sending veterans into the training business was a smart move. With a solid cadre of those, people they send to war will be well prepared for it. We did get some "nuggets," we called them, who had to be sent home, a mystery how they ever passed flight training. But our plan was better than that of the Japanese, who just kept sending their veterans out until they got killed, assuring that most of the replacements would be "nuggets."

At Fleet Air, I was told Admiral King wanted me in Washington for "an interview," but when I tried to give my B-24 to North Island, all civilian logistics folks at the time, they wouldn't take it, saying they had no record it existed. "Okay," I said, "I'll gas up and fly it home, my own private airplane." Then Crommelin stepped in, roaring, "I want that B-24!" So they decided to accept it after all, and it just disappeared, never showing up on the books. If I'd stolen a case of whiskey, hell would have been raised everywhere, but an airplane? Just "a paperwork foul-up."

Meantime, I went to D.C. for the "Interview." Convened on the 23rd of June, the room was admirals wall to wall. Only a message from Nimitz, telling King he should hear what I had to say, could have turned out that star-studded a crowd. But it was a testy situation. I learned from flying him all around in Panama that King expected me to speak frankly, that he avoided people who did not. I also knew few of the others did, especially in the Bureau of Personnel. I went at it head-on anyway. I covered the lack of trained people assigned to VB-106; the B-24's major equipment flaws we had to correct; how we lost the eight aircraft we'd lost; the "more than two hundred engine hours wasted on training flights, a drain more careful planning and supervision could have prevented;" the SPO-220 submarine business ("If we could rid the Pacific of some submarine experts from the Atlantic, we would be a lot better off"). I covered health and morale, the crazy "awards" routine; why I believed "the big-plane business has been neglected in the navy for a long time"; the flying hours ("far too many—either send out more squadrons or relieve us in six months"); and my beef with BuPers for not forwarding our letters to the Patella families.

Lastly I said, "A particular gripe I have is for enlisted men. We, all the wing commanders, have requested an air-crewman rating, not a machinist's mate, metalsmith, whatever. Of course, BuPers, which has opposed every advance, opposes this. So a man comes to us, a machinist's mate, second class, to handle the bow turret. That's all he'll do for a year. Doesn't know a valve from a piston, never will. BuPers says we can designate him combat air crewman but he still keeps his rating. Penalize him, not advance him, for doing his job? Send him to school, make a machinist's mate of him, if you want, when he quits flying. As it is now, we have a lot of senior rates being assigned to us who know absolutely nothing about aircraft maintenance."

I ended that with "A second complaint I feel even more strongly about is our enlisted men do a fine job but seldom was one in a B-24 given an award. If he shoots down an airplane, he should get at least an Air Medal but most are turned down. Our whole navy depends on the enlisted men. If we're not looking out for their welfare, we aren't going to get very far."

What the VBs did in the South Pacific was a chronicle of heroism, tragedy, laughter, boredom, fear, and devotion to duty; and I'd have praised them more if I'd known then what I learned later. In sum, they shot down 375 enemy aircraft, wrecked hundreds more on the ground, sank dozens of enemy ships—and would've done twice that if the navy had based us up forward as the army did. Yet, except in one memoir,[6] history's few pages on navy use of the B-24s in World War II cover only the use of navy-designated models, the PB4Y-1 and -2, against German U-boats. Evidently, nobody other than Nimitz' CINCPAC noticed us much, then or since.

But that was a story for the future. Meantime, the question-and-answer drill at the end of the debrief was as barbed as my remarks had been. It was a bitter session aggravated by my malaria acting up again. Alone later, King said I wasn't going back to the war. I was peeved no end. I'd done a good job, I argued, and wanted another shot. No, he said, after leave, I was to report to a new twenty-five-hundred-square-mile Naval Ordnance Test Station (NOTS) at Inyokern, California, as its first experimental officer, the person in charge of testing new military inventions. Capt. Sherman Everett "Ev" Burroughs was its director, but it was really run by the California Institute of Technology, he said. I was to give them any support they say they need and absolutely not tell any outsider what I was doing "except me or my vice chief, Admiral Purnell, but only if we ask."

At CalTech, he added, my key contacts, both Nobel Prize winners in physics, would be Dr. Richard C. Tolman, a fluid dynamics expert, and Dr. Charles Lauritsen, inventor of the first electroscope.[7] After less than ten minutes telling me all that, King had me taken to a doctor to treat my malaria. Then I headed for paradise: reunion with Lili, Mommie Floss fussing over us, and her maid cooking tasty food while I drank most of Daddy Pete's scotch whiskey supply.

Leave ended, I told Lili I'd send for her and the girls once I knew "what's what" at that arcane place in the Mojave desert, then caught a Naval Air Transport Service DC-3 all-night flight to Los Angeles. Droning along, not knowing just what my new job was, I thought of writing now-Vice Adm. Mitscher to ask if I could join his Fast Carrier Task Force Fifty-eight, which had just won a Philippine Sea battle his pilots happily nicknamed, "The Great Marianas Turkey Shoot." One reason I didn't was that "Uncle Pete" had been, in 1943, a strong, effective advocate for creating China Lake in the first place.

From Los Angeles, I drove to CalTech in Pasadena. It had contracts from OSRD (Office of Scientific Research and Development) to develop air- and ground-launched rockets and work on a "Manhattan Project" run by army Brig. Gen. Leslie R. Groves. To me, these CalTech folks were a very likeable bunch, instant friends and, in the science arena, awesome. They held enough Nobel and other science awards to paper a wall. The two in charge of my rocket-test agenda were Nobel astrophysicist W. A. "Willy" Fowler and his boss, "Charlie" Lauritsen, "designs fuses, can make anything work," Willy said. Dr. Carl Anderson, Nobelist for discovering the positron, my "rocket-launcher man," he said, was also a "makes anything work" guy.

CalTech's Tolman told me that Roosevelt had set up the National Defense Research Committee in 1940 "to help the army and navy develop war materials." FDR had named Dr. Vannevar Bush, president of Washington-based Carnegie Institution, as its chairman and Tolman as vice chairman. It was replaced by OSRD, created in June 1941, with Bush as director, the new "Office" and "director" hinting that he, as chief advisor to FDR on military science projects, had lots of "clout." After Tolman, I met Dr. "Ike" Bowen, the Wilson Observatory's boss and maybe the world's best optics man; Fred Lindvall, Carl Watson, and Leverett Davis, all tops in some technical field or other; and Fred Hovde, head of OSRD Section Three, "Our money man," they said.

On 4 August, a friend drove me out to "Inyokern," a derivation of an ancient Indian word meaning, "Place to pass by; don't stop here." As we pulled up over Red Rock Canyon, I saw why. Besides the tiny town, there was nothing but cactus, scrub brush—Death Valley was only fifty miles away—six quonset huts, bulldozers stirring up gritty dust, and a pair of buildings going up. I was shocked. Their short, eleven-year-old one-runway Harvey Field wasn't very useful, but the range was busy firing ground-launched rockets. "This," I muttered, "will teach you to mouth off to Forrest Sherman and the Bureau of Personnel." Coming off the ridge, I spotted bone-dry China Lake, so-named because Chinese immigrants from L.A. used to hike in to pan for gold when snowmelt off the Sierra Nevada filled it with water, briefly. The man who'd set up our test ranges said he'd had to pay them off. Why, I didn't think to ask and he didn't say.

At check-in, I met "Ev" Burroughs' technical director (navalese for "top civilian research and development guy"), Dr. L. T. E. "Tommy" Thompson, who had been a pal of Ev and of Deak Parsons since they were in BuOrd—Deak as experimental officer and Tommy as technical director at Dahlgren Proving Ground. NOTS' divisions, Ev said, were Science; Ordnance; Explosives; Navy Liaison; and mine, "in some ways, most important," Experimental Operations. His summary was simplicity, itself: I was to schedule and direct rocket tests and build a permanent ordnance R&D and test capability at China Lake. "That's your job," he said. "Go do it. If you need anything, just let me know." (Today, it is called China Lake Naval Weapons Center and, with some six thousand employees, is the navy's largest research, development, and test complex.) This, I told myself, is my kind of guy, a leader with a usually very productive ability to delegate authority right along with concomitant responsibility.

Once "Top Secret" cleared, Ev said, I'd also be working for Groves' Manhattan Engineer District and the folks on something called Project Y at Los Alamos, New Mexico. I'd be working with Drs. J. Robert Oppenheimer, Edward Teller, Luis Alvarez, Eric Jette, Neils Bohr, George Kistiakowski, Ernest Lawrence, John von Neumann, Enrico Fermi, and Cdr. Deak Parsons. "As he did you, Admiral King, a year ago last May, sent Parsons out to work for Oppenheimer, director of 'Site Y,' the Los Alamos Laboratory. By the way," he added, "of Site Y's half-dozen associate directors, Deak is the only one in uniform, the only one not a nationally eminent scientist."

Ev had asked for me, he said, because of my instruments and test-flight work at the Naval Aircraft Factory, because I could give a combat pilot's perspective to NOTS' rocket projects, and because I had a degree in nuclear physics from Moore. (Fact is, I did have the credits for a master's degree, but the war had come along to deny me time to write my thesis, "Capturing a Cross-section of Low-energy Neutrons against Certain Materials.") Besides, he said, he'd asked Deak's advice, and Deak, who knew that stuff about me and my opinion of Bureau bureaucrats, had said, "Get Chick."

It was a compliment, given Deak's own pioneering in navy-applied science but a pacifier too, really. Since late 1943 when Navy Secretary Frank Knox ordered NOTS set up, BuAer and BuOrd had fought over who should run China Lake. A hint to the intensity was Ev's being an Ordnance PG (Post Graduate School), a good one, I'd discover; while I was as popular at BuAer as a head cold. But when I was asked for, BuAer had said, "Oh, well, at least he's an aviator."

9

A Sailor's Start into the Nuclear Age

A basic rule in the R&D business is, "If a program manager doesn't control the assets, he's not running the program." BuAer, busy building its own rocket-test range at Pax River, could have frozen our program simply by not giving us any aircraft. (We did get them, all kinds of them, but the BuAer-BuOrd hassle was still so nasty in 1945, I called the White House [after Truman became president], where a two-star navy aide of Adm. Bill Leahy promised, "BuAer will support you.") As early as 1940, Deak Parsons had urged the NRL (Naval Research Laboratory) to make solid-fuel propellant and the Dahlgren Proving Grounds to build rockets. Finally in 1946, NRL began to do what he'd asked, but BuOrd's Dahlgren didn't. "We can't waste time on little rocket things," they sneered. "We're in the gun business." Their lack of helpfulness was one reason Deak endorsed the creating of China Lake.

Anyway, by 1943, Dahlgren had run out of room. Like the German rocket-test site built at Peenemünde in 1937, we wanted a range at least fifty miles long, laced with cameras and telemetry devices to track a rocket's speed and to record if it did or did not hit a spot in space preset by x (horizontal distance), y (height), z (hypotenuse) triangulation—the pendulum principle's trigonometry, basis for today's inertial guidance. China Lake's usually cloudless sky was a benefit and its ranges were isolated, like Peenemünde, so we didn't need to worry about killing curious spectators if a shot went haywire. And like Peenemünde, the remoteness would shield our work from spies and just-nosy officials with no need-to-know.[1]

We also were crosswise to a BuAer-BuOrd fight over who should make rockets to hang on aircraft. They and NRL wanted China Lake limited to testing, while Ev Burroughs wanted us to do R&D too. So he talked OSRD into giving us work and sold BuOrd's chief, Rear Adm. W. H. P. "Spike" Blandy, on us making navy rockets. Blandy probably was an easy sell. Soon to lead Amphibious Group One in 1944 and 1945 landings at Kwajalein, Saipan, Palau, Iwo Jima, and Okinawa, he was irked at BuOrd's slowness moving a weapon into production and at the long list by then of their failures—another reason for creating China Lake.

But what really peeved BuOrd and BuAer were the new kids in town, very bright civilians being paid to do what had been exclusively their domains. In that sense, we were indeed an "experimental center." I had range officers for rocket work but very few enlisted men. Mainly, NOTS was civilian contractors building laboratories, factories, housing, and hangars for our and the Manhattan Project's aircraft; making Harvey Field's airstrip into two ten-thousand-foot runways; and CalTech civilians in a navy base making air corps bombs, navy/marine corps rockets—all the costs paid by a combination of the Army Corps of Engineers, OSRD, and the navy.

And like Deak, I'd joined "the enemy," the scientists, not the Bureau bureaucrats who insisted that R&D "must be run the navy way, by officers." My real boss, Charlie Lauritsen, and I didn't agree. Nor did Oppenheimer and Parsons. In May 1943, on a train, riding together from Washington to take over Los Alamos, "Oppie" (as Deak called him) had sorted it out. His scientists would develop the nuclear "gadget," as they called it, for Deak to engineer into a bomb that military folks could handle.

Excepting Blandy and his 1944–1945 successor, Malcolm Shoeffel, to the navy letting a civilian scientist run the show was blasphemy. The "Brass" believed anything new was "disruptive of good order" but "new" is, of course, what scientists do. However, the answer to my saying that often was a nasty suggestion that I consider early retirement.[2] But when FDR had set up the National Defense Research Committee in 1940, converting it into OSRD a year later, he said its job, simply put, was to fund development of new, better combat equipment—such as the atomic bomb—and, as important, see that it got deployed. And OSRD had barrels of money. What it needed most were allies in uniform, preferably line officers with a science education. Indeed, Vannevar Bush, OSRD's director, had given the atomic bomb project to the War Department, the army, because he said, "Navy officers, especially some in NRL and the Bureaus, lack sufficient respect for or ability to work cooperatively with civilian scientists."[3]

Dr. Robert Pearson, the secretary of the navy's science advisor, had used that indictment to help him convince Frank Knox to launch China Lake. Pearson wanted "to break OSRD loose," he said, "from the bureaucratic mind-set of BuOrd and BuAer." His being in Fred Hovde's Section Three, the OSRD money tree, had led Knox to his decision. So had knowing that OSRD had never turned down a project recommended by CalTech, which wanted China Lake built. Pearson's indictment of the BuOrd-BuAer mindset was valid, in general, but the navy had some stellar exceptions.

One was Cdr. Frederick L. "Dick" Ashworth. He had not been sent to the usual sea-duty routine after Naval Academy and Pensacola graduation. Instead, because of his brilliant academic record, especially in physics, his assignments in the 1930s had let him stay in touch with technical people and, especially, nuclear research. Parsons had moved him to Los Alamos in 1943 as Parsons' right arm. Another guy he brought there, in 1944, was naval reserve Lt. Norris Bradbury. Bradbury had been with NRL earlier but also a physicist at Stanford University. At Los Alamos, Oppie was to him like chairman of the board to chief executive—Oppie more the theoretical guy, deciding what was to be done, and Norris deciding how to make it happen.

However, Deak Parsons was the best example of a naval officer willing, even eager, to work with civilian scientists. (When William S. Parsons entered the Naval Academy in 1918, his classmates made "Deacon" out of "Parsons," later shortening it to "Deak" Parsons.) Long before his academy classmate, Hyman Rickover, Deak Parsons was the first to back Dr. Ross Gunn, NRL's chief physicist, in wanting to develop a nuclear engine to drive a submarine, "make it a true submarine that never has to surface." (Interestingly, Gunn had gotten the idea in 1939 from a speech on nuclear fission by Enrico Fermi, an Italian physicist who had come to the States in January. Noting that German physicists, using Fermi's irradiation work [earning him a Nobel Prize in physics] had split uranium atoms in 1938, Fermi said we had to build an atomic bomb before the Germans did. But what Gunn heard was, "We can make a nuclear reactor to drive a ship.")

In June 1939, NRL's boss, Rear Adm. Harold "Ike" Bowen, gave Gunn $1,500 to research his idea—ten months before Fermi got $6,000 from Bush to build an "atomic pile." Gunn began working with Bush's Carnegie Institution where Dr. Philip Abelson had invented a thermal-diffusion way to separate from natural uranium the less than 1 percent that is fissionable U-235. (Most of a cube of uranium is U-238, which is not fissionable, but convertible to plutonium, which is.) And Deak Parsons, then at Dahlgren, began promoting and staying abreast of Gunn's work.

But further back than that, in 1933 when Deak was named BuOrd liaison to an obscure outfit called the NRL, he was impressed by its use of high-frequency radio waves to detect ships and aircraft invisible to the naked eye. And he was astounded at how few navy officers saw the dramatic potential in "radar." He went back to the project after sea duty, got the OSRD involved in 1941, and he and OSRD ended up getting credit for the navy having air-

and ship-borne detection radar and, on our big ships, radar fire control by the time Pearl Harbor hit us.

Deak also authored a proximity fuse of the type that both the British and Germans also were trying to make. As we'd known for years, antiaircraft gunners simply can't gauge an aircraft's speed or flight path rapidly or precisely enough to set a shell's mechanical fuse so it will explode on the target. So they just fired a cloud of flak, hoping aircraft would be crippled by flying into it. Deak and two men at Johns Hopkins' Applied Physics Lab—Merle Tuve, its top physicist, and Larry Hafstad—used OSRD funds to make and put into a 5-inch shell a tiny transceiver that would emit radio waves after firing—quite a trick, given the g force that hits it then. The closer the transceiver is to a plane, the stronger the pulse. When near enough to damage the aircraft, it detonates the shell.

Called the VT (variable time) fuse, it was rushed to production at Eastman Kodak in late 1942 while still being tested. And on a secret mission, Parsons took five thousand VTs aboard the light cruiser *Helena* at New Guinea, looking for a fight. He found one on 5 January 1943 when four Japanese Aichi dive bombers attacked a navy task force that included the *Helena* and eluded hundreds of antiaircraft rounds to damage three ships. As the Japanese flew off untouched, one ran past the *Helena*. Two rounds from one battery, now firing VT-fuse shells, sent the Japanese pilot to a fiery death.

When I arrived at NOTS in August, Deak was called the "godfather of China Lake" since our atomic bomb and rocket work, our two big projects, had been his idea. His creativity, energy, persistence, and what he achieved rank him, for me, the smartest scientist in uniform. In 1960, that informal title went to Capt. Levering Smith, technical director on the *Polaris* program, but I think Deak would have earned it by then, if he hadn't died suddenly, years earlier.

Most of that was in the future, obviously, in 1944 when my file of things to do was thick. After Ev's briefing and finding a bunk in a quonset hut, I flew to CalTech to start the security-clearance bit and rent a house for Lili and the girls near Pasadena. I'd room in the barracks during the week, I told her, and spend weekends with the family until our quarters were built at Inyokern. For the next eighteen months, as it turned out, I didn't see them a whole lot more than I had when I was fighting full-time in the South Pacific.

By September, I was cleared for Project Y. The reasons Brigadier General Groves hired me, he said, were that I could drive airplanes and wasn't in the air corps, the "Arrogant Corps," he called it. Marks against me were his hatred of my New York Yankees and his howling fits if Navy's football team beat Army. An able leader on the Project, from what I saw, he seemed to enjoy giving me the needle now and then, which, interestingly, created a friendly, frank rapport between us eventually.

At Inyokern, he had a few fifty-passenger C-54 transports called "Green Hornets" to move people and supplies around. He didn't let people at Los Alamos, Hanford, Washington, or Oak Ridge, or the other shops working on the "gadget" just go visit each other. He wanted to keep each one ignorant of what the others were doing. But he authorized me to fly him, Oppie, Parsons, and his other need-to-know types everywhere. Thus, I rather quickly acquired their perspective on what was happening all over, while the people we visited had no idea what I did.

A frequent flight was to Washington, D.C., to OSRD Section Three. They doled out money with no fuss, $2 billion during the war for just atomic bombs (ten times that in today's dollars), but required frequent reports of where we'd spent what. Often on that run was Johnny von Neumann, one of the world's smartest mathematicians, and his wife, Clarie—violating a navy regulation but after all, I told myself, this is war. And I discovered a funny fact about Johnny. In stormy weather, being bounced around, he was fine if I put him up front doing the navigation mathematics for me. But on a clear day, riding in back as just a passenger, he usually got airsick.

Soon after I arrived, China Lake got its first aircraft and I started a private project, seeing the next of kin of every man I'd lost in the war. I sometimes had to fly to obscure places. The visits were very difficult, telling strangers who I was and what had happened to a husband, son, or brother. But they were owed. A "Bureau of Personnel, U.S. Navy" telegram saying "We regret to inform you" is so cold and impersonal a thing. My other personal mission was reviving a habit begun before the war, taking graduate-level science courses at night. Since I would be flying to CalTech frequently for three years, enrolling there was easy. With rockets a NOTS high priority, my CalTech studies were mainly in ballistics and chemistry. In 1947 and 1948, working at Sandia on nuclear projects all day, my doctorate studies at New Mexico University were on atmospheric physics and nuclear processes. Then in 1949 when I was back working for the navy but still on the West Coast, I took a pair of doctorate courses on compressible and supersonic flows, and theoretical chemistry.[4]

But in what amounted to private tutoring, I learned more about nuclear physics from Oppenheimer, Neumann, Fermi, Ed Teller, Ernest Lawrence, and the rest at Los Alamos. From 8:00 to 9:00 A.M. each Monday on "The Hill," their name for what had been a boy's school, some guy would expound on a new discovery or wrinkle in science, in effect telling me how little I knew of what I thought I knew. It was a student's paradise, really, being with the brightest minds in the country, on the scientific frontier of physics, optics, chemistry, mathematics—everything.

Oppie was especially excellent. He had led the field in using Neils Bohr's quantum mechanics theories to evaluate molecular properties and, with Lawrence, analyze the structure of atoms. He had assisted when Lawrence invented the cyclotron for magnetic separation of isotopes. More than anyone else, Oppie taught me quantum mechanics; why all living things, including people, are radioactive; how Bohr's and Einstein's theorems were the basis for the atomic bomb and could lead (as we since have seen) to amazing gains in science, medicine, aviation and space research, solid-state computers, and more. I thanked him often, telling him that by teaching me quantum mechanics, he'd proved he could teach it to anybody.

By 1947, I had the course credits for a nuclear physics PhD, but neither time nor urge to do the required doctoral dissertation merely to tack a few initials behind my name. Poking into many technical fields, I was, my science friends said, a "tramp scholar." The University of Portland did give me a Doctor of Science degree, later, but Ernest Lawrence conferred what I rate my highest scholastic honor, a "diploma" ranking me a "Physicist, Third Class," to encourage me, he said, smiling, "to strike for First Class." Lawrence, Lauritsen, George Kistiakowski with his amazing corporate memory, and others were always teasing me, their way of saying they liked me, as I liked and admired them. But in the bureaus that made me "a rebel."

On the way to rent my family a house in August 1944, I stopped at Muroc Dry Lake, an army field (now Edwards Air Force Base) near NOTS, to fly a jet-powered P-59. It performed nicely, its lines much cleaner than piston-engine fighters, but its engine's very low thrust-to-weight ratio, typical of all the early jets, made it slower than an F4U, for example, so it never went to war. However, that flight did make me the twelfth navy pilot to become jet-qualified, thus a member of the Golden Eagles. (From Muroc later, I also flew the far better Lockheed P-80 jet.) I also earned a growl from someone in BuAer for my "sneaking off to fly an army plane." That was undeserved. It was a well-kept secret in 1944 and 1945, sure, but NOTS was pioneering by then in developing producing rockets for lots of others besides just naval aviation.

Lauritsen had started it, gone on "a rampage," it was said, in 1941, demanding top priority for missile development. At the time, we had only Dr. Robert Goddard's solo effort, dating back to 1911, and some OSRD work by Dr. Clarence Hickman. Lauritsen's campaign got a lift when Goddard was named BuAer research director in 1942. Germany's successful firing in October 1942 of its 46-foot-long, liquid-fueled, ground-launched V-2 rocket, carrying a one-ton warhead, helped too, ironically. So did its jet-powered V-1, first flown in December 1943, an unmanned toy airplane, really, both air- and ground-launched with a one-ton warhead. During 1944, more than eighteen thousand V-1s and V-2s were fired at England, Belgium, and France. Though less than one in five hit the intended target, they did a great deal of damage, psychological as much as physical.

Once again, Parsons was the first to promote, within the navy, the potential inherent in Goddard's work. Deak's one "course correction," so to speak, a prophetic one as it turned out, was in propulsion. Goddard, in fact everyone in the field, had used LOX (liquid oxygen) or some such cryogenic fuel. Obviously, stored in a ship or submarine, it created a quick way to self-destruct, so Deak had us work on solid propellants. These had been avoided until then because, once lit, a solid's burn rate usually shot right up to an explosion. What they needed was a cylinder-shaped solid fuel, a "grain," it was called, that burned essentially like that in a Fourth-of-July rocket.

The thrust in a solid propellant is a function of the burning surface. If there are cracks in a cylinder, or "grain," of it, the pressure goes up by the square of the increased surface—in other words, explosively. Thus, if the grain has cracks in it, it will explode when lit, blowing off its aircraft's wing. But in 1944, we didn't know how to cast large grains of the solid fuel we wanted, a form of nitroglycerine called ballistite, without getting cracks in it. So, against good old BuOrd's basic-safety rules, we used a heat press to extrude the ballistite cylinder. Still, at that stage in our learning curve, even extrusion didn't give us a grain that burned like, say, a cigarette. We had to put inhibitors on it to dampen the burn-rate pressure.

Similarly, making the cells was ticklish. Squeeze ballistite too hard, until its temperature is 130 to 140 degrees, and it will blow. That's why the army never adopted it for their weapons, and why BuOrd, which rated rockets a puny weapon anyway, didn't want any part of it. They were sure we'd kill ourselves. Not Lauritsen or his propellant man, CalTech's Dr. Bruce Sage. Using a design by a Commander Duncan (a very elderly man who, ironically, had the top-scientist job at BuOrd), they had us build the plant and extrusion presses at China Lake ourselves.

The design assumed that our "oven shop," as we called it, would blow up occasionally. So the process was set up to have a piece of ballistite ignite if it all was almost hot enough to explode. That told us to evacuate. We did, about once a month, with nobody ever hurt. We just put the press back together, the roof back on, and were at work next day. Initially, it was our Salt Wells plant but expanded later, doing fuse research in addition to propellant chemistry. At Lauritsen's urging, we named it the Michelson Laboratory to honor an 1887 Naval Academy graduate who was the first American to win, in 1907, the Nobel Prize in physics.

One other problem we had was that Deak wanted a way to hit a target without having to radiate electronics—that is, to have what we now call "smart bombs." The question was this: With the Earth spinning at nine hundred miles an hour, where is an airplane or missile in relation to the spots on Earth where it took off and where it wants to land? The answer, Deak saw, was in the formula for the period of a pendulum: the square root of its length over gravity. One translation, called an eighty-four-minute pendulum, says basically that if a pendulum can be stabilized and fooled into thinking it is the length of the Earth's radius, then by measuring a rocket's acceleration and velocity against it, a person can tell where the rocket is in relation to both launch point and intended target.

To do that required a gyro-type platform of accelerometers to record the acceleration length along all three course axes, x (horizontal), y (vertical), and z (diagonal), then integrate those twice to track in three dimensions the distance the rocket has traveled relative to its target. We had problems gauging the gyro's drift rate and the accelerometers' accuracy. A basic problem was that if the gyro-stabilized platform wasn't vertical, rocket acceleration and trajectory would have an error in it. It was the beginning, of course, of what are today's inertial navigation and guidance systems. Now, with Poseidon and Trident submarine-launched missiles, course correction can be done off just one star, and we have a laser ring-gyro so accurate that a missile, thrown sixty-five hundred miles away, will hit within three hundred feet of the target's bull's-eye. But Deak started it all. That man had so much imagination, was so smart.

Out on the firing ranges, we filmed our rocket successes and failures on state-of-the-art cameras made by CalTech's Ike Bowen—one of which took pictures at 875,000 frames per second. We filmed to see how accurate a rocket was at hitting an intended x, y, z spot in space, of course, but also how a weapon moved ballistically, how stable it was. In seeing that tests CalTech wanted were done, Oppenheimer taught me early on that test data, to a scientist, is not information. It's only words and numbers. Information, he said, is knowing something that will change the probability of something else happening after a test has told him what he didn't know before. I was always under pressure to give them data that would lead to information.

And both BuAer and the air corps were generous to a fault. If we didn't have a plane we wanted, they'd get it for us: the army DC-3, B-24, B-25,

P-51 Mustang, P-47 Thunderbolt; our own F-4F Wildcat, F6F Hellcat, SB2-C Helldiver, SBD Dauntless—any plane able to carry a rocket, we had. I flew and rocket-fired off a lot of them including the F6-F and its follow-on, the 435-mph Grumman XF7F-1. The Grumman was a beautiful airplane, I thought, but it wasn't very carrier-suitable, though the marines did buy a few toward the end of World War II.

Few outsiders knew we were in the rocket business, but China Lake's record from 1944 to 1946 was pretty good. We made the first ramjet-powered rocket, "Bumble Bee," and a 3-inch-diameter anti-surfaced-submarine rocket we named the "Mousetrap." We also cranked out hundreds of 5-inch, spin-stabilized Amphibious Landing Rockets (ALRs)—they never got a nickname—for launch from a barge. They were the result of seeing at Tarawa that even a two-day navy ship bombardment won't hurt dug-in defenses much. But with ALRs, once our big guns stopped, marines going ashore in barges could rain 5-inch shells on the exposed enemy. ALRs also could bust up the Japanese "kamikaze" attacks such as we'd first seen at the Battle of Leyte Gulf in October 1944.

We also developed two kinds of 5-inch-diameter air-launched rocket, one like the ALR and a better one we listed as an HVAR (High-Velocity Air-to-Ground Rocket) with a velocity of twelve hundred feet per second. At the time, BuOrd's Captain Moses, who opposed any new idea, was saying we were not likely to produce anything useful. So Charlie Lauritsen said, "Let's needle him. Let's call it the 'Holy Moses.'" We did and followed it with "Zuni," same size as the "Holy Moses," but with an improved ballistite, giving it a velocity of fifteen hundred feet per second.

We also built "Tiny Tim," an 11.75-inch HVAR with a bigger propellant grain than Moses'; and "Big Richard," a 12.5-inch HVAR of even larger grain than that. All Big Richard taught us, leading to its early retirement, was that the larger the grain, the more chances of cracks in its fuel cylinder, making it blow up on launch, of course. Tiny Tim was a different problem. Mounted under the aircraft, a 5-inch rocket was so small, its time-constant so short in leaving the aircraft that its ignition didn't damage the plane. But when we went to an 11.75-inch size, we had to put it on a rail, inboard. And when we first fired it, off a Curtiss SB-2C dive-bomber, it blew off the aircraft's tail, killing the pilot, Jack Armitage, the reason, sadly enough, that China Lake's airfield was christened Armitage.

That was the only tragedy we suffered in 1944 in rocket development and it wasn't the ballistite's fault. (One that was, indirectly, was the *Franklin*'s disaster off Okinawa prior to our landing there in March 1945, when two 500-

pound armor-piercing bombs drilled down to her hangar deck before exploding. That set off rockets hanging on her combat-ready aircraft, creating an inferno only ballistite can make.) CalTech's Carl Anderson devised a neat way to prevent another Armitage tragedy. He attached a lanyard to the rocket so that, once the rocket has dropped six feet off the rail, the lanyard jerks loose, igniting the rocket. All big rockets would be launched that way for the next twenty-five years or so. We test-fired Tiny Tim for penetration and velocity against concrete walls on our range since it was to be not just an anti-ship rocket but mainly to blast the concrete bunkers the Japanese favored for their island defenses.

One other forward-looking project started then was to build an air-to-air weapon. Dr. Bill McLean was project manager, Vince de Poix the project pilot. One mild glitch in that program was that de Poix wanted to go to the Olympic Games to compete in fencing matches but I wouldn't let him, nice guy though he was. Made him mad. (Later, he captained the *Enterprise* for one tour and became a vice admiral.) We had a lot of hot technical arguments in 1945 over how best to go air-to-air. We ended up taking the first steps toward what is today's Sidewinder.

Through it all, of course, we also trained fleet squadrons to shoot those weapons. In short, we helped turn our fighter aircraft into very lethal weapon systems, firing accurate 5-inch shells right at the enemy instead of scattering bombs dropped in his general area. By war's end, we were making $100 million a month worth of rockets; and our ships, aircraft, and troops were firing them at a rate as high as seventy thousand to eighty thousand a month. After the war, of course, China Lake and lots of other folks moved into the long-range guided missile era. Initially, however, a recurring event was visits by curious, unauthorized people asking why the navy was stuck out in a hunk of desert larger than Rhode Island. I recall a marine general landing at our airstrip once, and asking, "What the hell do you guys do around here? Don't you know there's a war on?"

"Oh," I said, "we do lots of things—like make rockets." After I showed him how we shot the barge-rocket off our Jeep vehicles, he left very pleased. As time passed, I found out that if we talked rockets, every visitor just assumed that was all we were doing. But this "Can't talk" stuff was raising a wall between me and much of the navy. One example was Adm. Royal Ingersoll's visit. Flashing his four stars at thirty-five-year-old Commander Hayward, who he'd been told commands all sorts of foreign-looking airplanes and some B-29s with a unique tube-shaped fuselage, he growled at me, "What do you use the B-29s for?"

"Admiral, I can't discuss that with you. It's classified."

Puffing up in all his beribboned glory, he glowered, "Young man, don't you know if it's important, I would know about it?"

"Yes, sir," I said—but he left without an answer to his question. None of us junior officers—Parsons, Burroughs, Bradbury, Ashworth and so on, boosted our image any with that kind of behavior. We had no option, of course, given Groves' and King's orders, reflecting a conviction that the atomic work was so vital to the nation's war effort. But that "Don't Talk to Anybody" rule was imposing a heavy personal burden on many of us.

To some extent, what launched China Lake started our atomic-bomb project, too. By 1939, inspired by Fermi's and German research, scientists everywhere were trying to make atoms fission. And in 1939 at Columbia University, Fermi, Leo Szilard, and two colleagues validated a theory for creating a sustained chain reaction. (Szilard, known as "an eccentric genius," was a Hungarian theoretical physicist who, to escape the Nazis, had fled to Vienna in 1933, on to London, and finally to America in 1938.) On 2 August 1939, urged by Szilard to do it, Einstein wrote to Roosevelt, telling him that Germany was trying to build an atomic bomb. Evidently, Einstein wasn't in a panic, however. Not until 11 October, hand-delivered, did his letter reach FDR.

And not until mid-1941, when OSRD was created and a Policy Committee of scientists and military officers set up to oversee OSRD projects, did the atomic-bomb project catch fire, inspired mostly by a very active British atomic-research program. War generally and Pearl Harbor specifically turned what had been a scientific challenge into an urgent military race. The DSM (for Development of Substitute Materials) Project, the atomic-bomb task's first name, all our R&D projects, in fact, suddenly were flush with cash. OSRD made Chicago University physicist Arthur Compton the head of the Uranium Advisory Committee, and he promptly consolidated all fission-related work from Columbia, Princeton, Chicago, and elsewhere into one site, his university. There, in what Fermi called "Chicago Pile One," the first man-made, controlled nuclear chain reaction occurred at 3:42 P.M., on 2 December 1942, in Chicago Pile One.

Earlier, in June, Groves' Manhattan Engineer District had been created to see that facilities were built and anything else they wanted was provided for the folks on the DSM Project (recoded Project Y by the time I arrived at China Lake and Fermi moved, a month later, to Los Alamos.) After the historic December event, a Met Lab–Du Pont team left for Hanford, Washington, to build a reactor with a million times the half watt generated by Fermi's Chicago Pile. By early 1944, our principal plants were Los Alamos;

Oak Ridge in Tennessee; Hanford; Monsanto's Mound Lab in Portsmouth, Ohio; China Lake; and Wendover Air Base, Utah—and we were developing two "fission" (meaning split the atoms) weapons. One, named "Little Boy," required 62.5 kilograms (167.6 pounds) of at least 92 percent pure U-235. (35 percent pure will run nuclear-powered submarines, a long-way-off project in 1944.) The other, "Fat Man," held 6.25 kilograms of plutonium, used because they knew plutonium would yield more energy than U-235.

Hanford was bombarding U-238 to extract plutonium 236, but what it sent Dr. Eric Jette's Los Alamos DP East plant was a liquid chemical "sludge." To shield themselves from deadly radiation, his people had to use remote control of his fast-neutron reactor to pull from the sludge and convert to metal its weapons-grade plutonium 49. (Jette called it "Clementine," from the lyric, "Miner, forty-niner, and his daughter, Clementine.") Oak Ridge was extracting U-235 from uranium. That uranium hexachloride, "yellow cake," they called it, like the "sludge" from Hanford, also had to be converted to a metal so it would fission. If it wasn't, hitting it with a detonator would be like dropping a rock onto a dust pile. The result would be not an explosion but a thud. Monsanto was making the very toxic polonium-84 device, Fat Man's explosion-initiator; Wendover was our B-29 bomber base; and China Lake was making Fat Man detonators, testing bomb components, and scoring on our ranges the results of Wendover training exercises.

Oppenheimer, Parsons, and crew did the Little Boy "Gun system" first because they knew Oak Ridge could extract U-235 and knew Los Alamos's DP East could convert it to metal. They weren't sure at the outset if irradiating U-238 would produce fissionable plutonium, nor how to implode it with enough force to make Fat Man fission. Making Little Boy work was easy. Any high school physics student today can explain fission, but in 1944 not many people of any age could. (High school physics texts said then, "An atom is the smallest unit of matter and can't be split.") As scientists knew, an atom is mostly empty space. Its largest mass, positively charged, is its nucleus, inside an orbiting swarm of negatively charged electrons and other particles.

When a neutron, which hasn't any charge, hits a fissionable atom's nucleus, it will split, releasing both energy and more neutrons to hit and split more nuclei. In mathematical terms, self-sustained fission occurs if the reproduction rate, k, equals 1 to about 1.01, the number of neutrons released by each fission iteration. Use a "sponge" like graphite, as Fermi had, to retain that tempo, and the energy released can drive a ship, submarine, or electrical power plant. Let the fission go exponential, from $k=1.01$ to a theoretical maximum of $k=2.5$, thousands of fissions per second, and it's a bomb. But the mass must be rushed through its subcritical range, k under 1. If not, instead of a huge explosion, it will just bust off chunks of itself. An easy way to reach critical mass quickly is to slam one subcritical hunk of U-235 metal into another, as Tommy Thompson and George Chadwick, an old ordnance engineer, did for Little Boy. They made a 39-inch-diameter, smoothbore tube, sealed at one end, 128 inches long so it fit a B-29's bomb bay, and a box-kite type of fin on the tail for stability as it fell through the air.

It had one 80-pound or so mass of U-235 in the nose and another in the tail, in front of a standard gunpowder shell in the tube's breechblock. Once fired, the shell shoved one uranium mass into the other at three thousand feet per second, a fact verified at the Los Alamos gun site where it was test-fired several times. If they were separated, the two masses were benign, but when slammed together, they became critical and fissioned. A lead shield in the nose, to be sure it went critical (and that we learned in postwar tests wasn't needed), made it a heavy bomb, 10,000 pounds. The big problem with Little Boy was that it required 168 pounds of U-235, which Oak Ridge, for unavoidable reasons, took a while to make. Indeed, a month's production fit in a small box and Groves ordered that regardless of amount, it was always to be shipped to Site Y by Santa Fe railroad, never by air. "Lord," he said, "if the plane crashes, we could lose one, two, three months work!"

Worried he'd not have enough U-235 to build a bomb any time soon, in late 1943 Deak restudied Johnny von Neumann's work, using shaped charges as precision explosives at BuOrd. The result was Fat Man, an implosion weapon whose contents were so secret that even use of the word "implosion" was classified. Its casing looked like a fat teardrop, 128 inches long by up to five feet in diameter with a five-foot-square fin on its tail. Inside its five-foot bubble were three balls. The innermost one, the detonation initiator, was polonium 84 wrapped in nickel-beryllium, polonium being so naturally radioactive that it's deadly to touch. Monsanto's Mound Lab assembled that ball by remote control. And it had a half-life of only 140 days so we had to keep replacing it all the time. (That problem eased a lot after the war when, on the hydrogen device, we used a lithium-deuteride initiator with no atomic half-life. Even adding tritium to boost a hydrogen bomb's yield avoided the polonium mess since tritium has a half-life of 12.5 years.)

Wrapped around the polonium was a metal globe, 6.25 kilograms of plutonium enclosed in a U-238 tub-alloy tamper. But the key to making Fat Man fission was 128 "lenses," so-called since they were focused on the bomb's initiator core to crush it, to release a burst of neutrons that would split the plutonium atoms. The tough part was that all 128 lenses had to blow within fifteen millionths of a second of each other for an instant, uniform implosion at all points on the sphere. If not, we'd just be hitting a grapefruit with a sledgehammer. Some juice would squirt out here and there, but the end result would be just a squashed hunk of fruit, not a flattened city. The detonator design used nuclear physics, chemistry, and high-speed electronics. Knowing that materials react at different speeds, the scientists picked Composition B, a TNT derivative, for the inward side of the lenses with a layer of baratol (barium nitrate and toluol) explosive on top of that.

In late 1944, Oppie set up a Site S division at Los Alamos under George Kistiakowski in Deak's bailiwick for Fat Man explosives and engineering. After Groves built a Salt Wells plant (as we called it) at NOTS for my CalTech guys, it took over much of the Site S work. The hard task was finding the right shape so that the lenses would put a near-perfect field of pressure on the core. Man-months went into covering a solid metal ball with baratol-Composition B, exploding it, and measuring its impact on the ball. But for all the hard trial and error, basically all we were trying to do was just build a very large condenser setup.

One expensive mistake fell out of all that. All things and people emit neutrons constantly. Afraid that in a bomb-drop a stray neutron in the air might hit Fat Man's plutonium, making it explode too soon, Ed Teller and Oppie had made a wraparound of boron, which absorbs neutrons like a sponge does water. We spent $10 million making and putting in that boron template, then decided it wasn't needed and never used it. But the root cause of our snail's pace in making a bomb was Hanford's reactor. Like Oak Ridge, and Fermi in Chicago, it used graphite housing to dampen the neutrons during separation. But the bombardment made the uranium expand too, so they could load in only a small pile of U-238 at a time. Too much and it would swell up hard against the graphite "can," making it a struggle to remove, especially since it was radioactive poison by then. Thus, to get one Fat Man plutonium ration, they had to bombard a mass for 130 to 140 or more days. Consequently, they simply couldn't produce Fat Man "sludge" or Little Boy "yellow cake" fast enough to make more than two plutonium bombs, one of them from U-235, in a year.

For the finishing process, Los Alamos had more options. Groves had built there three types of plants: an electromagnetic plant that just hurls atoms around to separate the heavier from the lighter ones, a chemical diffusion plant to do the separation chemically, and a gaseous-diffusion plant. I met Ernest Lawrence, who later would help me launch Livermore Laboratory, on my first visit to Los Alamos. In 1944, he ran its electromagnetic plant, which was, essentially, the cyclotron he had invented. Stranger, almost eerie, was the gaseous-diffusion plant at Oak Ridge where uranium atoms are hurled against glass barriers in a vacuum. Oddly enough, the U-235 isotope slips through the glass barriers, but the U-238 falls down. (By 1955, since it was the best method, Oak Ridge had three plants doing that type of gaseous diffusion and we'd tell visitors, "The only place with a higher vacuum than we have here is Washington, D.C.")

Groves couldn't have done today what he did from 1944 to 1945. By the early 1960s, the red-tape bastions were in place to require that we do a paperwork analysis on which of the three alternatives was best, rather than do it Groves' way (the smart way usually in my R&D experience): do them all to prove which is better. In any case, at that time no alternative could handle Lawrence's idea for doing the separation. He suggested using a "centrifuge method," as it's called today, which we know may eventually make gaseous methods obsolete. Certainly, it takes much less capital investment when we can do it, but back then we weren't able to figure out how best to do that.

And the driver behind it all was, once again, Deak Parsons. He was the practical link to the theoretical scientists. He insisted that the end result had to be hardware. Otherwise, he said, anything they did was pointless. His impact on them was why Groves loved him. In fact, rarely noted except by Groves, the key military players at Los Alamos all were naval officers. Beyond our little clan, the "gadgets" were cloaked in secrecy—and confusion. Each piece of the Manhattan District had a code name. Los Alamos, for example, wasn't Los Alamos but Site Y and Site S. A pesky problem was that the money spent on each piece had to be tracked separately. That's also why, after the war, they had so much trouble figuring out who'd done what when.

Inyokern was "Camel Caper," as in "Let his nose in under the tent and pretty soon he takes over." And we were. We were in bomber training, mainly to study the bombs' aerodynamics while at Wendover Field. The air corps under a different code ran the training itself. That 509th Composite Group had been picked by air corps Col. Bim Wilson and Groves who, given his opinion of army pilots, had demanded they be the best. Our "camel" scoring said they were very good. Commanded by Col. Paul W. Tibbets, Jr., a veteran bomber pilot, they'd arrived in October 1944, with fifteen B-29s modified to drop the 10,000-pound Little Boy and 10,000-pound Fat Man. In training, they used dummies we'd made in the shape and weight of the two bombs. They dropped these so-called "pumpkins" at isolated Karacot Valley, a spot picked because Fat Man had such a distinct shape that we didn't want it dropped where people could see it.

And Fat Man was a crazy bomb. Its length-to-width ratio being only two to one, its ballistic behavior was awful. Falling from the bomb bay, it tumbled like a barn door. We couldn't do anything for that, however. What science in that era said would make it implode had dictated its shape. So we put Fat Man in a wind tunnel to camera-record its drag pattern, then compared that data to the tracks we captured on photo-theodolites of its trajectories during the 509th's training drops. These trajectories were caused by its bulbous shape and its forty-five-second drop (a long time for a bomb) from thirty-five thousand feet. Then we used all that to feed corrections into the bombsight. We also had to gauge crosswinds' effect on its fall. And with the valley four thousand feet above sea level, we had to transpose those numbers to an airburst at sea level, the way to get the widest possible impact.

The airburst itself was up to the bomb's firing circuit: barometric clocks and switches and an APS-13 (auxiliary power source) tail-warning radar fuse. The bomb's release pulled out its bomb bay wires. That started its barometric clocks. At six thousand feet, their switches closed to start its APS-13 pulsing toward the ground. At 13 pulses, the lapsed time for the bomb to reach fifteen hundred feet, where we wanted it to explode, the fuse ignited Little Boy's gunpowder and/or the 128 lenses in Fat Man, instantly causing what Site Y and Camel promised would be a gigantic explosion.

10

A-Bombs and Turf Wars

We knew without a test that the Little Boy detonator would make its uranium fission. We didn't know if Fat Man's firing circuits would hold up under the g forces its "barn door" tumbling would generate in a thirty-three-thousand-foot fall. So a test of a telemetry-wired Fat Man, without nuclear content, was ordered. I'd heard that the 509th's "pumpkins," dropped in the soft sand at their test range, were sinking "to incredible depths" and I, with a talent for speaking when I should stay shut up, offered, "If you do the test at our NOTS range, my geology experts guarantee we won't have to dig down more than six feet for it."

Their drop from twenty-five thousand feet at China Lake of a "live projectile," lacking only the uranium component, was in December 1944. When that five-ton rock hit, it bored down ninety feet to the water table, sank thirty more as we rushed in earthmoving gear and ditch diggers, even carpenters to shore up the hole and prevent loose desert soil from collapsing in on them as they dug. Working through Christmas, making a hole big enough for a ten-story building, we took ten days to reach it. I'd been a bit off on the texture of China Lake's soil. Since we'd put telemetry on every piece of it, we got a lot of data as the bomb dropped. We still had to get it out to see if its fuses, barometric clocks, and radar all had worked in harmony. And they had.

A little perturbing to some of us was that Groves, before we got our bomb up out of its hole, told the Joint Chiefs of Staff in mid-December, "We have the A-bomb." Then we got hit by the Battle of the Bulge in Europe, a shock that had the JCS asking, "Should we use this weapon right now on Germany?" Roosevelt, Marshall, someone, decided, "No," thankfully. What good could one bomb do and, even if it did have some impact, what would we do for an encore?

As we used pumpkins to test firing systems, we also did the green-plug–red-plug exercise devised for the nuclear-armed weapons. Simply put, green plugs in the bomb shut down the circuits. Pull those out, put in red plugs, and that told everyone at a glance that, in space-age parlance, "All systems are go." The reason for the green-plug routine on Little Boy was that if the plane crashed on takeoff, we didn't want the bomb to "go nuclear," which it would if the crash lit off the shell's gunpowder. To prevent that, Parsons decided not to put the "gun" and its U-235 block in the casing until after takeoff.

We also wondered, if we had to jettison a Fat Man at sea, would hitting the water make its lenses "deflagrate," that is, ignite and detonate the bomb? So in January on one of eighty-seven flights I would make in it, I took a B-29 out of China Lake, carrying a Fat Man less the nuclear stuff. I intended to drop it from twenty thousand feet at a spot 185 miles west-southwest of San Clemente, off California's coast, where I'd put observers to see if the impact resulted in spontaneous combustion. But aloft, I flew into a problem. Because of being rushed to production while still in flight-test, early B-29s had lots of ailments. The worst was the cranky 3350 engine. If heavy-loaded as we were with our 10,000-pound bomb, it tended to overheat on takeoff, then catch fire at altitude.

And it happened. At nineteen thousand feet, an engine caught fire. I feathered that out but had to land at North Island. A navy pilot, civilian passengers, a Fat Man in a unique cigar-shaped aircraft—I knew we'd be as low profile as an elephant. First to greet us was John Crommelin, still head of training there. "Chick," he asked, "what the hell you doing in an army airplane?"

"Forget that, John," I snapped, "I've got a burned-out engine. The guys you have replace it will do only that. No peeking inside. No one else is to go near that plane. Nobody!" Well, my old Pensacola flight instructor was stunned. Years later, he claimed he knew what was aboard but I didn't believe him. Still don't. Back at Inyokern, I caught hell from General Groves. "Why the hell did you land there? Why not back here?"

"Because the plane was on fire!" I growled. I knew by then that his bark was worse than his bite most of the time, at least with me. Besides, the offset was that with help from him, I'd met nearly all the top scientists in this country in this century. We canceled the ocean-drop test then, since the chance of a detonation that way was marginal anyway. Besides, our atomic "gadget" attack was rushing to a climax. In February 1945, Deak sent Cdr. Dick Ashworth out to find a home for our 509th's A-bombs. Dick chose Tinian, a Marianas island taken the previous July, as were Guam and Saipan. Since November 1944, B-29s from there had been raiding Japan so often that Tinian was the world's busiest airport. Like a tsunami, the tide of war in the Pacific was now running strongly in our favor.

It had been, in fact, since before I left for Inyokern in June 1944. After taking Kwajalein and Eniwetok in February 1944 against surprisingly weak resistance, Nimitz had found Truk, once the linchpin of Japan's power in the central Pacific, nearly deserted. He'd bypassed it and other in-between islands, safe to do since we controlled the air over them, and went after the Marianas. By July, his marines and the army held those. That and Mitscher's victory in "the Turkey Shoot" forced Japan's "Hitler," Hideki Tojo, to resign. Then when Halsey found only a few Japanese ships near the Philippines, Nimitz and MacArthur decided to land at Leyte. King wanted to hit Formosa and Okinawa on Japan's doorstep, but War Plan Orange said and MacArthur insisted that we take the Philippines. Nimitz did say, once beachheads were secure, he'd leave the rest to MacArthur and send his navy north. Wars take two kinds of leaders, good planners and good fighters. Nimitz might not have led in battle as well as Halsey, but I don't think Halsey would have been the strategist Nimitz was—or as good at dealing with MacArthur.

MacArthur landed in the Philippines on 22 October and Japan's once-powerful navy threw what it had left into a counterattack. The three battles of 23 to 26 October 1944, known as the Battle of Leyte Gulf, involved 282 ships, thousands of aircraft. We lost a carrier and five other ships; Japan lost four carriers, three battleships, twenty other ships, and hundreds of pilots. Our combat edge was now huge. One clue: the first Japanese pilots in the war had averaged 700 hours of training. The 1944 replacements had only 150, explaining why Leyte Gulf was Japan's first use of the "kamikaze," or "Divine Wind," a phrase coined in the thirteenth century when typhoons drove invading Mongols out of Japan; but to us it was a confession that they'd lost the war.

Saipan-Tinian to Tokyo was a three-thousand-mile round trip with B-29 bomb loads at 30 percent of capacity to save fuel and, at twenty-eight thousand feet, too high for Japanese fighters to reach, since the B-29s had to go in without fighter escort. In March 1945, after three weeks of their bloodiest battle yet (5,931 killed, 17,372 wounded, nearly all the 23,000 Japanese defenders killed), the marines took Iwo Jima, 450 miles from Japan. By this time, P-51s could fly cover all the way to Japan, and bombers returning damaged had an emergency-landing airfield. Maj. Gen. Curtis LeMay, head of the Twenty-first Bomber Command, by then had gone to incendiary bombs and no longer needed fighter escort. (During the war's final three months, his B-29s dropped 160,000 tons of bombs on Japan, an average of 1,200 tons a day.)

In March, wary of its implosion scheme, Oppie set up Project Trinity at Los Alamos to detonate a Fat Man, and told Parsons to deliver the bombs to Tinian. On 12 April, FDR died suddenly and Harry S. Truman, now our thirty-third president, was told about "The Bomb" for the first time. In his memoir, *Year of Decisions,* he has Vannevar Bush telling him, "[It] is the biggest fool thing we have ever done. The bomb will never go off, and I speak as an expert on explosives."

On 8 May, "Victory in Europe" became official and in June, backed by Vice Admiral Mitscher's Task Force Fifty-eight (his chief of staff the destroyer ace Arleigh Burke), marine and army troops took Okinawa. At the end of June on a brief furlough, "Uncle Pete" came to see me. He loved to trout fish so I took him to a pet spot of mine, Red Meadows way up in the wild, high Sierras. He used his crazy flies, of course, needling me for using worms. "Go ahead," I said, "Have your fun. I'm trying to catch us some breakfast."

And there, he became the only outsider I told about the atomic bomb, the Michelson laboratory, Einstein, Fermi, the whole history. A quiet man, anyway, he listened in his patient way until I said, "So forget going back to the war. The war is over."

He didn't believe me, said the science types had addled my brains. "Chick, I have to get you out of here, get you an air group in my task force. You've gone just crazy."

"Forget it, Uncle Pete," I said, "Go home. Go fishing. The war's over."

But when Japan folded, he wrote, "Dear Chick, I'm old and gray. No imagination. The world belongs to young people like you. You were telling me the truth and I didn't believe it."

That month, I also logged more time in an F4U-4, our country's only fighter then, with a new high-altitude engine able to reach the thirty-two-thousand-foot altitude our B-29s were using to train for the A-bomb mission. Chasing them, I could photograph everything, then dive to follow the bomb down. Deak stopped by that month to check on their rehearsals out of Wendover and our rocket and bomb-lens progress at Inyokern, ending the month in San Diego to arrange for shipment of the Little Boy to Tinian on the cruiser *Indianapolis.*

The full-up test of Fat Man at Alamogordo had been scheduled for the Fourth of July. A few fumbles—one being Deak's 2–13 July trip to Tinian—delayed it until the 16th when, with Deak there, at 5:29 A.M. atop a one-hundred-foot tower, Fat Man was imploded. Its intense red-orange burst of light pierced thick clouds and shimmered across the sky, temporarily blinding observers ten miles away. Eight minutes later, its mushroom cloud peaked at forty thousand feet. For an instant, its roughly twenty-one-kiloton yield raised ground-zero temperatures ten million degrees, its shock wave vaporizing the tower and all life within a half mile. What Bush had said wouldn't work had become man's first nuclear explosion. It also left us with only two weapons, a Little Boy and a Fat Man.

Ten days later, the *Indianapolis* put a Little Boy on Tinian, and on 29 July, Fat Man's three pieces were hauled in separately by three otherwise empty C-54 "Green Hornet" cargo planes. Ordered not to drop the bomb unless they had visual as well as radar acquisition of the target, our B-29s spent a week rehearsing a Fat Man drop and a loading and assembly of Little Boy until a forecast predicted clear weather over Japan on the 6th. Meantime, LeMay, whose incendiaries had burned out huge chunks of several Japanese cities by then, asked Deak, "What are you crazy guys doing? We don't need this. We've got the war won."

Deak was to be the mission's "weaponeer," that is, assemble Little Boy and activate it, a task he had to do after Colonel Tibbets's B-29 was airborne. If it were armed earlier, a crash on takeoff—as four B-29s had, leaving for Japan just the night before—could ignite its gunpowder-shell, instantly exploding the bomb. As some guy in the outfit said, "If we must pulverize something, let it be Tokyo, not Tinian." Still, the exacting bomb-assembly had to be done quickly, below six thousand feet, as this was the altitude where the B-29 cabin, but not the bomb bay, was pressurized so the crew could work easily at a B-29's thirty-five-thousand-foot ceiling. (Obviously, if Deak were to open the bomb bay hatch above six thousand feet, instant depressurization of the plane could create a potentially lethal mess.)

Meanwhile, at China Lake, our family quarters finally finished, I moved Lili and the girls in from Pasadena, learning that Lili was pregnant again. The house was as nice as I'd seen for a guy of my rank, but flights to Cal-Tech, Los Alamos, Oak Ridge, D.C., and elsewhere kept me gone most of the time. Then, on 4 August, we set up a Research Board, a panel of scientists and engineers to put priorities on R&D and test projects. And we learned, on 18 August, that Ev Burroughs would be relieved by Capt. James Bennett Sykes, "strict about regulations," it was said. If so, I said, we will have war at Inyokern.

But before that, out on Tinian at 2:45 A.M. on the 6th, Tibbets's *Enola Gay*, named for his mother, took off on its seventeen-hundred-mile flight to Hiroshima. Minutes later, Deak had Lt. Morris R. Jeppson help him squeeze down to Little Boy and hold a flashlight for him as he worked, using the intercom as he did to report progress to Tibbets—first the block of U-235, then the projectile, then the green plugs, the weapon's "off" switch. At 3:15 A.M., his hands nicked and graphite-stained, he crawled out. At 7:30, he went back to replace the green with red plugs, arming the bomb, and turned on its radar to warm its barometric clocks. At 9:09, 32,700 feet over Hiroshima, Tibbets asked him to verify the target. When Deak did, he'd authorized the bomb's release.

Six minutes later, the *Enola Gay* bolted upward as its five-ton cargo left the plane. Forty-three seconds after that "a purple flash," he said, lit up Deak's protective goggles. Then a shock wave rocked the plane, then another, as the bomb's dirty mushroom cloud soared to twenty thousand feet, a white sprout in its center rising to fifty thousand. The airburst had come at under two thousand feet. Its shock wave, a report later said, flattened nearly all buildings within two thousand yards of ground zero, some as far out as four thousand yards. It also roiled up a torrential black rain that fell briefly on the city's entire western sector. For an instant, heat at ground zero was three hundred thousand degrees centigrade, starting fires that burned for three days, including some in areas fourteen miles from the detonation site. All told, two-thirds of Hiroshima was destroyed or damaged.

At 2:58 P.M., the *Enola Gay* landed at Tinian. The whole program had been unique: a naval officer on an air corps plane in charge of a weapon built by civilians working for an army general. The shame in it was what happened next. As Tibbets came down out of his aircraft, out strode four-star Gen. Carl "Toohey" Spaatz, commander, Strategic Air Forces, to pin the Distinguished Service Cross on Tibbets's coveralls. To witness the event, smile, and shake hands were LeMay, other air force brass, the press, a band—all that glitter and tinsel for the guy who'd been just the chauffeur, who knew zilch about how to do what a weaponeer does.

But when Deak dropped into view, hands grimy, shirt sweat-stained, he walked off alone, ignored. Later, quietly, off in a corner, an air forces one-star did present Deak with a Silver Star, about like getting a second-place trophy for winning a race. When Groves learned that, he snorted, "There was never any question by anyone but that Parsons was running the show. Apparently, the only person who did not get that right was General Spaatz."

And the navy, ever stingy about medals, was silent. Only after Japan's surrender did the navy give Deak a Distinguished Service Medal and promote him, in early 1946, to rear admiral. My science friends agreed there'd not have been an atomic bomb in 1945 without him, but how many Americans ever heard of Parsons? Not many. I decided navy policy hates a hero and he was a hero. Still, when I fussed at him, he had an answer I liked. "There's no limit to the good a man can do," he said, "if he doesn't care who gets the credit. And I don't care."[1]

On 9 August, Maj. Charles W. Sweeney, who'd flown the escort B-29 on Tibbets's mission, flew Capt. Fred Bock's B-29, *Bock's Car*, to drop Fat Man on Nagasaki at 11:02 A.M. (Prime target was to have been Kokura but fog and cloud cover blotted it out, making it off limits under the bombing rules-of-engagement.) Deak's pal, Ashworth, was weaponeer on that run. His Fat Man was a safer bomb, easier to handle than Little Boy. If the plane crashed, for instance, it couldn't explode, so Ashworth didn't have to assemble anything. With green plugs in the Fat Man on takeoff, he had a flight-test box with him in the cabin linked to its firing mechanisms. All he had to do, aloft, was check their circuits to verify that they worked, then simply replace the bomb's green plugs with red ones at under six thousand feet, before *Bock's Car*'s cabin was pressurized.

One Fat Man result was Groves raging at "Hap" Arnold and Lt. Gen. Lewis Brereton for talking as if they owned Fat Man, had made it, had done everything. They kept at it anyway, so he was forever mad at them, while the air force went to full power on its campaign to convince the White House and Congress that it alone should have a monopoly on use of the weapon. Also on 9 August, we learned about the *Indianapolis*'s tragedy. Around midnight on 29–30 July, she had been sunk by two torpedoes from a Japanese sub. Only 316 of her 1,196-man crew had survived, possibly as many dying of exposure, shark attacks, and lack of water as were killed in the torpedo attack. And her skipper, Capt. Charles B. McVay III, later was court-martialed for "negligence"—in my opinion, as terribly unjust as the event itself had been tragic.

When she'd left San Diego, King had wired Nimitz, "Leave the *Indianapolis* alone. It is on a special mission." So she off-loaded the bomb, then headed for Leyte Gulf by way of Guam; and CINPAC, obeying King's order, ignored her. Leaving Guam, she radioed Leyte she was coming, but an electrical storm scrambled the message. Her skipper not trying to resend it suggests that he felt she could outrun any unlikely trouble in that area. Then when torpedoed, she quickly capsized and sank, unable even to send an SOS since the hits had knocked out her electrical power. She'd intended to be in Leyte on 31 July, but no alarm was raised. Leyte wasn't expecting her. A patrol plane sighted debris on 2 August, eighty-two hours after she sank. Ships-and-aircraft search lasted until 8 August.

One thing I can't justify is King waiting until after the war to put McVay on trial for "not zig-zagging." I jumped King hard on that when he visited me at Sandia in 1948. It wasn't McVay's fault, I said. If the Philippine Sea Frontier people had tracked him, as they should, they'd have saved lots of lives. They're the ones who deserved to hang. And why, I asked, did he haul in the Japanese submarine commander to testify? On what to me was a key question, the trial record is moot: Did McVay, as was his right, demand a trial rather than accept a reprimand? The last captain I recall doing that was Alexander McKenzie of the *Somer*, tried and acquitted of murder at Newport, Rhode Island, in the nineteenth century. (James Fenimore Cooper was the court reporter.)

Nine days after Nagasaki, another destructive force, Capt. James B. Sykes, arrived to take command of Inyokern. At NOTS' birth, the plan was that BuOrd would take over at war's end, and they didn't waste any time. The Japanese were still two weeks shy of sitting down on the *Missouri* to sign the unconditional surrender when he showed up on the 18th. Just off being captain of the carrier *Bennington*, he brought with him an almost slavish demand that everything, I mean everything, was to be done strictly according to navy regulations. He was a regular tyrant about it.

My view on the subject of regulations had been formed mainly by "Uncle Pete." A good leader, he'd said, has five character traits: knowledge of and enthusiasm for his job; self-confidence, a trait that knowledge helps build; integrity, including frank, honest reports to superiors; and good judgment, which includes knowing when to intelligently disregard a regulation. Rules I'd been ignoring at China Lake forbade civilian women riding navy aircraft and any civilians riding navy buses. Common sense said that those CalTech folks, whose genius we badly needed, would stay home if they had to drive 140 miles to work. So I was giving them daily Beachcraft and Lockheed Lodestar flights between NOTS and Pasadena's Alhambra airfield (an old field too small for our C-54s), and taking them to and from their Inyokern work sites in navy buses.

I'd barely met Sykes when he snarled, "By what authority do you run a bus for non-navy personnel? I want that stopped."

Knowing he didn't have Manhattan Project security clearance, I could say only, "Captain, I wouldn't do that if I were you. It's a mistake. I'm not allowed to spell out the technical details but it has to be done."

"Well," he snapped, "I'm ordering it stopped."

So, I saluted, "Aye, aye, sir," went to my office, and called Deak down at Los Alamos. "Deak," I said, "China Lake has just been shut down." When he heard why, Deak called Groves, who had a huge investment in China Lake. Groves called Navy Secretary James Forrestal, and Forrestal called Sykes. "Captain," he demanded, "you let Commander Hayward run his buses and anything else he wants. Don't pester him over stuff like that. What he's doing is important."

From then on, Sykes looked under every rock for a way to fire me. Parsons got as furious as I was tense. The next two years were the most unpleasant time Lili and I had in the navy. Sykes and his ilk rated me "anti-

Navy," the "civilian outsiders' baby-sitter." It got pretty bad but didn't have to be. Lili was stuck there, unfortunately, but at the start, I was able to escape Sykes for a while.

Right after Emperor Hirohito announced Japan's surrender, I was ordered to go there when MacArthur did, he literally to run Japan, me to assess atomic-bomb damage for CalTech. I also was to measure residual radioactivity. We doubted there was any since, to induce radiation, the emitted neutron must have a free path to a metallic receptor. At the time, most Japanese structures were made of wood and paper, and the airbursts had been at fifteen thousand to eighteen thousand feet. At that height, neutrons, with a very short life anyway, can't induce much radiation. I was told to assess also the people with radiation problems such as lost hair.

I took along an American fluent in Japanese to be interpreter; a marine corps photographer, Al Uremovich; and Bill Penny, Britain's exchange-program guy at Los Alamos. (In a Churchill-FDR deal, Britain had been our A-bomb ally since day one, working with Tubelage, their code for uranium, and Clementine, ours for plutonium 49. Being excluded irked French leader Charles de Gaulle no end, a snub we'd pay for later on.) And, of course, we took along Geiger counters for the lingering-radiation business.

We flew over in a P5M Martin Mariner flying boat. The minefield our B-29s had made of Japan's Inland Sea was still there, so we were off-loaded west of Hiroshima on 5 September with some C rations and such and told to forage for our needs like today's big-city homeless. Just before the bomb fell, Hiroshima's 240,000 population had been augmented by a large influx of military personnel. The Japanese claimed that a total of 240,000 people died, 37,425 were injured, and 13,983 missing in the Hiroshima-Nagasaki raids. We said 69,000 injured, 66,000–78,000 dead at Hiroshima, 39,000 dead at Nagasaki. A third source said 23,753 dead at Nagasaki, 23,345 injured, adding, "but casualty estimates vary widely," the only for-sure fact in the tallies. We were delayed in checking that, however, because a typhoon hit as we arrived, shoving the city's hospital off its cliffside perch, messing up everything including medical files.

But it was amazing what we learned, collecting bits and pieces a month after the explosions, trying to reconstruct the time constant of the blast's overpressure, the radiation, and the burn-damage distances. I found one boy who'd been driving a wagonload of hay. He'd been perspiring, wearing a white shirt and black pants. From some seven thousand to eight thousand yards away, the blast had ignited the hay, but he wasn't burned except for a place on his arm where he had perspired. We also saw on a wall a sharp outline, like an X ray, of a man pulling a cart, but no shred of evidence of what, if anything, had happened to him.

In both cities, from the pressure it would take to squash or break off some objects we found, like crushed beer cans, I computed the probable yield of the bombs: thirteen kilotons at Hiroshima, twenty at Nagasaki. In one city, the blast also had knocked several feet off the top of a chimney. We measured the height of the chimney stub in order to calculate where and how hard the shock wave had to have hit the ground, then bounced up with enough force to sheer off the chimney at that height. (Normally, the force of a TNT explosion will go right around a lone chimney like that, not break it off.)

The Hiroshima bomb just cleaned the place out. Nagasaki's didn't touch hardly any of the city in some spots. The difference was in topography. Hiroshima is built on a flat, open valley where three rivers converge; but Nagasaki nestles in the Yurokomi valley, with hills and mountains all around to absorb much of the blast. For instance, the fishing village where we stayed, on the other side of hills outside Nagasaki—after the marines landed and ran us out of the Mitsubishi house, downtown—had no damage at all.

The Nagasaki bomb had burst right over a Catholic church. I chatted a bit with its Polish Jesuits, Japan's largest Catholic population, on a bank by their ruined cathedral less than a mile from ground zero. (The cathedral has since been totally rebuilt.) There, as everywhere else, we found no lingering radioactivity. We had expected that. What we hadn't expected was how cooperative people were. The only anger we felt, ever, was when part of Japan's Sixth Army came back to Hiroshima from China. They'd never been defeated in the war, and crossing their path was not very pleasant. But, other than that brief tension, the people in the bombed cities were very friendly, very cooperative. Maybe it was my interpreter, excellent in Japanese, explaining that we were there to help if we could.

Stafford Warren, a top radiation man in the future Atomic Energy Commission, headed the medical investigation team, documented all the cases and we followed them for years. But in 1945, Japanese in both cities seemed confused, wondering what had happened, or not sure the emperor told them what's what. Still, in most places except for the Sixth Army troops, I saw no anger toward us in any way by anybody, just a great relief that the war was over. And maybe afraid of what to expect from us. Tojo's crowd had told them we were vicious animals. What expunged that big lie was, I'm sure, our government deciding not to do away with their emperor.

Between the Hiroshima and Nagasaki stops, I was sent to Kure—a privilege, really—to release our prisoners of war, including "Pinky" Madsen,

Naval Academy class of 1928, captured on Guam in 1942; "Dub" Johnson, shot down in the Marshalls; and Tommy Payne, top aviator on the cruiser *Houston*. I greeted them with, "You all look fine," which, surprisingly, they did, and I knew Pinky well enough to add, "And I see you're sober for a change."

Lots of Japanese navy people lived in Kure. I met several. That got me invited to their naval academy in Itajima. I was a bit edgy when they saluted, then had me tour their naval museum. It had Tojo's nails and some of his hair, a hall for their Kamikazes, and a layout of where every ship in their Imperial Fleet had been sunk. I went all over with those people, lecturing in Hokkaido, at Tokyo University, in Watura Springs, and in Kanazawa. And they always talked very openly about the war. The whole drill was a big help when I had Anti-Submarine Warfare Force, Pacific, from 1963 to 1966, working on their new ASW Self-Defense Force's training. (By marked contrast, on our many postwar visits to Germany, including when Lili and I were honored guests at a big Luftwaffe party in Bonn, I met no one who'd say he'd been a Nazi. It was eerie. The Luftwaffe alumni acted coldly aloof on that subject, as if the war never had happened.)

We sent our Kure POWs home in one of our "Green Hornet" C-54s. Before they left, they turned over to me all the yen they had "liberated" from a Japanese paymaster, a big suitcase full. Then MacArthur declared that ten yen equaled a dollar but, to prevent shenanigans that had happened in Europe, he added that no yen could be taken out of the country. My chance at souvenirs gone, we did the Nagasaki survey, then went to Tokyo on our way home. There, ten-yen-to-a-dollar let me set up in style. I used some to refurbish what had been the University Club of Tokyo, then organized a big crap game to gamble away the rest.

In Tokyo, Adm. Ray Spruance (who'd become, in February 1946, president of the Naval War College in Newport, my navy career's birthplace) asked Bill Penny and me to join him on the *Missouri* for dinner—far, far better food than the best we could find, including C rations, in Hiroshima and Nagasaki. High winds blew that night. Tokyo Bay got rough as only it can, and returning to our Yokosuka dock, Bill got violently seasick, losing all that fine dinner. I told him, if this keeps up, it's the last time I take him to dinner. Groaning, he didn't seem to think my remark funny.

While in Tokyo, I met Paul Nitze, head of the Asia Strategic Bombing Survey team. The senior naval man on the team was Rear Adm. Ralph A. Ofstie, skipper of VF-6 when I was on the *Langley*. Also in that group was Cdr. Tom Moorer, class of 1933, whose aviation heroics in South Pacific combat I'd heard of while there. We'd hardly interviewed anyone in our work, but they'd interviewed a ton. Tom told me they'd also found in a cave near Tokyo a summary of Japan's war plan. The Japanese hadn't intended to take Australia. Once they held southeast Asia and Indonesia, they'd planned to offer us back the Philippines if we'd accept an armistice—and they thought, "because Americans have no stomach for war," we'd accept.

Groves hated my having to explain the atomic bomb stuff to Nitze's people. He didn't want us telling anyone what we did or how we did it. We were supposed to refer all such questions to his deputy, General Farrell, in Tokyo at that point. My big problem was that none of Nitze's team had an atomic-project security clearance. So, out of the 450-page report I wrote for Cal-Tech's damage-assessment people, we were allowed to give Nitze and John Kenneth Galbraith only the results of the blasts, but not tell Galbraith the yields nor even that Little Boy had been a U-235 weapon. Galbraith signed the survey report written by Nitze and company, got all the credit, and peeved me some, I confess, by not giving my team credit for all the damage information that was the best-read part of his report.

That itch aside, the report could, I thought should, have urged plans and policy types at home to assess how well their decision-making machinery had worked in the last months of the war. For instance, why did the Joint Chiefs tell the State Department and FDR, in February 1945, that it would take eighteen months to defeat Japan? I assume that's why, at Yalta, FDR offered to give away the farm to get Stalin into the Pacific war. He needn't have. He'd known in December we had an atomic bomb that worked. The day after Little Boy exploded, Deak wrote his father, ". . .there is a definite possibility [it] may crack them and end the war without an invasion. If so, it will save hundreds of thousands of American—and even Japanese—lives."

That was the conventional wisdom at the time, endorsed by Truman. But in Japan and later, I learned much that said this notion was long on emotion and short on facts. For instance, if intelligence folks had been as good as they should be, they'd have told Truman in April, three months before his Potsdam conference, that Stalin was no friend of ours. April was when the Japanese started trying to get Russia to tell us Japan wanted peace. The struggle was on then in Japan's highest echelons over whether to fight on, maybe committing national hara-kiri, or seek an armistice. The Russians knew that. We knew a lot about the Russians, we thought. Yet, we never knew about that. Anyway, when I got to Japan in September 1945, they didn't have one hundred thousand tons of POL (petroleum, raw or refined oil, and gasoline) to fuel their needs at home and their war machines in combat. They had a moribund industrial base. They were about out of food. Why in the world were we laying plans to invade? Why not just surround the place and let them die on the vine?

Clearly, our military intelligence was worth zero. It had swung from "ho-hum" analysis of the enemy at Pearl Harbor to a gross exaggeration of enemy power in 1945. But the most troubling question, critical in the nuclear age, is this: Why did we drop the second bomb? Command inertia was the reason, I think. No one stepped in to stop its being dropped, which it shouldn't have been. Indeed, when we had only three bombs, the scientists had asked for the Hiroshima one back to convert its uranium to plutonium for more weapons. Everyone has a story on why it was dropped. Groves reportedly said, "Look, I've spent $2 billion on this program. If it's not used in this war, they'll do nothing but investigate forever why I spent all that money." Interesting story but I doubt he said it. Congressional bureaucrats didn't scare him at all.

The problem, I think, was that our leaders didn't know how to stop a war, a scary thought in the so-called Nuclear Age. I did believe, back then, that the weapon would make a tyrant a little more leery of starting something. But deterrence is a plan with many pieces. You must have the weapon; know it will work; and have ample, survivable delivery systems to put it on target. An enemy has to know you have all those assets and believe you have the will to use them. And you must accept that, if you do use the weapon, you must annihilate him before he destroys you. If your answer to any one of those requirements is "no," deterrence is "zero." And of course, the bomb's mission is to deter war so, if it ever is used, it's a failure. It didn't deter. Deak and I were debating this when we had maybe nine atomic bombs, back before NATO (the North Atlantic Treaty Organization) was formed in 1949. (Even after it was, our 1950s so-called "Massive Retaliation" threat didn't seem to deter communists much from doing whatever they wanted.)

As I discovered when I returned from Japan in late November, lots of folks were mulling over that equation. However, most were chewing on a simpler question: "Should the military or civilians control the atom?" We were in the process, of course, of asking people to shift over from CalTech into the Civil Service, a whole different ball game. At the outset, the basic argument for starting NOTS was that it not be run like Dahlgren's rigid, rank-and-privilege bureaucracy. Now, the challenge was, could the navy run a laboratory like OSRD and Los Alamos had?

Adm. George Hussey, Chief of BuOrd; Hedrick at Dahlgren; and Sykes, an old sea-goer weak in the technical stuff, all insisted that NOTS "will be by the book." Of course, Sykes hated having our Research Board ("a batch of civilians" led by Technical Director Tommy Thompson) telling, not asking, him what to do; and my saying, "We must encourage these scientific types. They won't stay if you try to run this place like a navy ship." Two, Lauritsen and Warner Rhodes at CalTech even charged clear to Congress to testify on the whole business. However, most didn't want to fight the navy. They were willing to help set us up under Civil Service, then quietly retire. I was dead-sure we didn't want that. BuAer's Sallada, BuOrd's Malcolm Schoeffel, Parsons, Thompson, Groves, and Lauritsen all knew that's why I fought Sykes, and they backed me. 'Course, Sykes hated those guys. And me.

Then on 10 December 1945, I was made captain (T), for "temporary," and Forrestal gave me a Letter of Commendation with Ribbon and Bronze Star. The citation read in part, "(As ETO [experimental test officer] at NOTS) from 4 August 1944 to 9 August 1945 [his] extraordinary achievement in . . . organization and operation of [NOTS] . . . hazardous flights [on] experimental projects . . . skill, technical knowledge and initiative were [an] inspiration to those who worked with him [and] in large measure [led to] new weapons tested at the Station." I was pleased, of course, but knew much of the credit was due Parsons and my science pals. And Sykes may have felt like the guy warned by baseball pitcher Satchel Paige: "Don't look back. Somebody may be gaining on you."

By then, our new Michelson Laboratory was half built, a project outside Sykes' domain. At NOTS in charge of construction, we had Capt. Sandquist with a direct line to Capt. Pat Byrnes, head of BuOrd's Public Works section. I did have to do the architect phase to get an estimate (which, typically, is never what the bidders say nor what it ends up costing). But we didn't lack for money beyond the usual accounting track for the auditors. We'd argued for the lab well enough that Byrnes had approved it in 1945 with construction to be completed in 1946. It would double NOTS' capacity in fusing, propellant chemistry, the whole missiles-and-rockets spectrum. I wanted it named after Albert Michelson because he, an 1873 Naval Academy graduate, was America's first Nobel Prize winner in physics, a fact the navy had ignored until then.[2] I took a lot of heat for suggesting it but Dr. Lauritsen made it happen. He insisted, "Why don't we do that?" until everyone else caved in. (Michelson's daughters gave us all his experimental gear for a small museum and attended its dedication in 1948. Later, some jerk back East said that exhibit should be in the Naval Academy and took it away. Made me mad as hell.)

My work in Japan got me put in early December on Operation Crossroads, to test atomic bombs in July 1946, at Bikini, a long, necklace-shaped atoll twenty-five hundred miles west of Hawaii. They'd put in its big lagoon a fleet of big and small ships—German, Japanese, British, ours, seventy in all—to see what a nuclear explosion did to them. Groves refused to let an "Arrogant Force" guy run it, so Parsons, the tests' technical director, who thought the world of Rear Adm. "Spike" Blandy, got Blandy the command of Joint Task Force One, the Crossroads manager. (In the war, Blandy had set up a Navy Operational Test and Evaluation, T&E, team to make sure a weapon worked before he sent it to war. For a time in the 1970s and 1980s, the Pentagon combined it with R&D as RDT&E, a mistake. How something's made in R&D rarely is how it will be produced in the factory.)

At Bikini, CalTech's Dr. Roach and I were to measure yield of the bombs. Until then, they'd used the Beta-Feynman method. It figured yield from a "Bang" meter's radiochemical analysis of debris scooped from clouds after an explosion. That gave us the radioactive stuff, but its computation of yield assumed that the blast expanded symmetrically. We knew it didn't. (The dumb nut who wrote the computer model predicting global warming also assumed even distribution of atmospheric content. We know that's just not true.)

I proposed that we use Ike Bowen's high-speed camera at 875,000 frames a second for seventy-one microseconds to capture the cloud's expansion rate, and compute from that the bomb's yield. Parsons agreed, so we flew it in an old PB4Y-2 Privateer we had at Inyokern. For the Abel test, they were to drop a Nagasaki-type bomb set to detonate fifteen hundred feet above the *Nevada*. The ship that had taken me on my first academy summer cruise, she was now painted a garish, easy-to-see rust-bucket red. Ike's camera had a tiny field, so we put it on a tower just two miles from the doomed *Nevada*. To prevent irradiation of our film, we put atop the camera lead shields, held up by primer cord set to blow at seventy-one microseconds to drop the shields down over the camera.

The Able shot came on a soft, clear tropical day—and the "crack" air force B-29 bomber crew missed the target, dropping the bomb on a herd of Japanese AK cargo ships and our carrier *Independence* three miles away, out on the far side of the fleet of target ships we had anchored in the area and a near-blank for our camera. Forget bomb runs, Blandy said, and, for the Baker shot, we put the bomb in a caisson under the *Arkansas*. That twenty-one-kiloton blast lifted her carcass high in the air; blew up the nearby *Saratoga;* and sank or damaged most of the other ships, confirming what we'd learned in the war: the only way to sink a ship is fill it with water. And the lagoon now was a pool of intense radiation—no surprise, either. Salt (sodium chloride) irradiates easily, the particles having varied half-lives. Some degrade quickly, others last a long time—as also occurs in a ground burst, depending on the soil's metallic content. We wanted to do a third test but had no weapon. Even so, Bikini was the first use of high-speed photography to measure yield, a camera that explosive technological growth enabled us to perfect later. We also learned something about dead radioactive fish. We pulled one out of the water, laid it on a piece of film, and it x-rayed itself.

11

When in Doubt, Reorganize

Before I left for Bikini, Lili had added John Tucker Hayward, Jr., to our family. Ever since we'd lost Charles, we'd wanted a boy in the family. Prior to John's birth, our girls had called him our "Five-Star Final," and Lili, on the way to the hospital, had told me, "If this isn't a boy, we're going to have another baby." A bit stunned, I'd mumbled, "Oh? Well, okay." (A loose cannon on a pitching deck, growing up, he was an example of why some parents say all baby boys should be locked in a deep freeze until age twenty-one. Years later, he told me, "Well, no wonder I was spoiled. I was raised by four mothers.")

While in Bikini, I was reelected head of Inyokern's PTA (Parent-Teachers Association) because Lili told them I'd accept. It was a foray into politics begun in 1944 when a priority was creating a small city in the "Place to Pass by." Sandia folks could live in nearby Albuquerque but Inyokern was ninety miles from any large dot on a map. The houses built were a quick, cracker-box answer, really, but our major concern was schools. The scientific people we wanted to attract wanted their children well educated. We built a high school, first, since most of their children were at or near that age. Then they elected me head of our PTA partly because Lili didn't want the job but really because I could be their money machine. If we needed, say, a physics teacher, I could double that salary with a part-time job in my operation, improving by half our ability to lure good teachers to remote, arid Inyokern.

As a result, I was the only male PTA chapter president in the United States in the 1944–1945 school year, I was told. I was reelected, I think, because I allowed beer and cigars at our meetings out on the desert. It upset the national PTA some but assured that we always had a big turnout. That, in turn, got me invited to address the PTA National Convention held in Long Beach that year. In my hotel room when I arrived was a big bouquet of roses. That was before my speech.

Father Brady had inspired in me a strong commitment to education, both my own and the idea in general. I decided to tell them my opinion of PTA ideas for improving it. Excepting the national president's husband, I was about the only man there. To jar my dozing audience awake, I opened with, "This is a strange place for a naval aviator but this father of five children is sick and tired of a war leaving them for teachers nothing but frustrated old maids and gelded men."

That did it. I was given, coldly, a nice lifetime membership in the national PTA but no more roses in my hotel room, and back home I was not elected to a third term. Still, my inept choice of words aside, my point was that without good teachers and a strong school system, we would not attract top civilians to our work. And without good people, we can't get good results—in any endeavor. Similarly, in typically remote navy stations, we had to let our civilians use the commissary, the exchange, in spite of regulations saying otherwise. The navy was very adaptable in that area, letting common sense override the rules.

Except Captain Sykes. By September, his whip was cracking. Our weapons business was a bit schizoid back then. We had the "old school," typified by Dahlgren's Capt. Hedrick and Sykes, who abhorred the idea of "civilian outsiders" telling them what to do. My diary repeatedly labels "very forward-thinking" the few officers who wanted China Lake competing with the "old school," officers who knew it needed first-rate scientists who'd resign if we let Civil Service red-tape rule their work ethic, or a captain ignorant in R&D run R&D programs. That was the core of my war with Sykes. I always sided with the civilians. Deak backed me. And Groves said the navy was stupid to pick a Sykes mind-set to run China Lake. The feud was so nasty by late 1946 that when Deak came up to discuss a problem or watch a rocket test, Sykes refused to see him. My diary notes, "Ed Burroughs was marvelous, a fine leader, never made Admiral. I don't know why other than that the 'Old Fuds' don't like him. And they can't stand Parsons."

Nor would BuAer, BuOrd's leading critic, help us, acting as if China Lake were not on their watch. "Very discouraging," I wrote, "even appalling, the number of Naval officers who can't see we must tackle R&D like OSRD did in the war. If the cob-webbed antiques remain in charge, the Navy will open the next war with Task Force 58 just like the French began War II where they left off in November, 1918."

What lent urgency to it all was the fact that we weren't wrestling in a cozy little cocoon. One year after Japan's surrender, there were guerrilla battles in the Balkans; Red Army probes into Iran; Soviet puppets imposing harsh military rule in Eastern Europe; threatened communist political takeovers in

Austria, Italy, and Greece—all glaring spotlights on Soviet territorial greed against freedom-seeking East Europeans—and Moscow tossed off these events as merely "the inevitable workers' revolt against the intolerable yoke of capitalism." Truman maintaining our wartime spending level for atomic bombs implied that his people had little hope of reaching an accord with Stalin. "Lord help us and our children," I wrote, deciding, in spite of the Sykes-types, I'd "stay in the Navy a while to see which way the dice roll."

Sharply opposite Truman's resolve in the atomic arena, he had our military arsenal in free fall. On 31 December 1945 with fiscal year (FY) 1946 half over (in that era, 1 July 1945 to 30 June 1946), the army had 91 combat divisions; the marines, 6; the army air force, 213 air-combat groups (some twenty-seven thousand fighters and bombers); the navy, 1,166 combat ships including 16 carriers and 50 escort carriers. By fiscal year 1947 (1 July 1946 to 30 June 1947), the army had been shrunk to 12 combat divisions (1.6 million men) with only 2 at full strength; the marines, 2 divisions; the air force, 63 groups with only 11 fully equipped; and the navy, 343 ships plus a drawdown from 70,000 aircraft of all types at the height of the war to one-fourth that number. Four of our big carriers and most of our escort carriers were in mothballs. Military spending had plummeted from $43 billion in FY 1946 to $14.3 billion in FY 1947.

Up against that were service shopping lists for new hardware. The air force wanted to have seventy air groups, mostly by buying a new B-36 heavy bomber, a nag its fans claimed was a race horse. The navy wanted to buy 3,300 new planes a year to maintain a force of what had come down to only 14,500 navy and marine corps combat aircraft, and funds for six new ships, seventeen ship conversions. The army wanted to "modernize" with atomic artillery, battlefield missiles, new tanks, and so on. The total annual tab would have been $28 billion. Yet Truman was adamant about holding yearly "defense spending" to less than $15 billion. Fighting for a ration from that, we, mainly the navy against the air force and army, also were in a related, snarling debate over "Unification," a proposal to create a Department of the Air Force and put it, the War Department, and the Navy Department under the "general direction and control of a civilian Secretary of Defense."

The issue was a bag of snakes by 1947 but no longer over "whether," just "how to." Its fans, including Truman, justified it entirely on claims that "Unification" would cut costs, make the "military establishment" more efficient, and "end inter-Service rivalry." Ignored was a rush of new technology, ships and missiles with worldwide range, supersonic aircraft, electronic computers—all likely to make World War II tactics, doctrine, and even military organizations obsolete. Nor had there been any change in strategy or in presidential orders on what role each service would play in the nation's defense—the kind of basics that should drive a dramatic reshaping (which this was) of any organization. (A study of what future navy roles might be was ordered in December 1948 by Captain Arleigh Burke when he became head of the newly created OP-23, Organizational Policy and Research Division, in the CNO's office.)

Meantime, 1946 watched another emotional squall, a fret in Congress over who should control "The Bomb." My science friends insisted on civilian control, most of them as scared as Japan had been numbed by its wartime use. General Groves, for one, wanted military control as stated in the Senate's May-Johnson Bill. But Sen. (D-Conn.) Brien McMahon's bill, backed by Truman, became Public Law 585, the Atomic Energy Act. Effective 1 January 1947, a five-civilian Atomic Energy Commission (AEC) would control all things atomic, both military and peaceful uses of nuclear energy. The law also set up a Military Liaison Committee. Headed by a civilian chairman, its members flag-level officers from all the services, it was to tell the AEC how many of what kind of atomic weapons the military wanted. AEC review of that was to include comment by a General Advisory Commission of top scientists, also created by PL-585. Finally, all of the Manhattan District's empire was to become the AEC's on 1 January 1947.

And late in 1946, our duel with Sykes went to critical mass. To declare open war, the issue he chose was our Salt Wells plant. He began a memo barrage in mid-September, decrying its violation of BuOrd safety regulations—ironically, while I was in Los Alamos at a conference on what we could do to increase its productivity. Of course, if his bitching prevailed, he'd cripple our rocket program. I knew BuOrd's Hussey would not help us. He'd left Hedrick at Dahlgren for years, its consistently dismal performance a tribute to Hussey's bad judgment.

At the time, if I could fire up L. T. E. Thompson, Wallace Brode, Art Warner, and Bruce Sage to take a tough stand, I felt we could force the issue. But in two weeks of push-and-shove, I'd made little headway. For instance, Bruce Sage, though Salt Wells was his baby, wasn't fired up at all. I added it up for BuAer's chief, Rear Admiral Sallada. We had a good visit but he didn't sound steamed up enough to battle BuOrd. In October, Rear Admiral Schoeffel, who was to replace Hussey, stopped by. Sykes kept him away from Thompson, a snub so obvious it was funny—except that I began to worry if Tommy would stick. (In 1947, he joined private industry.) I did steal an hour alone with Schoeffel to lean on him. This billet needs a guy who knows what this job is and wants results, I told him, not a bureaucrat just telling us why we can't do something. The prize in this struggle, I said, is the future of navy research. He was attentive but said nothing. That night, I wrote in my diary, "It's time I draft a letter to Forrestal and buck it up the line through everyone, putting the facts as I see them on the table."

In December, Sykes was passed over for admiral, nailing me at the top of Mrs. Sykes' "hate" list. But life did pick up some after that. Groves got us money to make Salt Wells a permanent facility, and to finish building Michelson Lab and our modern housing project. And I bought Lili, my love, a mink scarf for Christmas, afraid she'd fret at the cost more than like the gift. (I got back a hug and a kiss.) And I got word I'd move to Sandia next year, a face-saving nuance my CalTech friends knew was just to avoid a shoot-out at high noon with Sykes. Simply put, BuOrd didn't want an aviator running any piece of China Lake.

On New Year's Day, I noted in my log, "I'm broke, as usual, but rich in wife and children. Europe is weak, arguing what to do with Germany. They need a strong Germany but remember how often it has invaded them. That and 'The Bomb' are giant problems in the new United Nations. Disarmament is high on the agenda of our Republican-controlled Congress. Will the 'Valor of Ignorance' prevail, the dark abyss of 1941–1942 forgotten? The world is now smaller than the United States was when Mr. Lincoln said we can not exist half slave and half free. Neither can the world, now." A footnote says, "If ever we are to govern wisely, scientists must have a say in what we do, not merely in how to do it most easily and economically."

In February 1947 Britain announced that its own war-drained economy was forcing it to pull its military forces out of and end financial aid to the Greek-Turkish theater. Truman's reaction, given to a joint session of Congress in March, was an American promise "to support free peoples resisting attempted subjugation by armed minorities or outside pressures." That could apply to just about anyone, but we all assumed it was to fill the vacuum the British had left in the Near East, as the Middle East was called then. What few people knew was that the unsung authors of the so-called "Truman Doctrine" were Navy Secretary James Forrestal and Vice Adm. Forrest Sherman, Sixth Fleet commander in the "Med" at the time.

That doctrine took a long time to write. In 1945, a Turkish ambassador to the United States had died, his remains temporarily interred in Arlington National Cemetery. Then in 1946, Soviet troops had massed near Greece, threatening to invade if communists weren't elected to run that country. They also were demanding that Turkey let them patrol the Dardanelles, their Black Sea access to the Mediterranean. They'd even invaded Iran, briefly. As a show of force without it seeming to be, Sherman advised and Forrestal bought sending the Turkish ambassador's body home on the *Missouri,* named for Truman's home state and site of the Japanese surrender six months earlier. Truman agreed but he and Secretary of State James Byrnes denied Forrestal's request that it be escorted by a task force—prompting Forrestal's friend Winston Churchill to tell him, "A gesture of power not fully implemented is almost less effective than no gesture at all."

The ship arrived in Istanbul on 5 April. Then during the next ten months, as the *Missouri* made "courtesy calls" in Greece, Italy, Algiers, and Tunisia, two carriers, seven cruisers, eighteen destroyers, and four auxiliary ships joined her, calling on a total of forty ports in all, being sure as they did that everyone watching, including Russians, knew that all the ships were well armed and had lots of marines on board. I also heard that, when one of our ships docked in Athens, its captain paid his crew in gold before giving them shore leave, using his sailors as a subtle way of asking the Greeks who was more likely to help them find peace and prosperity, us or the Russkies.

Then Forrestal drafted a policy proposal. His memo was delivered on 5 March to White House special counsel Clark Clifford, unsigned so Truman could take credit if he used it. A week later, Truman announced a summary of it as the Truman Doctrine, soon followed by a call to launch the Marshall Plan and create NATO (the North Atlantic Treaty Organization). It was the first time in the modern era, except in war, when U.S. military force was deployed in direct support of a foreign policy. As Sherman wryly observed, "There are a lot more things you can say here and do there if your fleet is there than you can say or do when your fleet is here."

That same month, I also was told Bobby (for Roberta) and Gail Burck, my intelligence guy in VB-106 on Cactus, finally had agreed to separate. She was a Ralston-Purina heiress, it was said. Her affair with a man named Dixon during the war had made Gail frantic. Now she'd said she preferred Dixon to Gail permanently. Their split was not unique, just one of many after the war. But that was it for Gail. He ended up committing suicide, another of the war's uncounted casualties.

In July, Congress passed and Truman signed the National Security Act, informally called the Unification Act, creating the National Military Establishment. Its amorphous name hinted at the bitterness of the fight for power within it. The act made the War Department's air force equal to the Departments of, now, the Army and Navy; legalized the wartime Joint Chiefs of Staff but without a chairman; created a secretary of defense with only "general direction, authority and control," not command, and two Boards and a War Council to help him do that; created a National Security Council to be the president's politico-military advisor (a Forrestal proposal); and created the CIA (Central Intelligence Agency). It seemed to me that all this massive shuffle did was create at the top echelon a whole new batch of committees. I also noted in my diary that day, "Unless we put the atom on a carrier quickly, 'Unification' may be marked in history as the beginning of the end of Naval Aviation."

Then I read Sykes' efficiency report on me. It said in sum, "He was more knowledgeable than I on technical things, but he was not concerned with orderliness or with regulations." The only negative ER I'd receive, ever, it forced a hard look, for the second time, at my leaving the navy. It didn't require genius to see what was in the navy's future: wars and threats of war everywhere, "Unification" causing turmoil and disunity, the rush of new technology. Yet merely in arguing that we had to devise answers to it all right now, I was being penalized, labeled as a "maverick," a "trouble-making rebel." At least that's how it looked to me.

Leaving Inyokern, I did take a little chop at "turf" warfare, a popular activity, it seemed, since passage of the National Security Act. In late July, a navy assistant secretary, BuOrd's Hussey, and Admiral Badger, head of the Eleventh Naval District, came to town. Why they hit on me, not Sykes, I only could guess. My rebuttal to their plan was, briefly, "The 11th Naval District should butt out! We must report to, be under the direct control of, BuOrd. Letting the District poke its nose in here would be a classic case of giving someone authority without responsibility, a malady we have too much of in the Navy, already." They left peeved but my view prevailed. Pleased, I guess, I wrote in my diary, "The reasonable man conforms to the world. The unreasonable man expects the world to conform to him. Therefore, all progress depends on the unreasonable man."

"Unreasonable" fit my stormy transition to Sandia in mid-1947. I wanted to go to sea. Deak wanted me at Inyokern. But BuOrd, Hussey and company, wanted me out of there. So, Groves and Parsons shot back, "Okay, fine, we'll take him down here in New Mexico."

I was drifting farther and farther away from the navy, casting a worse pall than the negative ER on my career future. On the other hand, Forrestal's navy finance expert, Wilfred McNeil, about to do the same comptroller-type work for him in the Pentagon, had said once, "Power flows where the money goes." If so, a job at Sandia had its merits because AEC had a virtually open-ended money flow from Truman.

Hussey picked "Red" Hean to relieve me, but Hean couldn't handle the pressure. He committed suicide in 1949. As I'd learned, being experimental test officer is a test of the man, too. One proof: the next four ETOs—Jack Monroe, Tom Connolly, Tom Walker, and Tom Moorer—all made admiral. Meantime, to stress China Lake's newfound value to the navy, Sykes was replaced by one-rank-higher Rear Adm. Wendell Switzer. A terrific "get it done" kind of guy, Wendy made the place hum, earning him a boost to BuOrd deputy chief later on.

The boss when I signed in at Sandia on 1 August 1947 was army General Montague. The base itself was field headquarters for AEC's Armed Forces Special Weapons Project (AFSWP), the R&D agency for atomic bombs. With a line officer and future admiral, Nev Schaefer, as my XO, my job was director of plans and operations for atomic warfare for the armed services. That and my being senior naval officer there meant we could stick our oar into most of Sandia's business—while I got quite an education in the warts and wheezes of a new mania, a stepchild of "Unification," that big programs must be run by a tri-service command as the Joint Chiefs of Staff were. Air force Col. Monte Canterbury was head of R&D, and the army's Corps of Engineers' Col. Gil Dorland ran the Engineers Thirty-eighth Battalion, the only team we had available at the outset to recruit and train tri-service bomb-assembly teams—the army and the navy—with the air force a reluctant bride about joining the group.

As that suggests, how to handle the nuclear era was, at that stage, a very complex question with few clear answers. A key one had come from the Riehlman Committee, named for its chairman, a congressman from New York. Tied to the Atomic Energy Act, the committee had authored Public Law 313 to divorce our technical fields from the Civil Service job-classification process. In that rigid drill, an agency wrote a job description, then a clerk in the Civil Service shop—who'd never heard of Einstein nor knew a spectrograph from nail polish—chose what grade, GS-1 up to GS-15, the job was worth. Back then, that number also was roughly the starting annual salary, in thousands of dollars, and most new hires began at GS-3 or -4, almost never at above GS-10.

Now, under "Riehlman," we technical folks would decide the grade a job deserved. Later, as the missile race with Russia fired up, we helped push through Congress creation of so-called "Super Grades," GS-16 to -18. To help us compete with private industry for civilian scientists and engineers, those did more than just create new top-pay levels. The law also allowed us to hire people directly into one of those top rungs, bypassing the standard Civil Service rule of promotion only by seniority. Riehlman, PL-313, was opposed by many in all the services. In the navy, BuOrd's Parsons and

Schoeffel, and BuOrd's R&D Division Director Arleigh Burke and a few others, won that debate. Today's system would not exist if they hadn't.

Meantime, one of the first tasks AEC had me tackle, besides writing technical job descriptions, was to turn Sandia and Los Alamos into permanent facilities. With 158 old B-17s and B-24s stored there, Sandia looked more like an aircraft graveyard than a research complex; and Los Alamos, all wartime temporary, needed a complete rebuilding. Army engineers did most of it but some was by Brown & Root. We had a lot of hassles because bids never matched final cost. They weren't inept or crooked. It was just hard to make a realistic bid with so many unknowns in the thing.

Groves was AFSWP's boss then, but his AEC parent ran the construction work, its goal being to double our weapons-production capacity. Otherwise, with just Los Alamos and Hanford to extract plutonium, we could build only two bombs a year. And Hanford was now so radioactive it was a ton of trouble. (Still is.) The University of California at Berkeley also was a factor. To Los Alamos what CalTech had been to China Lake, it wanted out of the weapons business in 1946. So the AEC had Donald Quarles (who'd become the Pentagon's top R&D man in 1953) and Merwin Kelly of AT&T's Bell Labs bring in Western Electric to form Sandia Corporation. Most of what AEC then built was put together conceptually by Quarles' outfit.

The Bendix tenant in an AEC-owned Kansas City plant replaced Raytheon as manufacturer of the bomb's firing circuits. The Mound Lab went from polonium initiators to making lithium-deuteride "triggers." A Rocky Flats, Colorado, plant (with today somewhat the same radioactive problems as Hanford) was built to do the final uranium-plutonium separation and bomb assembly Los Alamos had done. A Pantex plant was built in Amarillo to make shaped lenses and assemble the weapon, eventually putting both China Lake and Los Alamos out of that business. A second plant was put up in Portsmouth for U-235 separation. More and bigger facilities were built at Oak Ridge and Burlington, new ones at Savannah River, Georgia.

The best payoff, I thought, was Savannah River. The basic scheme for extracting U-235 or plutonium was to put the uranium in a canister and then to surround that with a "moderator" to dampen the neutrons bombarding the uranium, controlling the atomic reaction so it won't increase exponentially into an explosion. The original reactors had used graphite as the moderator, as Fermi had done. The problem with using graphite is that the uranium swells up as it's bombarded inside the chamber until it's hard to pull the material out of the reactor "oven," impossible if it's jammed up against the canister wall. That meant being very careful how much material they put in the reactor at a time, explaining, in turn, why it took up to six months to get enough plutonium for one bomb.

But Savannah River's moderator was "heavy water," deuterium, a hydrogen isotope with twice the atomic weight of ordinary hydrogen (hence the name, "heavy water"). It was easier and safer to run, with no "swelling" ailments. And in addition to plutonium, the neutron bombardment in "heavy water" also produced tritium with a half-life of 12.5 years, a big gain over polonium's 140 days. (We arbitrarily decided to accept that half-life, not hunt for a longer one, because we knew at the time exactly how much a bomb's yield would increase if tritium were added to its initiator.) After Savannah River was in production, by the way, its personnel said they'd detected, in their big pools of blue radioactive wastewater, evidence that Cherenkov's postulate (named for the Russian who devised it) was true. This postulate said if high-energy particles are injected into a medium whose radiation velocity is less than that of the particles, the resulting electromagnetic radiation will exceed the speed of light—refuting Einstein's thesis that nothing can travel faster than light.

That startling possibility was never researched further. Everybody, including Savannah River and ourselves, was behind schedule. Since that era, the whole bureaucratic setup has changed, but it's still pretty massive, still under civilian control. Even today, the Pentagon doesn't make any of the weapons. It states its needs and the Department of Energy has the weapons built, a weird dichotomy in drafting and Congressional approval of annual defense budgets. But in 1947, we had a capacity to make test vehicles and not much more, putting us way short of what the AEC Military Liaison Committee said it wanted stockpiled each year. We had only two "live" bombs—and had misplaced one of those—when AEC headquarters came out to count our inventory for the first time, in 1947. We had lots of "pumpkins" and one of the guys said, "Put the real one up front and tell them the rest are real, too." How few we had was "ultra top secret," never coming out publicly, but the truth is, during those first five years "stockpiling" was a joke.

Air force oracles were insisting that uranium always would be scarce and that they must have a monopoly on its use. After all, they argued, their strategic bombers were what would win any war. Echoed by swarms of politicians and media pundits, that defense-on-the-cheap idea was very attractive to an American public historically unwilling to pay much in peacetime on military forces. Ancillary darts tossed out included giving the air force what navy aircraft it could use, trashing the rest, and turning our ships into an army-escort service. Flimflam, and personal and professional insults spiced by "leaks" to the press, were rampant. In the bitching, the navy seemed always a poor loser to a powerful campaign the air force called "public relations."

A weird by-product was the Hiroshima "holocaust," as some called it, sapping enthusiasm even for projects offering dramatic gains in the quality of human life: nuclear power plants, medicine, and more. (Lawrence's Los Alamos cyclotron's first use was in cancer research.) On 17 September 1947, the first secretary of defense, Jim Forrestal, was sworn in to make the National Military Establishment official. My log notes, "This is a critical time for our Nation and for the atom. Yet, parochial Service ambitions are a marvel, Air Force the worst. And even in the Navy, those of us with the nuclear know-how are odd-man-out, most of our leaders ignorant in our field, blind to what new technology offers, what politics and 'Unification' will require. We must dump our old-thinking people, meaning a lot of younger-age people, too. The old world they want to regain is a Stone Age relic." Such opinions only convinced others that I was incorrigible, a "rebel against all things Navy."

As for the air force "scarcity" sermon, as new facilities came on line, the AEC offered to pay hundreds of dollars a pound for uranium, inspiring dozens of people with Geiger counters to scour our deserts and mountains, hunting the stuff. Soon we were up to our ears in uranium and production capacity. That sired another abrasive conflict. In a meeting I attended, LeMay said all he wanted, all he'd need, was a sixty-inch-diameter, 10,000-pound bomb with a twenty-kiloton yield. But Parsons wanted smaller, more efficient bombs, always asking, "How can we get them into the navy?" Army leaders, notably army Chief of Staff J. Lawton Collins and (in the 1950s) Maxwell Taylor, though they rarely said so on the public stage, agreed with Parsons.

One result, with Deak pushing the trend, was doing what I could to help build a 2,000-pound Fat Man with LeMay's desired yield—which eventually was in fact done, the Mk-28. It wasn't a big leap, actually. We'd been making prototypes at seven-tenths of scale since 1944: the Mark VI, a more slender copy of Fat Man (which still fell out of a bomb bay like a pregnant brick); at seven-tenths of that, the 45-inch Mk-V; the 32-inch Mk-VII at seven-tenths that, on down to a 12-inch Mk-12, even an 8-inch atomic shell. We also made a Little Boy-type Mk-IX but only ten since they took so much uranium. We had two ways to go, of course: LeMay's city-buster or small bombs for long-range missiles. A study Johnny von Neumann and I did said we'd likely have to go both ways. People not snared in the politics of it agreed.

A more pressing operations problem at the time was that a forty-eight-person team needed twenty-four hours to assemble a bomb. On the *Enola*

Gay, about all Deak had to do was put in the shell and powder bag and do the "red plug" work. Deploying a plutonium bomb was much more complex. Basically, forty-eight technicians had to do what the scientists had done, building it like a set of blocks: assemble the plutonium core around the initiator; put a uranium casing over that; then install the 128 lenses, the firing circuits, fuses, barometric switches, and the battery power source (which also had to be kept charged constantly). It added up to a very meticulous, demanding job.

Eventually, by early 1954 in fact, engineering design changes in the weapon and a mechanization of the assembly process made all that unnecessary, but in 1947 and 1948, no assembly team meant no weapon. Initially, we had just Dorland's 38th Engineering Battalion at Sandia to be trained, then deployed for bomb assembly. Eight of us, including Parsons, "Rivets" Rivero, army Col. Ken Nichols, and I, met to list the skills a team must have, then gave each skill a GS-rating number. The worst foul-up was that AEC let the air force, clearly against being in our program anyway, set up its own Training and Tactical Command. The TTC's first try at proving it could assemble and load the bombs on a plane, my log notes, "was sad, makes me mad, such a waste of time and people. Partly to blame is AEC headquarters ignoring the problem, what with so few weapons, but mainly the mess is due to Dorland's utter failure to devise a cogent training plan, then order TTC to live with it."

Dorland's Weaponeer-training curricula included instruction in the "FTB," a firing-circuit checkout device like Ashworth used over Nagasaki, but, since a bomb commander (BC) had to be aboard any aircraft carrying an A-bomb, I took that course. Its final exam was harder than the weaponeer's because a BC had to have not only a weaponeer's bomb-assembly knowledge of how it worked, but also know flight conditions that could affect his decision on when it could be or should not be dropped. In short, the weaponeer was an A-bomb mechanic, the bomb commander the engineer.

About then, I read Parsons' memo to the AEC General Advisory Committee, chaired by Oppie but not part of AEC itself. The memo was an excellent summary, I thought, of our ailments: the technical aspects, the pains of rapid program growth, the raw debate over who should do the atomic mission, the military risks of the bombs being in AEC's pocket and the bombers somewhere else, and Parsons' strong belief that the navy's first atomic priority ought to be building a nuclear-powered submarine. More on my mind then was being stuck in AEC, maybe for good. My 30 September 1947 diary got near-poetic as I sat one evening, staring past the New Mexico desert toward the Sandia range: "Breakers roll in across the miles of sand, hiss and subside, leaving a fringe of lace, ephemeral as moonlight on the hand. Salt grass forever ripples in this place of windy business and prolonged unrest. A gull goes over in a scud of spray; Tides wash each hollow where a foot has pressed and the dunes slowly, slowly drift away. Here I am, miles from the sea, but my thoughts of it seem constant. Will I ever return to the ocean's broad expanse?"

Of course, I still had Groves. In the Manhattan days when we first met, he'd told me what I could do and what not, and had people following not just me but a whole lot of us to see that we behaved. It was worse at the AEC with the FBI "trackers," we called them. If I went to, say, Los Angeles, my first call back in Sandia was Groves wanting to know what I was doing in the Roosevelt Hotel's bar at five in the afternoon.

"Having a drink, of course. What the hell else?"

"Well," he'd growled, "no loose talk."

While that history evolved in 1947 and 1948, I coped with a crisis or two myself. One was the seven-day, hour-a-day course on the atom I'd been asked to do for Washington's top echelon. I covered the Michelson-Morley experiment, Max Planck's constants, Einstein, Chadwick and the neutron, Neils Bohr, Fermi, Maxwell's equations, everything. My classes being generals, admirals, and department heads, I dared not flunk anyone. One student who seemed to be struggling was Gen. Omar Bradley. Then the army's chief of staff, in the snarling "Unification" debates, he'd proposed that the marine corps be abolished. The A-bomb, he said, had ended a need for amphibious landings. Yet he seemed simply unable to grasp how we got plutonium from U-238, made tritium, any of it. Neither could Adm. J. J. "Jocko" Clark, skipper of VF-2 at Coronado when I was on the *Langley*. In fact, Jocko quit after four days. "Hell, Chick," he said, "I don't have to know all that stuff."

My worst briefing brawl was on a visit by General Aurand, Department of the Army's deputy chief of staff, R&D, and navy Cdr. Pete Aurand's dad. The AEC's John Manley had tagged along. Groves asked me, in my business suit, to give them a tour. We were passing the "Clementine" fast-neutron reactor when Aurand said bluntly, "Why do you have to put a bomb in there? Our most important objective is to make a nuclear-powered airplane."

"General," I barked, "you don't understand a thing," then fired off scientific facts to prove it, no-rank Hayward lecturing an army four-star. To avoid a fistfight, I walked on. That plant did plutonium conversion, a deadly "hot" process in which radioactive isotopes might adhere to a person's clothes. If they had, that automatically set off showers we had in the exit aisles to wash them off. At the first shower, Groves whispered, "Aurand's a real pain. I'll give you five bucks if you'll douse him." Until then, I'd assumed they were pals, both being army,

I ignored Groves but, one aisle later, he pressed me on it, again. I skipped away from that, too, but as Aurand went through our last shower setup, a gushing waterfall drenched him—how, I have no idea. I was innocent. Groves shoved the money at me anyway. I shoved it right back. "Didn't do it," I said, "Keep your money." So insistent Groves and his aide drove out to our house that night and gave the five dollars to Lili.

Then in late 1947, BuPers said that on 1 January, my academy class' pilots would be demoted to commander. Aviation has too many captains, they said, and we'd have to stay commanders for at least two years. No admirals were demoted, I noticed. Once more, my class had been singled out for a hit. Once more I looked hard at quitting. Stuart Symington, former army assistant secretary for air, now secretary of the air force, in my Sandia class that week, saw that I was a captain one day, wearing commander's stripes the next. "What happened?" he asked.

"The navy demoted me, said it had too many captains."

Back in Washington, "Stu" wrote to Vice CNO "Wu" Duncan: "I'm told the Navy thinks it has too many Captains. I want to make the following of your people Air Force Brigadier Generals," equivalent to a rear admiral, lower level. My name was at the top of his short list.

Duncan promptly called me, snarling, "What the hell goes on here?"

"Well, the navy can't count, so I've taken a pay cut. With a family to support." The cut from $600 to $525 a month in pay was, in 1948, a month's groceries. His family Pensacolans, too, he knew Lili; had to know, with inflation soaring, that BuPers' off-the-wall order had hurt my family. I didn't tell him that, unlike the Sykes mess, now I had an option. General Dynamics wanted me to head their development of an air force liquid-fueled Atlas intercontinental (six-thousand-mile range) ballistic missile (ICBM), an attractive offer in a booming field. Besides the Atlas, in IRBM (intermediate range, fifteen hundred miles) R&D were an air force Thor, army Redstone and Jupiter, all liquid fueled; and a half-dozen navy and air force air-breathing cruise missiles.

After Duncan came a three-page letter from Admiral Sprague, head of BuPers, urging me to stay. Deak said it was my call. I asked Lili if she'd like to be a rich air force wife. "Forget it," she said, "You'll be treated like a leper in the air force and the navy will call you a traitor." I sent Sprague a note, giving him my opinion of his whole outfit. He sent a nastier one back but I felt better. The demotion was "typical navy" from way back. In 1920, when it cut Admiral Sims back to two stars while Gen. John J. Pershing, his counterpart, stayed at four, Sims retired. In 1927, Congress passed a law, restoring Sims' four stars. He was enraged. Arleigh Burke, a pallbearer at Sims' funeral in 1936, told me he was buried in his two-star uniform, wearing no medals he'd received after 1919, a rebel to the end.

Highlights in my January 1948 diary note a breakup of the final Big Four conference "with the usual name-calling by Russia's Molotov"; my brother's son being born; trips to San Antonio to squelch Fourth Army's push to run our AEC plants; and passing my "Einstein" course at the University of New Mexico. By then, Groves was acting as if he were indispensable to the AEC, but on 29 February, he retired. "We've come far in four years," I noted, "from a threshold of history to this gray day. Army Lt. Gen. Ken Nichols will succeed the old man, maybe give our outfit a future after all. Congress will buy The Marshall Plan," I added, "but Moscow won't. They'd have to adopt free-enterprise economics to get the aid. No way. The Plan is a gamble. A Stalin swipe at it could happen any time." That entry went in as Russia seized Czechoslovakia and hit Finland with demands that, if accepted, would make it a Russian puppet.

Operation Sandstone in April 1948 had General Hull as Task Force commander; Adm. Jim Russell, an excellent choice, and Dr. Carol Froman as test managers; the Naval Ordnance Lab's Greg Hartman to measure yields; and the University of Chicago's Dr. Shonka for the neutron-energy work. Parsons was very involved, of course, both on Eniwetok and on the seaplane tender *Albemarle* where we assembled the weapons. (In the midst of it, after a swim with "Rivets" Rivero and Deak, I jotted in my diary, "It's Easter Sunday. Hope my family is having fun.")

Each test involved very innovative design ideas. For the one later most publicized, the X-Ray shot, we levitated the plutonium in an air pocket, away from the U-235, and added tritium to the bomb's initiator. Those two new ideas were to increase the bomb's yield, and they did. When X-Ray, atop a two-hundred-foot tower, was ignited at 6:17 A.M. on 15 April 1948, it delivered a thirty-seven-kiloton yield; a second shot forty-nine kilotons, more than twice what hit Japan in 1945. More interesting to me was the Zebra shot. Ed Teller's analysis said that if a lithium-isotope core is hit with two million electron volts (MEVs), the neutrons that the core emits will create tritium, helium, and an energy reaction of fourteen MEVs. To do it, timing of the reaction sequence was critical, so he nicknamed it "The Alarm Clock." That Zebra shot got an eighteen-kiloton yield, basically proving that we could make a low-weight bomb with very high yield. We also proved that lithium-deuteride was as effective as polonium, extending from days to years how often we had to change an A-bomb's initiator.

My log that week adds, "Trieste, Germany, Finland, Korea, war is a runaway train, less than two years away, I think. Don't know where or how it will come but ominous clouds and lightning flashes tower on the horizon." (The Korean War would erupt in June 1950.) One sour note at Eniwetok was all the air force "boy-colonels" there, a hint to how much political ground the navy was losing at home. Another was at dinner with Adm. "Rinky-Dink" Denebrink in Hawaii. His limp words told us he was in the phalanx of navy types who'd rather let naval aviation die than fight the air force. On a side trip to Kwajalein, I saw retired Johnny Vest, by then an old man, and ill. Not being selected for admiral had taken its toll.

And I heard that Vice CNO Arthur W. Radford had raged at BuAer over our captain-pilot demotions. It implied that naval aviation was downsizing, a move the air force had urged—along with wanting an okay to offer air force commissions to Naval Academy graduates. Radford was sore, too, at Adm. Louis E. Denfeld, CNO by January 1948, for not, as Nimitz had, fighting back on that or on anything. It all was head wind against his drive to put a nuclear-weapon delivery capability on our carriers. As he'd told Parsons, Ofstie, and Mel Pride, the navy's best pilot-engineer, "We must do it. The very existence of naval aviation is at stake."

12

Pushing the Envelope

Creating a navy nuclear-strike capability began in 1946 and 1947 when Radford was OP-05 (deputy chief of naval operations for air). Mitscher sired part of it. OP-05 himself in late 1945, he was offered the CNO job but picked Eighth Fleet skipper instead, taking along his wartime Task Force Fifty-eight chief of staff, Arleigh Burke. Before he did, he sold Forrestal on building the largest ship afloat, a sixty-five-thousand-ton (over eighty thousand tons displacement, full-loaded) "supercarrier" to be named the *United States*. Its one-thousand-foot top deck would be flat like the old *Langley*'s, with no starboard-side island. This would limit an atomic bomb's shock-wave damage, designers said, a debatable premise; and would allow for safe launch of aircraft weighing up to one hundred thousand pounds, almost twice that of my B-24s in World War II. Like we patrol-bomber guys then, he argued now: Existing carrier aircraft can't lift tons of bombs or fly in bad weather, can't or at least won't fly at night, and don't have a bomber's range.

Still, persuading Forrestal and CNO Nimitz didn't make it a done deal. In the psychological warfare called "the budgeting process," their fiscal year 1947 (1 July 1946 to 30 June 1947) funding request already was before Congress, and their "Ships" plan for FY 1948 already included a call for funds to modify our largest carriers, the forty-five-thousand-ton (sixty-two thousand, full-loaded) *Coral Sea*, *Midway*, and *Franklin Delano Roosevelt* for nuclear operations. The supercarrier couldn't get into the cycle until FY 1949. Amending the FY 1948 plan to put it in might have been justified by a crisis, but the only one evident at the time was the attack at home on naval aviation. (Largely because of an assault on Berlin begun by Moscow in mid-1948, Congress in late 1948 voted to build the supercarrier, a small victory, we thought, against the "anti-navy" onslaught.)

In any case, from 1946 on, building the carrier-based big-bomber force evolved along two parallel, interactive lines. One focused on hardware; the other, on hiring and training able people. In both, we were "pushing the envelope," as pilots say. In hardware, getting big carriers left the question of what plane to put aboard. Mitscher wanted the same one on all our carriers. The sound, businesslike plan would have been to build and deploy one, a task that might take five years. That was too long a wait, given the political heat, Truman's miserly defense budget, and dramatic air force success convincing people that its bombers were all the military they needed. CNO Nimitz had agreed, "No way Congress will fund what we'll need for a nuclear war until we first show we already have the capability. We've got to go with what we've got."

Radford asked me, right after my return from Bikini, if we had a plane that, off a deployed carrier, could take a Fat Man far enough, nominally fifteen hundred miles, to hit Soviet military targets. We did, I said, and proving it, in effect, on 29 September 1946, a P2V-1 "Truculent Turtle," weighing 85,500 pounds, carrying 8,396 gallons of fuel, left Perth, Australia, to land fifty-five hours later in Columbus, Ohio, setting a world record, 11,236 miles, for nonstop, nonrefueled flight. For two years after that, Radford talked to me a lot on the subject. But he'd started "pushing the envelope" in June 1946 when BuAer ordered three North American Aviation (NAA) A-2As, navy-designated the AJ-1 ("A" for "attack," "J" for the builder).

Two wing-mounted twenty-four-hundred-horsepower Pratt & Whitney (P&W) "Wasp" piston engines plus a forty-six-hundred-pound-thrust Allison J-33 turbojet in its tail gave the AJ-1 a five-hundred-mph speed, 40,800-foot ceiling full-loaded at forty-three thousand pounds—triple World War II carrier aircrafts' weight. A *Midway* hydraulic catapult could launch it or it could take off on its own, its seventy-five-foot wingspan clearing by some thirty-two feet the flight deck's island. First flown in July 1948, its range was only eight hundred miles carrying a 5-ton Mark-Six (Mk-VI) Fat Man, but seventeen hundred miles with an Mk-V weighing just six thousand pounds. With the same twenty-kiloton yield as Fat Man, the Mk-V was an early reward from our atomic seven-tenths-of-scale work. The NAA's AJ-1 project engineer was Ray Gayner, BuAer's project officer was Cdr. Roger Woodhull; but I worked mostly with Capt. John N. "Mother" Murphy, head of BuAer's Armament Division.

I became sort of a utility infielder on it, advisor to Murphy on how to put the A-bomb in it, and acting as test pilot and flight instructor. But my frets over it began in 1945 when I test-flew a weird Ryan FR-1 fighter with a radial engine up front, a turbojet in its tail. That's why Parsons sent me to a 15 October 1946 look at an AJ-1 mock-up, BuAer program review; and this "observer" observed that both piston and jet engines in one airframe is a nutty idea. Make it all jet, I said. I was ignored. "Forget it," said Rear Admiral Lonnquest, the AEDO (Aeronautical Engineering Duty Only) Chief of BuAer. "We'll never see jet-powered aircraft on a carrier."

That's why he put turboprops in its follow-on, the A2J. He was a mule in a track meet, everyone else racing to build jet-powered aircraft. When one of his only two A2Js crashed, killing the pilot, that killed his program. Still, he had a point. Carriers all had straight flight decks back then, and jets did a high-angle, high-speed landing approach that easily could miss even the last of a carrier's four arresting cables. (We called them wires; now, I'm told, they're called traps.) If it did, it would plow into planes stacked astern, full of fuel and live bombs. Also, a jet's reaction, shoving it to full power, was much slower than that of a piston engine, a potential disaster if the landing signal officer is a tad late giving a wave-off to a jet pilot coming in at one hundred knots.

But the AJ-1 needed its jet for unaided carrier takeoff and for the same reason Norden was making a new AN/A SB-1 bombsight. World War II strategic bombers flew in set high-altitude formation, a tactic assumed in their bombsight's design. But to evade enemy defenses, the AJ-1 was to fly at cruise speed, low altitude, until in the target area; then kick in its jet to jump at top speed to a high-altitude run over its target. Using radar and optics, the SB-1 was to aid navigation as well as to provide all-weather bombing capability during all that maneuvering. (In use, the SB-1, far better at what the old Norden Mk-15 had done, didn't work well in the new flight plan, but gains in radar and related technology soon made it a relic anyway.)

Also worrisome was the fact that, to stay under the AJ-1's specified forty-three-hundred-pound gross takeoff weight, its crew was only three people: pilot, copilot-weaponeer in the right-hand seat with no flight controls; and a radioman-navigator. Another fret was that all its five hydraulic systems—landing gear, flaps, bomb doors, brakes, and so on—ran on three thousand psi (pounds per square inch) pressure, the effect of that provoking me to dub it "the flying hydraulic leak." But the main reason the AJ seemed a poor answer was that its jet fuel line ran the length of the plane. A break anywhere in it could start an explosive fire, probably killing the crew.

Radford's problem with it was having to wait until 1950 to do our show-and-tell. I told Ralph Ofstie, OP-05 then, that a modified Lockheed P2V could already do it, and in early 1948 he sold Radford and Adm. Louis Denfeld, CNO since January, on the P2V. Then Radford, our top aviator, ordered fourteen P2V-3Cs for Composite Squadron Five. It took a crew of eight, had two thirty-two-hundred-horsepower Cyclone high-altitude (thirty-three thousand feet) engines on its wing. As a hydraulic catapult was unable to pull its seventy thousand pounds of gross weight, it had to take off on its own, using eight JATO (Jet Assisted Takeoff) rockets on its tail. Into the 28-knot wind (the speed a *Midway*-class carrier was required to make even with only three of its four screws turning), a P2V at full power ran halfway down the nine-hundred-foot deck, fired the JATO, and instantly went to a 150-knot takeoff speed—its one-hundred-foot wingspan clearing the starboard-side island by only some ten or so feet. Worse, it couldn't ever be much of a nuclear weapon. The Little Boy, all it could handle, had such a poor ratio of U-238 needed to plutonium produced that the AEC planned to make no more than ten Little Boy–type bombs, ever.

The day Radford ordered the P2V-3C, he called me, with Parsons in his office, and told me to create Composite Squadron Five ("composite" saying it had more than one aircraft type), to be based at El Centro, California. A 146-person group, including air and ground crews and bomb assembly teams was to create VC-5 ("V" for "heavier than air," "C" for "composite") with nine AJ-1s and three P2V-3Cs. Then, learning from VC-5, we'd create VC-6, then -7, and so on—one for each *Midway*-class carrier. Radford wanted VC-5 operational by 1951, he told me. With our first AJ-1s not due to come aboard until late 1949, that would be a squeeze.

I'd expected Radford's call. By late 1947, his frequent updates to me on the subject had me sure I was headed there, but I was surprised, too. Parsons, then head of CNO OP-36 (Atomic Warfare), had to agree to his decision. Lots of people had leaned hard on Deak to pick Dick Ashworth, AEC Military Liaison Committee executive secretary at the time, a strong leader well liked in the business. But he lacked the necessary flight time, Deak had said, and Ofstie, I was told, said, "Well, our 'Nuggets' (new, young pilots) can't handle this. Get Chick."

Other than my ten thousand flying hours, my having flown all types of aircraft was not unique for a 1930s Pensacola graduate, but was in 1948 with our new-hatched World War II pilots all having been trained in only one aircraft type. Anyway, once Deak accepted me, I made Ashworth my XO (executive officer), intending, as we later did, to have him command VC-6. But, as the next few years proved, Radford, Parsons, Ofstic, the AEC Military Liaison Committee, and all my pals were just making it easy for me to make a lot of enemies.

Meantime, the do-it-now mandate aside, Parsons in 1946 persuaded Radford to build a new aircraft designed specifically to carry the Mk-V and -VI, have a fifteen-hundred-mile range, and be able to fly day or night in bad weather. The contract eventually, in 1949, went to Douglas Aircraft for a twin-jet, swept-back-wing (a radical design back then) aircraft designated

A3D-1 Skywarrior. One of my jobs as Composite Five skipper was to work on it in its mock-up stage with Douglas Chief Engineer Ed Heinemann who, in his thirty-year career, would design more good aircraft than anyone else in the business, I think.

Initially powered by two Westinghouse J-40 jet engines, each with seventy-five hundred pounds of thrust, it had a 72.5-foot wingspan, seventy-thousand-pound full-loaded weight, and 1,150-mile range. Its 620-mph speed and 40,500-foot ceiling were at least twice, and its weight five times, that of a carrier-based World War II torpedo bomber. Still, to lighten it, Ed left out ejection seats, and the cockpit had just one set of flight controls, the pilot's. Its three-man crew would have to be well trained to get out of that bird all Ed had designed into it. And he and I didn't want the J-40 engine, a dog anyway, its main asset being the eagerness of a Capt. Ramsey in BuAer to get Westinghouse into the jet engine business. The fight was nasty and I made another enemy, but we did sell the folks upstairs on using P&W J-57s with ninety-seven pounds of thrust, giving the A3D-1 a range beyond 1,500 miles with a seventy-eight-hundred-pound Mk-VI implosion bomb. And as the bomb's size and weight came down, it of course would do even better.

Initial A3Ds arrived in 1952, but the first of fifty-two more we bought, at some $5 million each, weren't deployed until 1956, a delay partly caused by carrier skippers refusing to let "that damn monster take over my whole ship." And in 1960, it got a traveling companion, the A3J with a gross weight of nearly eighty thousand pounds, fifteen-hundred-mile range, and Mach 2.1 (1,385 mph at forty thousand feet) speed. Given the Mk-V bomb's crazy aerodynamics, a fear was that if dropped at that speed, the bomb might waggle right back up into the aircraft tail. So in the A3J, the bomb was set in a tube between its two jet engines to slide backward out the tail.

But their carrier-suitability risks were pretty well solved by about 1955. Carrier flight decks were changing from straight to angled for far safer landing and takeoff, and they were acquiring two steam catapults, each able to hurl heavy aircraft aloft easily. The LSO (landing signal officer) had been replaced by a "ball-light" electronic device to tell pilots if they were or were not on a correct glide path to touchdown (all British inventions, incidentally); and new arresting cables were able to survive being hit by a sixty-thousand-pound aircraft landing at 140 knots. By 1956, I was taking an A3D onto a carrier's angled deck on 92 percent power, using the stabilized "ball-light" mirror on the deck, calibrated to the glide path, to guide me in, kind of like playing a video game. If the lighted ball was in the mirror's center, I was fine; up, I was coming in too high; down, I was too low. Miss the wire and I just went up and around again; and as soon as I caught the wire, I whacked off the power. But all that stuff was ambitious dreaming in 1948, when Radford told me to create and deploy VC-5.

With hardware decided, my top priority was people. Radford told BuPers to let us have anyone I wanted. Since VC-5 would be the first big twin-engine aircraft put on a carrier, I chose half carrier-qualified pilots, half patrol-plane people. I was told that getting big-plane guys carrier-qualified ("CarQual," they called it) would be our hardest job. In fact, those pilots adapted quickly to carrier operations. Much harder was teaching the single-engine types to fly in bad weather. And the VPs ("V" for "heavy aircraft," "P" for "pilot") obeyed the LSO ardently, but the hot-rodding single-engine jockeys tended to ignore him. A couple of crashes cured them of that. Parsons, Charlie Martell, Joe Jaap, and "Rivets" Rivero, all in OP-36 for a time, monitored my picks, advising, not deciding—though Jaap himself took VC-7 when we couldn't find any other qualified guy willing to do that. In short, I was able to hire very capable people for what I called our Atomic Squadron.

Some of my brief notes read: "Dick Ashworth, weaponeer at Nagasaki. Bernie Strong, dive-bomber, one of two who wrecked the carrier *Shohakaku* in the battle of Santa Cruz. Bill Romberger, torpedo pilot, helped sink giant Jap battleship *Mushashi*. Eddie Outlaw, fighter-type, loud, brash and good. Kelly Harper, Operations deputy, level-headed, multi-engine type. Steve Morrison, has flown only an F6-F fighter, smart, hard worker, ability to laugh at himself great for morale. Bill Shryock, Dave Purdon just out of Test Pilot school, top-grade pilots and officers." And my list of junior copilots is headed "all outstanding." Some Composite Five types—Jaap, Jim Dare, Outlaw, Bob Baldwin, Tom Connolly, and Ashworth—would end up admirals.

We also had to pick leaders for the three twenty-four-man assembly teams, one for each carrier. A team's forty-eight-hour assembly and aircraft loading of an atomic bomb had to include a mechanical section, electronic section for fusing, ordnance section for the implosion lenses, and a nuclear section. I needed people expert in all those fields. What they, mostly enlisted men, had to do was what scientists at Los Alamos and Inyokern had done, put together a Fat Man's components like building blocks. I chose people like Commander Dobey, gunnery officer on the *New Jersey,* for one, and Cdr. Steve Carpenter. And I was amazed at how easy it was to find good people all the way down to jgs, (many, like Dobey, also eventually becoming admirals).

Given the evolution planned for VC-5, then -6, -7, -8, and -9, recruiting, begun in March 1948, ran on (at least with me involved) until late 1949. For instance, I was not able to collar Lt. Cdr. William Scarborough until July 1949, at the end of his year at Ohio State under Admiral Holloway's program to get college degrees for aviation cadets. "Willie" had enlisted in 1935, and was an aviation chief ordnance man with a ship aviation unit when he started flight training at Pensacola in 1938. A lieutenant, junior grade, by 1942 with a patrol-bomber group in the Pacific in World War II, he'd come to Inyokern in 1944 as NAF operations officer and pilot on my rocket projects. He didn't mind at all my saying he'd have to spend five months at Los Alamos, learning the weaponeer business, before he could go flying again. In short, he was my favorite kind of guy, a bright, ambitious enlisted man who'd earned his way upstairs.

My batting average wasn't 1.000, however. Because he was one of the navy's best at devising carrier-squadron tactics, I tried to hire Pete Aurand, son of the general I'd seen drenched at Los Alamos. (Though we became good friends, I never learned if he, like Arleigh Burke and Tom Moorer, had tried to enter West Point before ending up at the Naval Academy.) But Pete turned me down. He wanted to skipper VF-5A, our first squadron of North American FJ-1 Fury straight-wing jets commissioned in March 1948. An "X" FJ-1 had hit Mach 0.87 in 1947, making it, briefly, our country's fastest fighter. It did lead to one of the 1950's best jet fighters, the swept-wing F-86 Sabre; and his unit, renamed VF-51, was our first jet-fighter unit to operate off a carrier at sea. But not for long. In May 1949, VF-51 was disbanded, its FJ-1s given to naval reserve units. I wasn't the only guy hitting potholes in the road that year.

The people business, tracking P2Vs, AJ-1 upgrades, and A3D work wasn't the full range of my VC-5 duties in 1948. Another was the *Midway, Coral Sea,* and *Franklin Delano Roosevelt* (which was to be the *Coral Sea* until Truman renamed it after FDR died) being modified to store and assemble the bomb before loading it on the bombers. The equipment and training on it all assumed an Mk-VI, Fat Man–type 10,000-pound weapon. Flight and hangar decks had to be reinforced, and a stronger drive put in the elevator for aircraft far heavier than any carried before. Parsons' OP-36 boys had the Bureau of Ships write the specifications and compute costs; then he had me approve the package, making sure all the interfaces really interfaced. Then BuShips' Captain Trescott, in charge of the ship modification itself, had me watch that closely, actually walking through the entire process from assembly to aircraft loading of the bombs to verify that it all worked.

And when Radford became CINCPAC in April 1949, he wanted a staff briefing on VC-5 every two weeks, it seemed. So in 1948 and 1949, I was a moving target, flying to all three carriers' shipyards; to the P2V-3C, AJ-1, and A3D factories; to Washington, D.C., and Hawaii; and, in 1948, to AEC sites too, since I was still AFSWP director of plans and operations for atomic warfare—and the air force was in a hot argument with everyone else over their wanting me relieved by an air force general. Upstairs, some officers, Radford the most prominent, were in a hassle with the air force, Forrestal, Congress, and the media, a roles-and-missions debate newspapers called, variously, the "B-36 bomber vs. the Supercarrier" or "Revolt of the Admirals." If the air force won that one—and it was way ahead on points at the time—Composite Five would be dead.

Of course, all that commotion was, for me, just an add-on to my AFSWP work at Sandia where we had a snake pit's worth of problems. One was at our New Mexico bomb-storage site where we had by then six bombs, with nuclear components for forty or so more in the pipeline. This caretaking challenge flared up in early 1948, before Radford's VC-5 phone call. My log's excerpts note visits to storage sites, "with the General [Montague] and Vice Adm. Gingrich, the AEC man in charge of security, smart guy, excellent choice for his job. But it's obvious the vaults are not adequate what with the polonium initiators having to be changed so often." It also cites "very bad air conditioning, safety features in vaults 6 and 7," and Col. Gil Dorland's failure to check our little stockpile regularly or keep continuing temperature and humidity records, mandatory for knowing even such mundane-seeming facts as when the bombs' batteries, their ignition-power source, had last been recharged.

I also had a long session about then with Parsons, Rivero, army Gen. Ken Nichols and company in Washington to update assembly-team skills needed (in my case at the time, aboard ship) to build and maintain a plutonium bomb, the initiator, plutonium core, and all the other components—all of it a meticulous, demanding job. (The Sandstone tests at Eniwetok in April— the tritium initiator, alone, a big factor—began to change all that. By 1953, reengineering and design changes had made pointless all the assembly-team stuff. In short, what we did from 1948 to 1952 was like teaching people to drive a stick-shift car while automakers were building automatic transmissions.)

My diary hints at flying's therapeutic value to escape the brawl, noting happily, "Over the top at 20,000 feet in an XC-97, five hours, 15 minutes, Albuquerque to Washington! Good trip, cold, to get a portable bomb-assembly building," an oversized Quonset hut, "to take back to New Mexico." The trip did open a wound over my demotion I thought healed. Says my log, "The injustice is what rankles. Why can't the navy just fire a man? FDR fired Kimmel. Can you name another Admiral?"

Then in April, I missed by six feet making Lili a widow. I was in the right-hand seat of a new XC-97 Stratocruiser, an air force junior pilot at the helm, departing Bolling Field with another bomb-assembly hut for Los Alamos.

Whatever the kid was doing, flying that C-97 wasn't it. I grabbed the controls, lucky that Anacostia's Naval Air Station runway, almost an extension, really, of Bolling's, was empty. We barely staggered over the Potomac River's Fourteenth Street bridge. I'd been scared before, in the war, climbing into an untested aircraft. This time, even my bones shook.

But April's headliner was Rear Adm. Dan V. Gallery, assistant CNO (guided missiles), at the time. A visionary like Parsons, he was admired, at least by me, for his dogged persistence, pushing the navy to develop ship-launched rockets and missiles, and Earth-orbiting satellites for radio communications and navigation. In January 1948 he wrote Denfeld a memo, the gist of which was, "If we develop the proper planes and tactics, the Navy can become the principal offensive branch of the national defense system . . . we should start an aggressive campaign aimed at proving we can deliver the Atomic Bomb more effectively than the Air Force can."

Some ghost, the usual source, "leaked" Gallery's memo to widely read columnist Drew Pearson in April. Pearson's column on the memo, salted with sneers at the contents, started a firestorm in Washington, of course. John L. Sullivan, Forrestal's replacement as secretary of the navy, was struggling to gain a truce with Symington in the annual defense-budget battle, and he was as enraged as the air force. Sullivan sent Dan a "letter of reprimand," after making Denfeld co-sign it, then publicly denied he shared Gallery's view. What he did believe, he suggested, was that there were valid roles in the future for both the air force's pet, the B-36 bomber, and the "Supercarrier." That ducked the real issue: which department and programs deserved what share of Truman's annual $15 billion budget for defense? Said Symington, "It was like throwing a piece of meat into an arena and telling three hundred hungry tigers to go in after it."

Trying to get in there first had spawned the so-called roles-and-missions debate. Briefly summarized, those holding army and air force reins wanted missions defined by weapons each had and their use confined to an arena: the "ground" army, "airborne" air force, "seagoing" navy. An echo from the 1920s Morrow Board nonsense, it meant, of course, no more marine corps, and naval aviation limited to ASW (antisubmarine warfare). Unfortunately, many admirals agreed with the "area of operations" business. In his memo, Gallery had scolded them too: "The assumption that the next war will be fought like the last one is basically wrong and if we stick to it, the Navy soon will be obsolete." The few navy leaders such as Radford, Ofstie, and Burke who agreed with Gallery wanted military objectives to decide who did what missions.

While they hacked at those issues in Washington during all of 1948, we wrestled with flaws in the atomic stockpile's security. However, getting that act organized was hung up, awaiting a decision on who should be doing it. Public Law 585 gave the AEC custody of the weapons. Yet in wartime they were military weapons. In short, the AEC had authority without responsibility for the end result, a setup found frequently in our federal government. The Military Liaison Committee visited Sandia a few times, trying to settle the custody question. And we—Parsons, Ashworth, myself—were in an incestuous sort of rotation: on the Committee one day, ruling on a military request's merits; next day writing those requests for the navy. And behind it all was the air force wanting a weapons monopoly, the navy saying that must not happen.

On 27 April, as that war raged, I returned from Eniwetok and on 10 May flew to Inyokern for the Michelson Laboratory dedication and to see Charlie Lauritsen receive his Presidential Citizens Medal that we'd all worked so hard to bring about. It stirred a flood of memories, including my own work at China Lake well done, I told myself, in spite of BuOrd. I also recalled the troubles and ill feeling Sykes had provoked. By this time he was out of the navy, encouraged out by an offer to reward his resignation with a retirement promotion to rear admiral.

And on 14 May, the United Nations decreed the new state of Israel. Within eleven minutes after that, Truman had us be first to recognize it. British troops promptly left Palestine and war erupted, Arabs attacking Jews all that day. Earlier, hoping to encourage the British to stay a while to prevent the otherwise predictable bloodshed, Forrestal had begged Truman to hold off until we could put a task force in there. Well-known radio-newspaper columnists Walter Winchell and Pearson savaged "anti-semitic" Forrestal for "trying to squash" Israel's creation. My diary for June 1948 notes, "This new thorn in the UN's side may fester until it kills the patient. Truman's move is smart politics here. We have a large bloc of Jewish voters. But it will offend many of our Arab friends. We'll see guerilla wars and worse in Israel before, if ever, there's peace—short of Armageddon. Once more, the British have left us holding the bag."

On 2 June I was in San Antonio on another "Joint-Service" drill, part of the bomb-custody battle. Of this army engineers, navy district, air force meeting's waffled results, I noted, "The idea seems to have caught on that, just because a decision may affect someone, he has a right to help make it." It was spawning a rash of "Joint" committees, such panels being, as many know, a way to delay doing anything. In mid-June, I was told that Composite Squadron Five's base now was to be Moffett Field, south of San Francisco. AEC security demands can be met more quickly, I was told; and Palo Alto is much nicer than El Centro, I thought, with a bonus—Stanford University for night school. They also said I'd be promoted to captain in January to make me the only captain in Composite Five. If a private company ran the rapid-fire demotion-promotion merry-go-round the navy seemed to favor, I said, it would go broke within two years.

Next day, Gallery, an outspoken backer of VC-5, took me up to see the Grumman folks in New York. "They have a new approach," he said, "on a way to deliver our beloved 'gadget.'" That told me two things: Dan had started to talk like Parsons and Oppenheimer; and Sullivan's reprimand had bothered him about as much as water does a duck. But Grumman's flip-chart serenade was an offer to build an atomic-powered aircraft, its power-to-weight ratio so high from using lead shielding to contain engine radiation that it made absolutely no sense. I cut their scheme to shreds. (Nuclear-powered rockets are a different matter, of course. Unmanned, most of their flight time is in isolation, shot through the atmosphere in a suborbital trajectory.)

I returned to Sandia just in time to take Lili to the hospital. She had to be operated on to remove a breast cancer, surgery as life-threatening in that medical era as the disease. I was angry all month at having to be somewhere else when I should be with her. It told me VC-5 would be my last job for the navy. I was tired of moving, of making my family a tribe of nomads. John Jr. was a hellion near impossible to control. The demotion was a second reason. The clincher was Lili. She'd be in the hospital a month, recovering. She'd be a healthy girl after that and I wanted to give her, owed her I thought, a stable home and normal life. No one believed I'd take that step at the time, but what could the navy offer to make me stay in? Did it even care?

The Communists lost the election in Italy, a tiny victory for us in what Churchill had titled, "A Cold War." But on 24 June, after months of harassing our road and rail traffic to West Berlin, the Soviets halted it totally. Too late, hindsight whined, "Why did we ever agree to let only them command access to Berlin?" By July in a knee-jerk reaction, our people on-site were airlifting food, coal, and medicine to 2.5 million West Berliners and our thirty-thousand-man Allied garrison. That squelched Soviet propaganda telling Western Europe we'd abandon them all if pressured—drivel postwar leader and future French President Charles de Gaulle, for one, was inclined to believe. It also spurred the creation, in 1949, of the North Atlantic Treaty Organization (NATO).

I didn't think just an airlift could sustain the city but it did, brave pilots flying usually in bad weather, often landing only a mile apart. If one had been shot down, we'd have been staring at World War III, but clearly Russia didn't want a little spat in Germany to go that far, either—though in 1948 neither the Republican-controlled Congress nor Truman saw that as the issue. In May

1949 Moscow gave up and lifted the blockade, but the airlift went on until 1 October. The RAF had been in it, and navy Skymasters—my academy roommate, Harry Badger, in command of one squadron—had hauled 25 percent of the tonnage in ten flight hours a day per plane, mostly at night or in bad weather—a feat our publicity folks ignored. More amazing to Paul Hammond in his essay "Super Carriers and B-36 Bombers" was that "The Air Force did not unleash its powerful public relations talents. It deferred to the State Department which deserved no credit, at all." Was it, he asks, a fear that praising its transports might smudge its "gung-ho" strategic-bomber image?

A nastier question to me was where, when it all began, were tools created by the National Security Act: the CIA that was supposed to warn Truman it might happen; the National Security Council, presumed author of policy to deal with a crisis; the several Pentagon offices to turn that policy into military plans and operations? Nowhere. Instead, a few State and Defense officials, after a Sunday meeting run by Forrestal, sent Truman a vaguely worded set of politico-military options, deploying bombers to England, for example, but said nothing about an airlift option. Our track record before that, in 1948, hadn't been much of a blue-ribbon performance, either.

Meanwhile, on 23 July 1948, a Monday, Lili came home after more than a month in the hospital and I left for Washington at 3 A.M., Tuesday, to give Secretary Sullivan a midmorning answer to "misquotations in the Hearst Press." Normally, I'd have flown up on Monday, as coldly callous a thing to do as I could imagine in that situation. Say, "Hello, Lili, goodbye," in one breath? No way. The CNO had told me to see Sullivan in D.C., an emotion-charged city at the time, but I was peeved. I could have briefed him by phone. I'd given a speech to an Institute of Aeronautical Sciences group in Los Angeles—my job had me always getting "pinged" on to give a speech to someone. Asked there if Russia had an atom bomb, the Hearst reporter said I'd said we had a system that can detect an atomic explosion by its reflection off the moon, a crazy idea, of course. I suspect the reporter hoped, by saying I'd said what I hadn't said, he'd get me angry enough to blab a secret or two while telling him so. Make me fly all the way back to D.C. just to explain that puny incident face-to-face? Phooey.

By late 1948, that poke at me and the big one at Berlin were not all that had gone awry in the Pentagon. Since his "Truman Doctrine" speech had noted the Communists' war in China, the Yugoslavian Tito-Soviet face-off, Soviet pressure on Greece and Turkey, fighting in Palestine and in French Indochina, and the "saber-rattling" in Germany, most people had assumed Truman was issuing a call to rearm. But in January, he asked Congress for only an $11 billion FY 1949 budget to buy fifty-five full air force groups, fifteen "skeletonized" ones (giving it its desired seventy [mostly bomber] groups), 10,922 navy aircraft, and a five-hundred-thousand-man army. And in March 1948, he asked Congress only to enact the Marshall Plan, resurrect the Selective Service draft (both of which it later did), and enact Universal Military Training (UMT). Congress eventually did approve the seventy-group air force, "to be built up incrementally," and killed UMT to pay for it as Republican Senator Robert Taft fussed like he had before—"There'll be no more war."

Common sense argued in 1948, as it still does with as limited an impact, that deciding what to spend on military forces should be the result of a logical three-step process. The trigger should be a "threat estimate," a forecast distilled by the National Security Council from political, military, and economic data gathered by the CIA, State Department, Pentagon, and others. From that should come presidential policies and programs to neutralize, or failing that, defeat the threats. Then the Pentagon and the Joint Chiefs of Staff should decide how much of what kind of military forces will enforce effectively what the president has decreed. The final step is simply to compute the R&D, procurement, and annual operating costs of those forces.

A policy that could have guided all that planning and funding was first proposed to Washington in February 1946 by George F. Kennan, counselor to our embassy in Moscow, who repeated it in an anonymous article, "The Sources of Soviet Conduct" in the July 1947 *Foreign Affairs* magazine. Basically, he urged the White House to launch a "firm containment" policy against the Soviet Union. A Communist-run economy cannot sustain itself, he said, and thus a communist form of governance can't, either, unless it continually expands. Containment should be backed by military force when necessary, he said, though he implied that didn't mean anything near the kind of war footing and combat forces mobilized for World War II. By 1949, the State Department, at least, was pursuing and promoting a policy to do something new in our military history: fight what it called "limited wars."

In the Pentagon, the roadblock to preparing for "limited war" was in Truman's defense budget. In early 1948, he'd put a "cap" of $14.4 billion on the military budget for FY 1950. Adding a force structure for fighting "limited wars" to that needed for "strategic nuclear war"—itself, a hot navy-air force argument—resulted in an aggregate Joint Chiefs of Staff budget agreement at $30 billion. Meetings with Forrestal and other maneuvers whittled that down to $23.6 billion. Finally, in a compromise suggested by Forrestal, they whacked that to $16.9 billion, a result that surprised even them. And Truman hung onto his $14.4 billion "cap," insisting, as did most in federal power then, that any more "will bankrupt the country," a confession of economic ignorance if nothing else.

The trouble with either low number was, as the Joint Chiefs indicated, it meant being able only to launch a massive bomber counterattack and hope that would halt a Soviet advance long enough to mobilize for all-out war. A very minor footnote was that the air force, more popular than the navy both publicly and in Congress, and vastly better at "public relations," had peddled successfully a message that, as some pundit said, "resonates with America: 'We can deliver a bigger bang for the buck.'"

Truman's budget "cap" had hurt in other ways. For instance, his recognition of Israel and it then quickly coming under siege by military force from both inside and outside its borders, trying to destroy it, should have energized Truman's Doctrine. Yet he'd refused to send in either a navy task force or army troops, or even air force bombers that would've been useless anyway. Adding it up in August, still thinking the Berlin airlift was not enough, I wrote in my log, "Virtuous motive, bound by inertia and timidity, is no match for armed, resolute wickedness; a sincere love of peace no excuse for muddling millions of humble folks into total war. The cheers of weak, well-meaning assemblies soon cease to echo, their voices cease to count."

An alibi for my literary flair is that it was a presidential election year, rife with political hyperbole. New York's "tough on crime" Governor Thomas E. Dewey and California judge Earl Warren led the Republican ticket, Truman and Alben Barkley the Democrats', and Henry Wallace, a fan of Russian collective farming fired by Truman as secretary of agriculture, was running as a third, "Progressive Party" candidate. The Republican nominees were from our two most populous states, and the "smart money" and the new art of opinion polling all said Dewey would win easily. Listening to the campaign noise, I wrote in my diary, "These are perilous times and politics may get us in trouble, again, as it has previously on the international front."

August 1948 was a chock-full month, the 4th marking one year since I'd left Inyokern. And the air force–navy feud still raged with no end in sight, kept alive, I thought, mainly by the lunatic fringe on both sides. One crazy piece of it was LeMay insisting that we not build any atomic bombs with a yield of less than twenty kilotons, while guys like me were reciting the well-worn adage that "Accuracy will win out over yield every time." Eventually, I was sure, sometime, Congress would resolve the roles-and-missions business, maybe by law. The same was likely, my diary notes, for a change in the weapon-custody rules. There was no basic disagreement at my level on the need for it, but who will make that milestone decision and when was unclear. I did write one other note, at that point: "I think the Navy is fighting a losing battle in not letting us transfer to the Air Force," an aberration proving I must have been tired that day.

In midmonth, Cdr. Dave Young, an ordnance Post Graduate, visited me. He'd been picked as my relief in the bomb-assembly and storage business at Los Alamos. We'd need a good man there to help get us operational, but he acted as if this would be a three-year hold on his career. That and his chatter told me he wasn't strong enough nor his views "Joint Service" broad enough to do the job well. The day he arrived, most of my VC-5 VPs were in Sandia, taking the bomb commander's course, convincing me that one place VC-5 would not fail was in the "people" department. (On the 18th, Jacob Lamokin, Soviet general consul in New York City, was deported for being party to the Kockensina espionage case, the first public hint that we had Russian spies in our midst.)

Air force Gen. Hunter Harris, I learned then, would replace me at Sandia, but not report in until November. That meant another commute on my agenda, one between VC-5 at Moffett and my job at Sandia. As a result, I flew with General Montague on Labor Day weekend to inspect the atomic-storage setup, back on Tuesday, back to Moffett on Friday, 8 September, for the squadron's commissioning, then back to Sandia, leaving my XO Dick Ashworth in charge until, "if ever," I wrote, I can take command.

In politics that month, the pundits still called Dewey a shoo-in, and Drew Pearson wrote a new acerbic criticism of Forrestal, accusing him of "openly" helping Dewey's campaign by giving him an "unauthorized, classified" briefing on our military programs. Democrat Forrestal quickly, publicly reported that he had, in fact, gotten Truman's permission first, adding that he believed a secretary of defense, because he is that, must avoid political campaigns; and that Gen. George C. Marshall, whom Truman had called "the greatest living American," felt the same way.

The end of September, I was in Norfolk on the *Midway*, doing a final check, then a run-through of the script for an exercise at sea to test the ship's modifications, to be sure they worked for what a team had to do to assemble and aircraft-load the weapon. I also saw Jimmy Flatley, operations guy for our Air Atlantic Commander, Adm. Felix Stump; and my Coronado-days squadron mate, "Dusty" Rhodes, "who," my diary says, "hasn't changed a bit in all these years except to lose his hair." I also noted on that trip: "Since Radford, Ofstie, Stump, all seem to call on me, we have prepared a briefing on the entire program to give all Fleet Commanders and senior people in the Shore Establishment."

In mid-October, I was told I'd join VC-5 in January and that my class would go up for selection when the General (promotion) Board met on 29

November. Just hearing it irked me. Added to my peeve, army Secretary Kenneth Royall, that week, sent ex-President Herbert Hoover's Commission on Government Organization a proposal to absorb naval aviation into the air force. Dan Gallery fought back in a speech on 31 October, setting himself up as a major navy public opponent to giving the air force a monopoly in the nuclear business. I perked up in late October when Admirals Pride, C. Turner Joy, and Lonnquest were in town to take my course on the atom and its production problems. Leader of our VF-3 squadron when I was on the *Langley*, Mel Pride was now one of the navy's very best, both in flying and in aviation technology. My admiration was doubly so because, like me, he was a bluejacket, machinist mate, second class, who, in his case, had made admiral.

November went by like a whirlwind. Truman, to the stunned amazement of the pundits, beat Dewey to win the presidency by an easy margin, and, at the urging of Vice Adm. J. D. Price, DCNO (Air) by then, Col. William R. "Soupy" Campbell joined VC-5 to begin building a marine corps atomic-delivery capability. Composite Five's first P2V-3Cs were delivered that month, and I flew one to London to brief the commanders of U.S. Navy forces in Europe and the Sixth Fleet on what VC-5 meant. Flying home, I had to stay aloft nineteen hours, moving around to evade a hurricane. Finally, on 2 December, I was relieved as AFSWP Plans and Operations—in a change-of-command ceremony done while I was out of town, earning Harris a big "demerit" from my Sandia pals and giving me a bellyful of this "Joint" stuff for a while. We rushed East to spend my leave in Pensacola. The children had a great time with "Gramma" Floss and "Grampa" Pete, as did Lili and I, including a lovely Christmas. But the General Board's silence had me saying, if I don't get promoted, I'll likely be out of the navy by June. I'd been far too long at a desk job, mostly on the civilian side of the house. Then just before Christmas, they told me I was to be a captain, again. "For how long this time?" I wondered.

13

This Is Not a Boat, Hayward

I had to be in Moffett on 2 January, so I left Florida after Christmas, driving older daughters Shelley and Marion in our black Cadillac, Lili and the little ones to fly out later. After stops in Shreveport, Louisiana, and Midland, Texas, we reached Tucson on New Year's Eve. I was having a drink that evening in the hotel bar with several lovely young ladies, all seeming eager for male companionship, when Shelley and Marion walked in. Quickly, my covey of ladies vanished. Naturally, the girls told their mother who got very upset, if a wry smile and shake of the head can be called angry. By then, I'd rented Lili and company a house in Palo Alto. It cost more than I could afford, really, but staring at bankruptcy so my family would have a nice place to live was routine by now.

Then in February, VC-5 took its first hit. On a NAA test flight west of Los Angeles, the tail section broke off one of the only two AJ-1s built so far, the one I'd intended to fly after they did. The instant crash killed both North American test pilots aboard, Al Conover and Chuck Brown. BuAer grounded the aircraft, saying NAA's fix on that problem would delay to early fall the AJ-1's delivery date to VC-5. So, after the Senate confirmed my promotion to captain on 19 February, our party at home was a bit subdued.[1] And in Washington, Radford's team was in two big fights, one to prevent the navy being ordered out of the atomic-bomb-delivery business, another to pry out of the AEC an allocation of the weapons themselves.

Thinking we wanted them only for sea warfare, the air force, hawking its "monopoly" theme to everyone in sight, said no nation, certainly not Russia, had a navy big enough to justify the expense. Besides, it argued, the Bikini tests proved one atomic bomb can sink an entire task force. They hadn't, of course. The Able shot had sunk five ships, damaged nine, but they'd been anchored like fish in a barrel, not moving under a fighter screen. Radford's answer to me was, "VC-5 can make or break our program. You must get operational as quickly as practical."

He, Ofstie, Parsons, and Adm. Felix Stump (then the DCNO, operations) all seemed to want me to oversee promoting the whole program to put the navy into the atomic-bomb business and not just be VC-5's skipper. So we had to battle both VC-5's engineering flaws and also uniformed and civilian leaders in the Pentagon, White House, Congress, even in the navy. Still, I groused, "It's hard having to go straight to Radford and Forrest Sherman (after he became CNO in late 1949) for my orders with no one telling people senior to me in between what I'm doing and why. Ofstie says he'll watch out for me and Parsons says just go ahead and do what has to be done."

Our air force opposition sounded off on 1 March when a modified B-29 landed in Texas to complete an historic, round-the-world, nonstop, aerial-refueled flight, and Gen. Curtis LeMay thundered, "We could drop an A-bomb anywhere on Earth!"

I chose myself to fly VC-5's reply since, "If you want a good outfit, you have to lead." (That idea wasn't mine, alone. In World War II, one reason our best squadrons were best was because their leaders—Mel Pride, Sears, Jimmy Thach, Flatley, and all the others—never asked their men to do a mission they hadn't already done themselves.) I flew a P2V to Norfolk, had it lifted onto the *Coral Sea* on 3 March; and, at sea on the 7th, we roared off her windy deck at 74,100 pounds, our "bomb load" a Little Boy "pumpkin"; flew two thousand miles to the AEC's Salton Sea test site near El Centro; then flew back to land at Patuxent River, Maryland. And on the 22nd, I led a nearly as long three-P2V flight, taking off in forty-five knots head wind, using 750 feet of deck to get airborne with no strain. Next, Dick Ashworth flew off the *Coral Sea* to Panama, then Alaska, and back to Moffett, making headlines as he did. The following week we ran two more flights with several senators and congressmen invited to watch.

But alarms blared that month when Truman told Forrestal to resign so he could nominate Louis Johnson to be his secretary of defense. A West Virginia lawyer-politician, Johnson had run the Democratic Finance Committee, personally raising most of Truman's presidential campaign funds. Forrestal was shocked. He'd been led to believe he'd be asked to stay. Besides, he'd said often, party politics must not decide who becomes secretary of state or defense. Two days later, Forrestal collapsed and went into Bethesda, Maryland, Naval Medical Center, suffering what the Pentagon called "an illness not unlike the combat fatigue suffered by soldiers in war."

Then Truman said he was scrapping the "high Defense budget Forrestal had talked me into." The new FY 1951 (July 1950 to June 1951) number was $14.6 billion. After setting aside $1.6 billion to "stockpile scarce raw materials" (whatever that meant); pay a military salary increase Congress had passed; and build a new guided-missile test site at Cape Canaveral, Florida, the services would have only $13.5 billion left. And for FY 1952, Truman said, that figure would shrink to $11 billion. Johnson had asked, "Can we argue?" and Truman answered, "No."

The Joint Chiefs, who once had berated Forrestal for not getting them a $15 billion budget, didn't say a word. After a few gyrations, each one said it might squeeze by on $9 billion but that the others could do nicely on $2 billion each. Meantime, I flew an aircraft from the *Franklin Delano Roosevelt*, off San Francisco, to Panama and back to Moffett, a repeat of my four-thousand-mile trip off the *Coral Sea*. The air force answer was a ninety-six-hundred-mile B-36 flight with forty-two tons of bombs, telling the press that seventy Soviet cities within that range would be "annihilated if total war comes." To many of us, their city-busting plan was odious, a dereliction of duty. Our job is to defeat the tyrant's military force. Cities would be irrelevant in a nuclear war quickly over, once started. Assuming we'd one day return to acting military, I countered their stunt by flying with a fake Little Boy off the *Midway* in the Atlantic to Alaska, circling a "target" twice, then landing in Florida, an eight-thousand-mile round-trip.

The air force erupted, of course. We were wasting money, they said, trying to grab the strategic bombing role given exclusively to them by the Joint Chiefs and Forrestal. Radford, with help from Ofstie on the Military Liaison Committee, muted that bitch with some fine essays. A key point was that PL-585 forbade storing our atomic bombs on foreign soil. So they could be sent at Russia only from the States by air force bombers on six-thousand-mile, unescorted, air-refueled flights or from our carriers. To back Radford, I invited Ernest Lawrence and his industry pals aboard the *Roosevelt*. In giving them a VC-5 rundown, I did note, not very subtly, "Our country can get in lots of trouble, assuming airpower is just bombers. Airpower's job is to obtain and maintain control of the air. That includes over the sea lanes, our gateways to the rest of the world. If the navy can't do that, the air force won't be able to do its job."

On 18 April, I was at Lockheed, urging them to put a tail hook on a P2V-3C. I'd gotten Radford's okay for it in March. He didn't want to tell Congress that a P2V, after a mission, couldn't land on a carrier but would have to ditch alongside. I did as he wanted but couldn't see much point to it. Even if the AEC gave us all their Little Boys, one flight and the war'd be over for ten P2Vs whether they got their feet wet or not. That same day, at the Newport News shipyard, keel-laying began on the USS *United States*. Then on Saturday, 23 April, a thunderbolt struck. With navy Secretary Sullivan in Texas, Johnson, with Truman's approval, canceled the "Supercarrier."

Asked by Forrestal in March, the Joint Chiefs had okayed its construction. Now, Johnson asked them, again. Now, under Truman's new spending cap, the army and air force chiefs said they hadn't endorsed it in March, merely accepted what "navy man" Forrestal already had decided. CNO Denfeld let pass their sudden change of mind, dropping his popularity in the navy to near-zero. Later, Johnson said, "Truman sent me to the Pentagon to take charge and this issue gave me a chance to do that."

Predictably, navy's Sullivan and Under Secretary W. John Kenney resigned, enraged. In his three-page resignation letter, Sullivan noted that Johnson, a week earlier, had sidestepped the subject twice, then announced his decision while Sullivan was away. "I am very deeply disturbed," Sullivan wrote. "(This is) the first attempt ever made in this country to (stop) development of a power weapon," one already approved by Truman and Congress. "(Worse), the consequences of a Secretary of Defense drastically, arbitrarily changing and restricting the operational plans of an armed service without consulting that Service . . . are far-reaching and can be tragic."

With that, the so-called "Revolt of the Admirals," mainly but not exclusively aviation guys, was on. To organize for battle, the navy's Pentagon focal point was its CNO Organizational Research and Policy Section, OP-23, headed by Arleigh Burke. Set up originally to study what ways the 1947 Unification Act could or should affect existing navy policies and organizations, in the "Revolt" business, it was a defensive outfit at first. "You can't defend yourself," Arleigh said, "until you get into the other guy's camp and find out what he's doing."

The other guy's big gun was Convair's B-36, with up to a fifteen-man crew and a 357,500-pound top takeoff weight, the largest bomber ever built. Powered by six thirty-eight-horsepower P&W engines pointed aft, it was to cruise at 240 to 300 mph at forty thousand feet, above the reach of that era's piston-engine fighters, and carry a 10,000-pound bomb ten thousand miles. Its evolution riddled with mismanagement, by its first flight in 1946, it was obsolete. Even Forrestal had wondered, "Has the air force really faced up to the urgent strategic and tactical problem of whether or not the B-36 can survive against modern radar defenses and the new jet fighters armed with rockets?"

In 1948, with the B-36 still unable after two hundred flights to do what was specified for it in 1941, nearly half the top air force people wanted to end that program. Even the WSEG (Weapons Systems Evaluation Group), set up by Forrestal to provide "impartial" studies of a weapon's effectiveness, had panned it. Noting that its 45,200-foot ceiling was 10,000 feet higher than a B-29's, WSEG had concluded that "in all other respects, its ability to reach Soviet targets is no better than that of existing bomber alternatives." A swept-wing, six-hundred-mph jet-powered B-47 with a four-thousand-mile range, 40,500-foot ceiling, twenty-thousand-pound payload, had flown in December 1947; an eight-jet Mach 0.95 B-52 was coming along; and now air force ICBM folks were competing for funds. Still, in early 1949, Symington said, "We must go with it. We can't wait" (like Radford on VC-5), and air force spokesmen bragged, "Until the advent of ICBMs, the B-36 will be outmoded only by a better airplane," adding that "Russia has no adequate defense against this bomber."

Navy "leaks" in Washington told newsmen that was a lie. The Soviet Mikoyan-Gurevich MiG-15 jet fighter can go above forty thousand feet, they said, as can our Grumman F9F Panther and McDonnell F2H Banshee. And, they said, a rocket-armed Banshee, at its climb rate, could be on a B-29's tail within minutes after takeoff. To prove it, they hung an aerial photo of Washington in the National Press Club bar. It had been taken by a Banshee from fifty thousand feet ten minutes after takeoff, they said, and challenged the air force to a B-36 versus a Banshee duel. The Joint Chiefs, including Denfeld, promptly shot that dare down, saying it wouldn't prove a thing. Snapped back some anonymous naval aviator, "Just like Billy Mitchell didn't prove a thing twenty-five years ago when he challenged the battleship admirals to a duel."

Symington fired Johnson a note, "Are we back in the old rut of two Services doing what they are told and the third what it thinks best?" And on 5 April, HASC (House Armed Services Committee) Chairman Carl Vinson shot off a public salvo, saying, "I want it clearly understood, if persons in the Armed Services or their employ continue to pass statements to the Press calculated to deprecate the activities of a sister Service and at the same time jeopardize national security, (HASC) will step in with a full-scale investigation."

On 22 May, Forrestal dove out his sixteenth-floor hospital window, killing himself. His virulent critics in Washington's intramural wars were abashed. My diary notes, "What a sad end to such an excellent career. I guess Walter Winchell and Drew Pearson are happy, now, but, in years to come, I think we'll see he was right on the Israeli matter. French astronomer Laplace's quote seems very apt: 'What we know here is very little but what we are ignorant of is immense.'"

Then on 24 May, Francis P. Matthews became secretary of the navy. An Omaha lawyer with no government experience, he said at the outset that all he knew about the navy was that he'd owned a rowboat, growing up in Minnesota. "Rowboat" Matthews quickly showed he was not our top man but a sheriff hired to do whatever Louis Johnson wanted him to do. He accepted without a murmur Johnson's order in July, backed by Omar Bradley, to cut navy air and fleet size 50 percent. I jotted in my log, a cut that big could bust me back to commander. "If that happens, my Navy farewell address will be titled, 'I gave up the ship.'"

Meantime, Dan Gallery made headlines, again, with an article in the 25 June *Saturday Evening Post*. Titled "Admiral Talks Back to the Airmen," it was an improved version of the speech that got him in trouble back in April. Matthews wanted to fire him but navy Under Secretary Dan Kimball told Matthews, "in very salty language," it was said, "Since he named no one, personally, you can not do that to a man for merely exercising his Constitutional right to free speech—unless you want to look in public like a damn fool."[2]

In July, the North Atlantic Treaty, signed in April by the foreign ministers of Belgium, Canada, Denmark, France, Iceland, Italy, Luxembourg, Norway, the Netherlands, Portugal, and Great Britain; and by Secretary of State Dean Acheson for us, was ratified by the Senate. (Greece and Turkey joined NATO in 1952, West Germany in 1955.) Counting all our military cutbacks, my diary wonders, "What will we do to back it up?"

Congress took a big step in August, mostly in Forrestal's memory. He had urged them to amend the 1947 Unification Act to give him, he told Senate Armed Services Committee Chairman Millard Tydings, "sufficient authority to control effectively conduct of our military affairs." Congress did as he asked. The "Establishment" became the Department of Defense, and the secretary of defense now had "authority and control" over it, not just "general direction" as earlier.[3] Now, only the secretary of defense, not those of all three services, had a seat on the National Security Council. Also, his three assistants became assistant secretaries, he got a deputy secretary of defense, and a chairman was established for the Joint Chiefs of Staff.

Also added to the original act was a Title IV section, its powerful nuance hardly noticed at the time. It required uniform budgeting and fiscal procedures for all Department agencies. In effect, that meant the secretary of defense, not the separate services as before, would disburse Congress-appropriated funds, making him second only to the president as the military's chief executive. Wilfred McNeil, just promoted to assistant secretary, comptroller, said of Title IV succinctly, "Power flows where the money goes."

And on 29 August, a B-29 equipped for "long-range detection," a first step toward the U-2 spy plane developed from 1953 to 1956, found highly radioactive particulates in air samples gathered on a flight between Japan and

Alaska. Analysis of their strength and of wind patterns concluded that the Soviet Union had exploded its first atomic device that day in Kazakhstan. It amazed so-called experts who'd said it would be years before they'd be able to do that. But we in the business knew Russia had competent physicists and probably were well informed on our program too, since so many Americans at the time were sympathetic to the Soviets.

From before that event, beginning in late May, I had been fussing with North American to fix all the "trouble and nothing but" we were having with the AJ-1's hydraulic systems, flight controls, turbosuperchargers, the two fuel systems, and the jet engine. Many of the ailments would have been preventable as early as the design stage, except for two facts: NAA was on a "rush" schedule, expected to deliver AJs to VC-5 while the plane was still in flight test; and NAA had big problems of its own. Drastic postwar cuts in military programs had shrunk the company workforce from forty-nine thousand to five thousand, its engineers now spread over several air force and navy projects. By 2 August, noting a rash of near-fatal flight-test accidents, I wrote in my diary, "We shouldn't accept it until it's right. The risks with reversal of control in the wave-off configuration is the worst thing. If not fixed, it will kill people"—as it did, twice, almost before that entry's ink was dry.

Radford's deadline said I had to accept it anyway. So on 17 September, I picked up at the Downey, California, NAA plant the first of twenty-eight AJ-1s on order, flew it to Moffett to let my guys poke around in it, then to New Mexico for weapon fitting, "a quiet, easy trip," my log says, "to get a feel for the fuel consumption." To keep "need-to-know" visits minimal, we did not involve Patuxent River Naval Air Station, our leading flight-test facility by then. After New Mexico, I tested it, running the flight test program out of Moffett.

Pax River was no stranger, however. We finally pried loose a pair of P2Vs with a tail hook on them, so, from 1949 to mid-1950, I landed a P2V one hundred times at a sink rate of ten to eleven feet per second (which worked nicely) into an arresting gear on a "carrier deck" painted on a Pax River runway. I also did a few touch-and-goes off the deck of the *Roosevelt*. Then in mid-1950, Kelly Harper, my XO since January 1950 when Dick Ashworth became VC-6 skipper, tried one. A novice at it, he came in at too high a sink rate, 13.25 feet per second. That made his P2V bounce when it hit land, forcing an "in-flight engagement," meaning he caught the wire while airborne. That jerked him nose-down onto the deck, mashing his P2V's front end. BuAer promptly told me not to let anyone but me do that exercise. (As a C-130 transport proved years later, with angled flight decks [which we didn't have then], any well-trained pilot can land that size aircraft on a carrier with ease if weather and seas are not extremely rough.)

During those tests, I saw a lot of Capt. Fred "Trap" Trapnell, head of the test-flight work at Pax River. The first naval officer to attend my Sandia nuclear-orientation course, he was close to Mel Pride in aeronautical engineering knowledge, a first-rate test pilot, perfectionist in flying technique, and other than maybe "Little Bill" Davis, the ablest senior naval aviator of that era. A bachelor, he had his own F7F and didn't want to do anything but fly, anytime, anywhere. So I didn't mind at all, nor did he, my asking him to come see me at Moffett. I picked his brains a lot in the VC-5's early days on P2V, AJ, and A3D flying qualities when I was doing engineering on them. To lure him away from his F7F long enough to make him a rear admiral, Stump gave him command of the *Coral Sea* for a tour. Later, he took Dave Young's slot at Sandia.

Meantime, in September 1949, my brother, Dick, came by, bound for Korea. We put him on a Martin "Mars" flying boat, his destination no worry to me then, but a big worry just nine months later. Also in September, a verbal storm shook our nation's capital. Some zealous navy junior officers and a civilian, Cedric Worth, "leaked" libels to the press, alleging that there were B-36 cost cover-ups by Convair and the air force; that former Convair board member Johnson was buying B-36s to up his Convair stock's value (stock he'd sold upon becoming secretary of defense); and that he and Symington were cooking a deal to buy lots of B-36s so Symington could take over Convair and make it "The General Motors of the Air." (It came close for a time, later, as General Dynamics, but neither man had anything to do with that.)

Instantly, Carl Vinson ordered a "full-scale" HASC investigation and a Navy Court of Inquiry was convened; Johnson jumped up to slap the navy for "waging a campaign of terror against unification," and Matthews indicted "insubordinate officers engaged in surreptitious disclosure of information to persons not entitled to it." Within two days, John Crommelin, at the time a captain serving on the Joint Chiefs' staff, was proving what we'd said at Pensacola: "good aviator, terrific flight instructor, but with a knack for making people mad." Afraid the Court of Inquiry might not hear his testimony, he invited a herd of newsmen to his home to give them copies of his prepared statement. Then, in his typically forceful way, he elaborated, calling unification "in its present form" a terrible mistake, claiming the navy was being "nibbled to death by the Joint Chiefs and Secretary of Defense," and "The Air Force strategic atomic bombing concept is militarily unsound, perhaps morally wrong," "the B-36 an inferior airplane."

Headlines forced Vinson to reopen his B-36 hearings, now as twelve days of October hearings on "Unification and Strategy"—after Radford had urged him to do that. In eight of those twelve days, the navy presented its case. For openers, Matthews claimed the whole furor was fueled by "spoiled" naval aviators more "keenly disappointed" than the rest of the navy by budget cuts that had affected all the military. Next, said the HASC report later, "Almost the entire high command of the Navy appeared—three of four fleet admirals, six other admirals, three Marine Corps generals, one vice admiral, three rear admirals, supplemented by a number of captains, commanders and civilian technical experts," a tribute, said author Paul Hammond later, "to how thorough Navy staff work can be at its best."

CINCPAC Radford came in from Hawaii to lead off. His basic point: "The B-36, itself, is not so important as the theory of atomic-blitz warfare it symbolizes (which) as a threat, I do not believe will be an effective deterrent to a war nor win one. (Moreover) the unusual, unjustified ways used to push the B-36 to its present status have prevented progress toward mutual trust, understanding and unified planning, (the essence of) unification."

Up next was Trapnell, still at Pax River, who detailed how "an unescorted B-36 can be intercepted and shot down easily." Next, experts in radar, night intercepts, jet-fighter capability and Capt. John H. "Savvy" Sides on antiair rockets elaborated on Trap's theme. Ofstie, from being on the Joint Chiefs' Bikini-test evaluation group, tore apart strategic bombing, showing why it wouldn't deliver a military victory nor support national policies. Adm. William "Spike" Blandy, now CINCLANT (commander in chief, Atlantic), spelled out why funding left the navy unable to carry out its assigned missions. OP-23's Arleigh Burke emphasized, "Our country will always need to command the sea," and that justified the new supercarrier. Jimmy Thach recited the future need for fast carrier task force operations and amphibious landings. Blandy, Burke, and others, not being aviators, undercut Matthews' claim that only navy aviators were the source of all this "insubordinate bickering."

Our cleanup hitter was CNO Denfeld who finally had been persuaded, too late, to be not a mediator to the rest of the Pentagon but the navy's spokesman. Pouring funds into the untested B-36, he said, was starving both the army and the navy of essential resources; a twenty-five-year lead in carrier and amphibious warfare know-how was being squandered on false doctrines of war; the army wants to abolish the marines and the air force wants to abolish carriers. In sum, he charged, the navy "has not been granted a full partnership in defense." Matthews' reaction was to storm out of the hearing. Vinson's HASC mostly avoided the root cause of all this acrimony, the defense budget.

Not avoided was Matthews. He got Johnson's and Truman's okay to fire Denfeld, flying in Sixth Fleet commander Forrest Sherman to replace him on 1 November, making Sherman, skipper of VF-1 in my *Langley* days, first naval aviator to be CNO in peacetime. Matthews also took Burke's name off a list for promotion to rear admiral, charging that his OP-23 had orchestrated the attack on the B-36, a credit undeserved. But Vinson had promised revenge on anyone who punished witnesses for saying what they honestly thought, so Truman put Arleigh's name back on the list.

John Crommelin didn't fare as well. Initially, Dan Kimball issued a press release endorsed by Radford and Denfeld, saying the "official Navy position is, Capt. Crommelin merely stated his personal views which is his Constitutional right." A surprise to Matthews, that press release boxed him in. But later, when Crommelin was to get a slot requiring rear admiral rank, a flag we thought he deserved, Matthews stepped in. Saying that would look like a reward for publicly opposing Johnson, Matthews put him on "indefinite furlough." So Crommelin retired.

As Sherman was named CNO, my P2V was put on the *Midway*, Capt. Wallace M. "Miles" Beakley its skipper, Tom Moorer the air boss, and we sailed to a spot one hundred miles off the Florida coast. The cream of the navy's political foes would be aboard that week: Johnson; Omar Bradley who, in the HASC hearings, had called us "Navy fancy-Dans who always have opposed unification"; Symington; army Chief of Staff J. Lawton Collins who'd replaced Bradley and wanted the Joint Chiefs to have command authority like, he claimed wrongly, the German General Staff had in World War II (and put my navy out of business); newspaper mogul and air force admirer William Randolph Hearst; and some others. They'd been invited to watch a firepower demonstration by a two-carrier task force, a convincing display, we hoped, of why we needed a big carrier that could carry lots of attack and bomber aircraft. After the show, I was to fly them to Washington.

They seemed impressed with the firepower stuff, I thought. Meantime, my crew chief, J. E. Shadburn, had our P2V offset a bit to give its one-hundred-foot wingspan more clearance of the *Midway*'s starboard-side island. As I buckled Johnson into the copilot's seat, Shadburn whispered, "Skipper, don't worry. If this thing goes over the side, I know who'll get out." Some guys aboard ship did want me to dump the flight but I wasn't quite ready to wipe out the entire Pentagon top command. Still, warming up the engines, I told Johnson, "Mr. Secretary, our wingtip will clear the island by only about six feet down the deck. So if our right-side engine conks out, the navy will have a flush-deck carrier whether you want it or not and we'll be right in the middle of it."

Johnson frowned; Symington laughed nervously; Bradley, a solemn guy anyway, glared. (When he told that story after he became a U.S. senator from Missouri, Symington would say, "I knew Hayward's reputation, knew he was just crazy enough to do that on purpose.") With that, I revved to full power, released the brakes and, after we'd rolled some 150 feet, kicked in the JATO. We roared down the deck, Johnson white-knuckled, squeezing his seat arms, as we lifted off. Zipping past the bridge, I saw Beakley's face pressed against its glass; and once I was airborne, he sent me a message, "Never fly off my ship like that, again!"

We'd scared him too, I guess. My passengers let off at what is now Ronald Reagan National Airport, greeted by the usual press mob, I gassed up and left for Moffett Field, poor weather making us work hard on the gauges all night. We did a ground-controlled approach into Moffett and, as I taxied up to the parking line, the VC-5 duty officer, ran out shouting, "The NO is on the phone, wants to talk to you, immediately!"

What in blazes, I wondered, have I done now? Sherman told me, bluntly. "Don't I have enough trouble with the secretary of defense," he snarled, "without you getting in the act?"

Then he spent fifteen fierce minutes explaining the merit of my learning to keep my mouth shut. I was able to deduce from it that Symington's pals on *The Washington Post* had printed his telling them what I'd told Johnson. Sherman never again mentioned it to me but it was for a while a hot topic in Pentagon coffee breaks, and Symington hung in his office a picture of that takeoff, I guess to show his air force friends he'd flown off a carrier.

By mid-December, VC-5 had only six of the nine AJ-1s authorized, and I got so blind drunk one night in New York, I took a hard look, next day, even worried about becoming an alcoholic, typically meaning an irresponsible liar. One break was that there were no navy types around to see me make an ass of myself. That stupid behavior said the job pressure must have been getting to me more than I'd sensed. It didn't ease off any in the first half of 1950 when technical problems galore and accidents limited our AJs' flight time to only about twenty-five hours a month.

Sherman came by in January and I flew him from Albuquerque to San Diego in my AJ, a good trip. Our chat convinced me he'd do a good job. Like Denfeld, he was a peacemaker, but he listened to Radford and Radford was the man most responsible for getting me the tools needed to put atomic weapons aboard ship. But Sherman was alone. All the other big guns in the Truman administration opposed him. They wanted something for nothing. He understood that, but he understood Radford too. But Lili shook me terribly that spring. "Nothing has been the same since the war," she said. "You fought all those years to stay alive and return only to find ourselves in an ever-widening circle of strangers." I agreed, though I'd refused to admit to myself until then that only the excitement of the atom had kept me from leaving government to join private industry and its relatively "big bucks" paychecks.

The first four months of 1950, Radford visited me once, and I made eight trips to Hawaii to brief his staff on VC-5 progress. Our story went over very well except with old "Rinky Dink" Denebrink. Each time, Radford stressed the need for early VC-5 deployment. The briefing he gave in March said our next crisis probably will erupt in the Pacific. But the Pentagon was transfixed on Europe. So Sherman, with Radford agreeing, ordered VC-5 put in the Med; and, on 1 April, we began to re-base in Norfolk, which meant three months of moving aircraft, equipment, ourselves, and our families east, little of that time spent developing VC-5 itself. Nor did anyone have definite plans on what we'd do in the Mediterranean.

In mid-April, we spent a week on the *Coral Sea,* at sea, testing our aircraft atomic-bomb-loading skills. Their new installation aboard ship was excellent but I did say I hoped we'd not have to put her in a navy yard for extensive changes with every change in the weapon. During our tests, the Russians shot down a Privateer, a navy version of the B-24, which they said had invaded their airspace. All we knew was we had a plane missing on a flight from Wiesbaden to Copenhagen. No one ever did discovered what really happened.

In late June, Bill Sampson was killed when AJ Number 17 crashed at Bedford, Virginia. He was on a carrier-suitability test flight, trying to go nonstop from Muroc, now Edwards Air Force Base, California, to the *Coral Sea,* not a smart decision in a new aircraft at that stage. Meantime, Johnson, defending his $13 billion FY 1951 budget, one-third less than the services had requested, said, "This budget is adequate to defend the nation in any situation that may arise in the next two years," adding, "If Joe Stalin starts something at 4 A.M., the fighting power of the United States, led by the air force, will be on the job at 5 A.M."

But low budgets indicate low intentions. As my family and I headed for Norfolk, three North Korean communist army columns led by one hundred Soviet tanks roared over the thirty-eighth parallel into South Korea at 4 A.M. on 25 June, driving Republic of Korea (ROK) and U.S. Army defenders out of Seoul, south toward Pusan. And an unnamed admiral reportedly growled, "Louis said we could lick the Russians. He didn't say anything about the North Koreans."

Truman pushed through the United Nations a resolution to use military force, ours to be most of it, probably, to "restore the territorial integrity of South Korea." Surprising was the Soviet Union walking out before the UN

Security Council voted on the resolution. If they'd stayed, they could have vetoed it. Truman also began drafting World War II veterans, the only way we had of deploying a combat-ready force quickly without moving troops out of Europe, while Omar Bradley, focused on Europe, bemoaned, "This is the wrong war at the wrong time with the wrong enemy." By August, with MacArthur now UN commander, his army dug in on a defensive arc around Pusan, it looked likely we'd be driven off the peninsula.

And on the 13th, my AJ Number 18 crashed to delay carrier-qualifying tests yet again. I decided I'd have to tackle the qualifying flights myself. Then in mid-August with Sherman wanting VC-5 deployed in September, me thinking it impossible, I flew to London, Nice, Lisbon, and Morocco, to brief the commanders of all U.S. Navy forces in Europe on VC-5's status, and to survey our area of operations. Back in Norfolk on 21 August, I did the first-ever AJ-1 landing on and takeoff from a carrier, the *Coral Sea*, which had sailed three to four hours out to sea to make the test realistic. As Shadburn and I were preparing to leave Norfolk, Adm. Felix Stump, then COM-NAVAIRLANT (Commander, Naval Air, Atlantic), strode into our hangar. As stern and humorless as Admiral King, I thought he'd come to wish us luck but he had other plans. In a friendly (for him) tone, he said, "Chick, I'm going with you."

"Admiral, you can't. This is a test, one we haven't done before."

Growling, he said, "I'm COMAIRLANT, this is my airplane, and I'm going with you."

So we went in AJ aircraft #122600, Eddie Outlaw following in 122601. Whistling by the carrier, I radioed that I had COMNAVAIRLANT aboard. A long silence finally was broken by the ship's skipper himself, "Trap" Trapnell, saying softly, "Chick, repeat, please."

"The Commander, Naval Air Forces Atlantic is in my right seat," I answered, thinking, 'Oh, man, what happens if I get him killed.' The AJ already had killed a lot of people and didn't have ejection seats. Trapnell was shook too, for the same reason, but couldn't say much either. However, landing and takeoff went smoothly. Then, on my second flight that day, waiting for me at the carrier was Jack Floberg, navy assistant secretary for air, who insisted he ride back to Norfolk with me. My warning him this was a dangerous test flight bothered him no more than it had Stump. But that flight too was uneventful, after which my boys spent the day using what AJs we had, seven by then, to land on the carrier, a total of ten flights.

On 15 September in a move called "military genius," MacArthur landed the First Marine Division and an army division at Inchon, Seoul's port city, 150 miles behind enemy forward lines. To avoid being trapped, the communists fled north. After three bloody months trying to hold onto a piece of it, UN forces cleaned the enemy out of South Korea in just fifteen days. (The Joint Chiefs, in July, had denied MacArthur's request to do the end run because they didn't want to divert forces they believed soon would be needed in Europe.) So much, we said, for Bradley's claiming amphibious landings would never be needed again. Our carriers' image got a boost too. Offshore, carrier-based aircraft were in the air over targets five times longer than the fifteen minutes their fuel load allowed air force jets flying out of Japan.

That same month, I was told November was now VC-5's date to deploy and, after it did, I'd go back to AEC, meaning an end, maybe, at age forty-two, to my active flying. I was sore, of course. I'd always loved flying. Now the reward for it was to be tossed back among "an ever-widening circle of strangers." In my diary, I noted that there must be a lot more navy people than I realized who want to see VC-5, and me, fail. Truman fired Johnson that month, hiring George Marshall in his place, and we launched four "bombs" off the *Roosevelt* with no mishaps, dropping them through an overcast on a radar-identified target.

In mid-October, we were sent to Guantánamo for two weeks deployment training with the *Roosevelt*. Our last day there, Lt. Cdr. Dave Purdon and his enlisted chief aviation radioman, Edward R. Barrett, were killed in a crash on takeoff. Only Lt. Edward A. Decker, the bombardier-navigator, got out—how, even he didn't know. Their boost system failed and, when that happens, the controls lock, making aircraft control impossible. It was my fault, really. I should have been more hard-nosed nasty, demanding that NAA redesign that whole setup. Worst of it was, my boys had just begun to accept the aircraft. Now, everybody wanted to quit. That same day, VC-6 had an AJ go down, burning. All hands bailed out safely but VC-6 seemed to blame me for it all though I'd argued from day one against that damn jet fuel line. And my diary notes, "NAVAIRLANT really gets me goaty. Maybe I'm getting too sensitive, hearing a jeer in every remark, seeing fingers pointed at me that probably aren't."

Those disasters grounded the AJ again. Now I was sure the AJ needed a huge rework, including all-jet power, if it ever was to be a safe airplane, but I didn't tell my guys. They were mad enough at me already. We did enjoy two events in December. One was Navy's football team, winless all year, beating Army 14-2, Army's first defeat in twenty-eight games. The other happy event, in Norfolk waiting for the green light to head for the Med, was that I, who didn't rate them, got quarters on the base, opposite the Breezy Point Officers Club. Admiral Stump said, "We have atomic weapons here. I want you where I can find you in hurry if we have any problems with them."

But one night in January, our AJ Number Two caught fire in a jet-engine start at the Cherry Point marine base. That, hydraulic leaks, and other malfunctions on the AJs, delayed until 7 February 1951 my six AJs and three P2Vs starting our transatlantic flight, the first carrier-based squadron to do so—flying because the *Midway* and *Coral Sea* already had deployed to the Med in 1950, and the *Roosevelt* wouldn't let us aboard ship for the crossing. Sherman wanted us to go nonstop but I talked him out of that. We took three days, refueling first with an overnight at Bermuda, nostalgic for me, then the long run to the Azores through the North Atlantic's typically lousy weather that time of year, then into Port Lyautey (now Kenitra), Morocco.

With the PL-585 ban on storing our atomic bombs on foreign soil, ours had come over on the *Roosevelt*, the first ones to leave the United States, a display of navy flexibility the air force did not have. We also had an AEC civilian on board to affirm their custody of the weapons. But Lyautey was home base for us because the Med Task Force commander, Capt. W. B. "Little Bill" Davis, a future admiral, refused to let more than three AJs on his carriers at any one time—"take up too damn much deck space," he said. Consequently, VC-5 as a combat outfit was sort of half pregnant, our aircraft stuck on land, our weapons at sea. The *FDR* was his flagship, Pete Aurand her air-group commander, and on 26 February we began shuttling AJs on and off the ship, doing a daily mock drill, loading an atomic bomb on the AJs, then doing high-altitude, high-speed bomb runs with and without fighter escort. As my guys expected by now, I did the first one. Coming into the *Roosevelt*, I used authority most of its crew didn't know I had, calling down, "If you'll put your boat into the wind, I'll land aboard."

Davis exploded, "This is not a *boat*, Hayward, this is a *ship!*"

In February, we did the first-ever catapult of an AJ off a carrier. Then in March we lost an AJ when its tail, set afire by its jet, fell off. The crash killed the pilot, a crewman, and Lt. Cdr. Clyde Fairless, Jr., a flight surgeon. My boys' faith in the AJ died. The irony was that "Doc" Fairless was there just to take his first ride in an AJ. "How can I face his wife, Dale?" I asked and told my people not to use that jet engine except in an emergency. All the flying ran us out of spare parts quickly. This time I went right to the top, to BuAer and to Emmett O'Beirne who'd taken Joe Jaap's slot in OP-36. I was fussed at for bypassing the system, but got results. I was unable, however, to get the carriers to park the AJs aboard ship for extended periods. One result, one that weakened their value as a weapon, was that they weren't set up to do full aircraft checks or topside repair aboard ship.

When I was not in an AJ, my P2V took me everywhere: London; Paris; Beirut; Marrakech; Fez; Rabat; Tangier, where, in the open market, I met a foreign-exchange "banker," Moses Parieno, who gave me thirty thousand "dracula" (drachmae) to the dollar, on a personal check—the official rate being fifteen thousand to the dollar; Athens, with a layover in Malta, a flight total of ten hours in a fierce head wind to impress on me how wide "Mare Nostrum" is; and Palermo, for the gruesome sight of the dead monks hanging in the catacombs. On my trip to Rome, the latest of many by then, I was told that if I ever got tired of flying, I was qualified to be a licensed tour guide there.[4]

Meantime, in addition to the *FDR*, four AJs used the *Coral Sea* beginning 20 May for a week of simulated bomb runs. On the 25th, three AJs went aboard the *Oriskany* inbound at the Strait of Gibraltar, the first AJ landings on an *Essex*-class carrier. And on 8 June on a run out of Kirtland Air Force Base, New Mexico, another AJ crashed, caused by a fire in the jet engine section. The NAA guy bailed out but the pilot and radioman were killed. I was gone by then. In late March, I told Stump that VC-5 was now operational, and I was immediately shot into Washington, D.C., to head, as of 1 June 1951, the AEC Military Applications Division's weapons R&D section. I left Cdr. E. E. "Eddie" Outlaw in command of VC-5 to await my relief, due to arrive in July.

My crew chief being Chief Petty Officer J. E. Shadburn is why I survived that AJ flying coffin. I'd first seen his work at Inyokern, which is why I had him in my PB4Y-2 at Bikini and on my plane in VC-5. Aviation's finest crew chief, he kept my AJ in as perfect a flying shape as humanly possible. Still, that VC-5 tour had been painful. Ashworth told me that until he read the history, he was not aware of the pressure I was under. The worst was that we were on the edge of change: steam catapults, first angled flight-deck carrier, first all-jet A3D bomber in 1952. If they'd let me wait for those, we'd have had in 1955 what we got in 1951—without having killed a few pilots in the meantime. My guys hated me for that. I had to recite my 1925 Calvin Coolidge bluejacket adage often in that awful stretch: "Nothing can take the place of persistence. Talent will not; nothing is more common than unsuccessful men with talent. Genius will not; unrewarded genius is almost a proverb. Education will not; the world is full of educated derelicts. Persistence, determination . . . has solved and always will solve the problems of the human race."

14

Turf Wars: Some Angry, Some Deadly

In Washington by 3 April, following a debrief on the AJ's ailments and a swing by Pensacola to hug my family, I flew to New Mexico to see what currently was on our atomic-bomb development agenda. What I heard eased my peeve at being pulled out of naval aviation and back into working with AEC laboratories. At the Yucca Flats, Nevada, site opened in 1950 and 1951, they planned several tests, ranging in bomb size from large to small, one of an eight-inch artillery shell with soldiers testing protective gear and defensive tactics against it. In one test that month, using a tritium initiator, they achieved, as they knew they would, a forty-seven-kiloton yield, more than double the Baker shot's yield in Operation Crossroads.

After the Soviets' 1949 atomic blast, Truman had raised the AEC's FY 1950 budget by $93 million to $725 million. Now, they said, AEC's budget for FY 1951 through 1953 would total more than $8 billion. Our Savannah River "heavy water" plant was being enlarged. Research, mainly by Ed Teller, had us making a liquid-hydrogen "H-bomb," a fusion weapon one hundred times more powerful than a fission weapon. Atomic warheads were being designed for ballistic missiles. Hyman Rickover had a team building the world's first nuclear-powered submarine. Its keel to be laid in June 1952, if it worked, I asked myself, why not nuclear power for ships too? In short, atomic energy seemed the best place in science to be.

On 11 April while I was in New Mexico, Truman announced he'd fired MacArthur and made Lt. Gen. Matthew Ridgway, head of the Eighth Army in Korea, the UN commander. Truman's reason, he said in his memoir, *Years of Trial and Hope*, was that MacArthur had challenged publicly "the policy of the government in open insubordination to his Commander-in Chief." He then spelled out how Defense Secretary Marshall and Joint Chiefs Chairman Bradley came to support him. Marshall was ambivalent, at first, he said, but Bradley agreed with Secretary of State Dean Acheson, "You should've fired him two years ago." Years later, army Lt. Gen. Ken Nichols, who was in that 9 April 1951 meeting, told me that Bradley had opposed the idea at first, and agreed only after Truman said, "Let me fire MacArthur and I'll give you the atomic bombs."[1]

PL-585 gave Truman that authority but he ignored all that in his memoir. Even so, the AEC-Pentagon custody battle did plod on for a couple of years. Finally the military got the weapons, but under a control system where only the president had the authority and means to order them used. My brush with it all, which I didn't realize at the time, was hearing that our AEC civilian on the *FDR* left the ship right after MacArthur was fired.

The commander in chief can fire any of us he wants, of course, but Truman's argument was a bit shaky. What "government policy" did MacArthur openly oppose? At first, the UN "police action," as it was called, was aimed at freeing South Korea. As the invader's army fled north after Inchon, the goal became restoring all of Korea to its 1944 status as a single country. Then the Chinese Communists jumped in, after MacArthur had said they would not, driving thinned-out UN forces well back into South Korea and occupying its capital, Seoul. By spring 1951, our troops had stopped that advance, retaken Seoul, set up defenses along the thirty-eighth parallel, and the "policy" reverted to what it had been at the outset. Then MacArthur announced he was ready to discuss a cease-fire with the enemy, a task Truman had ordered be done by Acheson. Just as publicly, MacArthur said we must attack Communist China's cities with atom bombs and have Chinese Nationalists on Formosa invade the mainland, adding, in a letter to Republican Congressman Joe Martin that was promptly released to the media, "There is no substitute for victory."

Since most Americans agreed, the political eruption over MacArthur's being fired was enormous. The popularity of "a man not afraid to make a decision," Truman, sank into the sewer. Truce talks did begin in July 1951. As those dragged on, we saw American troops hit a position at the front to bust up an impending enemy attack, then retreat: a deadly turf war, basically, just for a bargaining chip to toss on the negotiating table at Panmunjom. Public support for the war went south, forcing Truman to say he'd not seek reelection, a preview of what would force President Johnson's retirement sixteen years later.

In March 1951, before all that commotion started, Ethel and Julius Rosenberg had been sentenced to death for espionage. One damning piece of evidence was Julius' having given Col. Alexander Feklisov, his Soviet KGB contact, a copy of Parsons' proximity fuse on Christmas Day, 1944. Its success was well known in Congress and we did discuss it in my AEC Division, but I must assume he gave Feklisov only a sketch of it. We didn't have any fuses at Los Alamos.

The events leading to the Rosenbergs' conviction had begun in 1948 when, in England, Klaus Fuchs confessed to spying for the Soviets, implicating a guy named Harry Gold and Ethel's brother, David Greenglass. To cut his sentence to fifteen years, Greenglass fingered his sister and her husband. Another guy caught in the net was Theodore Hall, a Harvard chemist in Kistiakowski's Los Alamos Division from 1944 to 1945, doing ordnance design on both bombs. He and Fuchs had given the Soviets rough outlines of what we were making. Though I don't recall meeting him, Greenglass was a machinist in our explosive-castings shop at China Lake who gave Rosenberg a sketch of the lens mold, key to making Fat Man work, eight months before we dropped Fat Man on Nagasaki. In trips to Capitol Hill, I did meet Fuchs, who seemed very upset with anyone from Los Alamos. The Brits gave Fuchs fourteen years in prison in 1950. Gold got thirty. Hall was never charged, and some Americans who probably were Soviet spies never were exposed.[2]

All that was just background noise when I entered my AEC office in Washington on 21 April 1951. Gordon Dean, AEC Chairman David Lilienthal's successor in 1950, was, I discovered, a fine person. We got along nicely. Officially released from VC-5 on 1 June, I already was head of a drive to build a second facility like Los Alamos. The site chosen was the Naval Air Station at Livermore, California, because it was near where a lot of bright scientists worked and the government already owned the property.

But to get Livermore Laboratory approved was a nasty turf battle, not as deadly as the one in Korea but just as hard on the nerves. Simply put, the air force wanted its own setup at its Wright-Patterson Field in Ohio. Let them, I told the AEC, and it will be a typical red-tape–obsessed military–Civil Service bureaucracy whose only product will be just more 10,000-pound Fat Man bombs. It was a touchy point in what was, by then, a four-year-long debate over which way to go: the air force "Big Bomb" way or the army-Parsons "tactical" weapons route. Of course, the "city-busting" folks just wanted something for nothing. Nobody'd have to storm a beach any more. Just fifteen men in a B-36 were all we'd need. But Korea had blown away *that* myth in a hurry.

Much of our battle was over the weapon itself, really. Air Force R&D Gen. Roscoe "Bim" Wilson and his cohorts, Bill Borden and Ken Michels on the Congressional Joint Committee on Atomic Energy staff; Teller; and David Griggs in the Rand Corporation, a nonprofit "think tank" wholly beholden to the air force back then, all wanted a huge-yield, "dirty" hydrogen bomb and, we learned, the Soviets did too. Oppenheimer was against it, the

Wilson-Griggs crowd said (not noting that Lauritsen, a lot of the rest of us in the business, were too.) His influence was why Norris Bradbury's Los Alamos was, they claimed, "too slow, not innovative" and needed competition from an air force–run laboratory. It was a critical challenge to the AEC, one I thought the air force would lose since PL-585 said making nuclear weapons was solely AEC territory.

Actually, Los Alamos had a good program going for the air force, as I knew well since one of my jobs was to review it with Norris and his people, going over the military requirements, then justifying their funding to Gordon Dean. In fact, Livermore wasn't needed because Los Alamos had done a bad job. Rather, Los Alamos was so tied up in air force work that Livermore was needed so we could fill other requests too. But the endless talk was fierce from June 1951 to June 1952. It got so I hated to go to the office. Few of all the meetings tackled basic issues: Will the new lab be able to test its devices at Yucca Flats and convert them to usable weapons? Who will decide each lab's priorities, and defend both their budgets in Congress? I also learned that whoever disagreed with the air force, and I surely did, instantly became one of "the enemy."

As the debate's noise level rose, the army, sparked by Lauritsen and Oppenheimer, put out a report titled "Project Vista," showing what small tactical nuclear weapons could do. An implied objection to making only huge-yield bombs as per the air force "bible," it enraged the air force. Maybe because I was tired from playing thirty-six holes of golf the day before I said it, my quip, which peeved Teller, was, "Accuracy always pays off over yield, be it in golf or atomic weapons."

Then, Dean had me write an essay on Livermore's merits as a way to generate competition and make small tactical weapons. I ended up writing eighty-three staff papers on why it was needed, why it must be in AEC to succeed, why the navy wanted it. AEC Commissioner Tom Murray was an early promoter of most of them. But my pal, Ernest Lawrence, was who worked with me hardest on them. With his persuasive help, we finally got the Joint Committee on Atomic Energy to endorse one of my papers. Meantime, when the University of California, Berkeley, was picked to be for Livermore what CalTech had been to China Lake, I had Berkeley's Wally Reynolds and Don Cooksey help me put together a budget for the lab's first year.

I got no help from Los Alamos, some there claiming that Livermore was a slap at their work, others snorting that it was unlikely ever to be much competition. But the lab helped us even before it was chartered. The Naval Research Laboratory had done all our test-site instrumentation before. Now,

Ernie Krause at NRL tried to stick us with a fat fee for the "Mike" test due in November at Eniwetok. So I gave the job to Berkeley even though Livermore wasn't a lab, yet, and we put a Sandia unit in to do for Livermore what it was doing for Los Alamos. And, helped by Lawrence, I defended in Congress a $52 million first-year Livermore Lab budget that Joint Atomic Energy Committee member Henry Jackson pushed through Congress. By June 1952 Reynolds and I had established, honoring Ernest, the Lawrence Livermore Laboratory.

Lawrence was named Livermore's chairman and wanted Herb York to be its director, which made that a done deal, of course. The other top people were unknowns in the weapons business: Harold Brown, a twenty-seven-year-old physicist to head the Theoretical Division; Johnny Foster, the Explosives Division. York, Foster, and Brown each would become the Pentagon's top R&D man, eventually, Brown also serving as secretary of the air force, then secretary of defense during President Jimmy Carter's administration. But in 1952, they were just young, creative, and energetic.

The hardest "sell" Lawrence and I had was Dr. Teller. We wanted him at Livermore but, at first, he wouldn't go there nor back to Los Alamos, either. He was mad, said they weren't working hard enough on his ideas. So Ernest and I went to Berkeley several times to meet him in the Claremont Hotel's bar, then off to Trader Vic's for a meal, trying to convince him, using a steady stream of martinis to weaken his resistance—and learned he was mad because, "The AEC won't let me blow up my own bomb." So I had Gordon Dean come out, listen to the beef, and had him say, "Ed, if you want to build and blow up your own bomb, go to it."

Now, I was really in the cross fire, meeting with both Norris Bradbury at Los Alamos and York and Lawrence at Livermore to decide what projects to do where and how. The AEC didn't tell them what to do. We weren't smart enough for that. We just ruled on their proposals. In the process, Livermore made the AEC's only "dud." It tried to use a special type of carbon, just proving that carbon is a nonstarter for an atomic bomb. Nonetheless, Livermore did all the warheads for the Polaris missiles and later was warhead architect for the *Trident* submarine's D-5 missiles. So they ended up with a big chunk of the business.

Livermore wasn't my only fret, back then. By 1952, my personal debt had climbed to $3,700 with no way to pay it unless I could get promoted to rear admiral, an unlikely event since I'd never commanded a ship. To escape the worry, I spent June 1951 on leave in Pensacola, living off family charity, really, insistently given, gratefully received, with Mommie Floss and Daddy Pete spoiling the children as usual. In July, as the Korean war waffled, I flew to California to hire Dr. Oppenheimer as an advisor to my weapons shop, amazed at how much he'd aged in four years. Then Admiral Sherman dropped dead in Naples. I wanted Radford to be CNO but his role in the "Admirals' Revolt" ruled that out. "Politics will govern," I wrote. "Whoever doesn't know that has rocks for brains." On 3 August, Adm. William Fechtler, an amphibious warfare veteran, was made CNO, a boost for the marine corps in political Washington.

August reports from Europe said Eisenhower's efforts as SACEUR (supreme Allied commander, Europe), trying to make a cohesive NATO team out of twelve squabbling brats, looked well nigh impossible. That had the intelligence folks asking: Will 1952 be the year Russia strikes in Europe? Will our atomic bombs persuade them not to? Still, I noted, "Their use of surrogates like North Korea is clever. It avoids a direct confrontation with us on the battlefield." By October, off the results of Operation Sandstone's "alarm clock" test, the AEC finalized plans to detonate a hydrogen device in late 1952. On 22 October, my diary gave equal billing to, "The Yankees won the World Series," and "The World still seems rushing to World War III." Three days later, the Russians shot down a navy P2V in the Sea of Japan, claiming it had violated their air space, a lie, probably. Then on the 29th, we successfully tested seven Mk-7s, a seven-tenths model of the Mk-5. Each one delivered a thirty-four-kiloton yield.

Navy crushed Army, 45-7, in December, my log says, after Army's star players had been kicked out of West Point for cheating on midterm exams. My final 1951 entry, 31 December, records, "We will try our Mike shot on Presidential election day, 2 November. It's so 'Top Secret' even the label, 'a radiation-implosion device,' is classified. A two-story-high device, it will use liquid hydrogen, deuterium, a U-235 spark plug, lithium, plus a fission device on top to ignite it. Very complex, it will require building a cryogenic plant at Eniwetok. An Ohio State chemistry-physics professor has that job."

In the spring of 1952 in a P2V, I flew to the Med to inspect for the AEC, and for myself, how well the carriers were doing in the atomic-bomb delivery business. Deploying VC-5 had put the navy in the business. Until then, we'd been working on it for the air force. Of course, we assumed, if we did have a war in the Mediterranean, it would be with the Soviets. Our first answer, VC-5, still had hurdles to jump in 1952: aviators, tied to the World War II syndrome, who just wanted better-performing copies of what they'd flown in the war; carrier captains trying to keep my few aircraft from "clogging up the flight and hangar decks."

To get those guys on board, we told them to look ahead to the A3D, built for angled-deck carriers that would begin to come into service in 1952. (In

1955, as I recall, they had a veteran carrier pilot try to land an A3D into the wire on the straight-deck *Bon Homme Richard*. He hit the ramp up to the deck instead, shattering the aircraft and killing everyone. So, it was ruled, "No more A3D landings on a straight-deck carrier, period.") The *Hancock*, first carrier to get a steam catapult that could launch easily an A3D's eighty-two-thousand-pound, full-loaded weight, made her first launch in June 1954. Delivery to the fleet of the A3D itself didn't start until March 1956. But one guy who volunteered to go with me on these "sell the program" visits was Dan Gallery. As well as I, he knew an A3D would have more punch than almost anything else. When a carrier skipper groused at him, his blunt answer was, "If you don't have an atom bomb on board, what can you do with those little firecrackers you do have?"

But as late as 1962, when I commanded a task force in the Med, navy guys were leery of the A3D. The best news I got, telling me they were coping with it, was that they'd given it a nickname, "the elephant." Reporting "an elephant on the porch" meant the rest of the flight deck was closed. My pal, Pete Aurand, skipper of one of my carriers in 1962, took particular delight in asking, as he did often, "Chick, I've got a sick elephant on the front porch. What do I do now?"

Meantime, on the 1952 trip to the Med, I treated myself to a visit with Pope Pius XII. I'd met him in 1938 when he was Cardinal Eugenio Pacelli, the Vatican's secretary of state, when I was asked to fly him and some other Catholic dignitaries from Boston to Kansas City, a navy good-neighbor gesture done often back then. He had a strong interest in science and, surprisingly, a good grasp of atomic theory, so we had a pleasant chat during the flight. At the end, he said if ever I was in Rome, "Please come see me." A year later, the college of cardinals elected him Pope Pius XII and in 1948, when I was in the Med to tell the Sixth Fleet that VC-5 was headed their way, I went to see him—not hard to do, I learned, since, his secretary said, "Much more than any pope before him, he's made a habit of granting audiences to individuals and groups frequently."

My 1952 arrival was on his seventieth birthday as a crew, excavating under St. Peter's tomb, hunting relics of ancient Rome, dug up what they assumed were St. Peter's remains. That's organic, I told the pope. You can carbon-date it to learn when he died. Carbon 14, I explained, is a radioactive isotope of carbon with a half-life of 5,750 years. We're all radioactive and have a lot of carbon 14 in us, but not after we die. It's a complicated process, I told him, but basically, we can count the number of carbon 14 atoms in one of the skeleton's bones, compare that to what he would have had when he was alive, and his time of death is whatever that ratio is as a factor of 5,750.[3]

Pius XII was enthusiastic. He gave us the bones and all kinds of material to test. I wasn't involved in that but did do some research. Folklore says, when Nero's persecution of Christians began in 64 A.D., Peter fled. Meeting Jesus Christ on the way out, he asked, "Domine, Quo Vadis?" Christ said, "I'm going in to be crucified, again," and Peter said, "No, I'll go back." It is said that behind a tiny church at a spot called Domine Quo Vadis ("Lord, where are You going?") where they met, in two Appian Way stones are two large footprints, allegedly those of the Lord. However, Martin Luther said St. Peter never went to Rome, so whose remains are they? I don't know. I do know that whoever is buried there died in 68 A.D.

Back home, the army general (an engineer) who ran my division, reporting directly to Gordon Dean, took me with him to that month's regular meeting of the five AEC commissioners. Our weapons-test work was under me and, he explained, on the agenda was a British request to explode their atomic bomb at Eniwetok. Winston Churchill and Bill Penny, my ally in the Hiroshima bomb-damage study, now Sir William Penny, chairman of their AEC-type outfit, met with us in the morning. Churchill soon left, returning at 8:00 P.M. when only Murray, Harry Smythe from Princeton, some AEC staff people, Penny, and I were still there. We sat up until 3:00 A.M., drinking brandy, Churchill giving us one of his big cigars. Penny pocketed his for a souvenir. I smoked mine—and my children later growled at me for not saving it as Penny had.

We had a bit of debate, in the staid British style, when we told them PL-585 required that we'd have to measure their bomb's "alpha," the speed of neutron generation. That, in turn, can reveal quickly the bomb's elements, design parameters, and yield. They didn't want anyone else to know that. So the proposal died and they blew up their bomb, alone, at remote Monte Bello island off Australia. That proved, in effect, that the know-how to make an atomic bomb can't be hidden from the rest of the world. So, we said then, as the active material becomes plentiful, the weapons and nations that have them will proliferate. Stacks of "arms limitation" treaties seem still today to please some people, but there's no way to keep a "have-not" country from getting nuclear weapons if it feels it must have them. The traditional military answer is, of course, to build a defense. So far, "treaties" have won out over "defense."

All that spring of 1952, each passing day in Korea made it appear that MacArthur was right in wanting to take the fight to the Chinese. And the turf war over atomic weapons was making me a rabid anti–air force, anti-Pentagon man, threatened much as John Crommelin had been, asking myself, why not just shut up, go along? My diary says, "Career be damned. I'm obligated to do my job. Persistence, Coolidge said." I also noted, "If Ike beats out Robert Taft, 'Mr. Republican,' as the Republican candidate, he'll win the national election in a walk."

Mommie Floss died on 30 April, as much a painful loss to me as losing my own parents. She'd been like a mother to me, an integral part of my whole family's life. Daddy Pete took her death hard, losing all interest in his tugboat business. Lili decided she'd have to stay in Pensacola, with the young ones, to care for him. So I gave up our house in northern Virginia and moved into a D.C. apartment, taking only Marion with me since she was about to graduate from St. Mary's—with a far better high school record than I'd ever achieved. Luckily, I could borrow a navy fighter for weekends in Pensacola all summer. In August, the Republicans nominated Ike; the Democrats, Adlai Stevenson, who seemed bright, well educated, a far better public speaker than Ike—but the rumble everywhere, from most people, was "I like Ike." So did Democrat Hayward.

At Princeton in June 1951, after evaluating the April-May 1948 Sandstone tests, Gordon Dean had approved putting a test, code-named Mike, in Operation Ivy, due to be held in late 1952 at Eniwetok. Norris Bradbury and Graves, the test director, picked election day because politicians and journalists all would be busy somewhere else that day. During September and October 1952, I flew to Eniwetok to check progress on our thermonuclear device being erected on Eugulap, and on the cryogenic plant being built nearby (run by an Ohio State professor, Dr. Johnson), the main reasons "Mike" was costing over $100 million to do. The expense was unavoidable at the time. Liquid hydrogen, at minus 300 to 400 degrees, boils over as it moves from its ortho- to para- stage unless run through a cryogenic plant to reliquify it.

On those trips, I also discussed with those folks where the AEC was headed and what its objectives ought to be in this business. A good deal of what I said reflected what Deak Parsons and I had talked about as World War II ended. We didn't believe this atomic "ultimate-weapon" stuff nor the claim that it had made the world's navies obsolete. We'd talked the "spectrum of conflict," as physics types usually do. That was before the IRBM and ICBM starts, but even then we saw many ways to deliver an A-bomb. We also wanted it to go to sea, not necessarily to sink ships but to increase the ways to deliver it on an enemy target and to complicate defending against it. The air force said destroying fifty or so Russian cities would win the war. Didn't we still need to put troops there to enforce the surrender, to fight enemy forces left largely intact while we blew up some cities? But the basic question was, and remains: Once started, how do you stop a nuclear war?

Back when only we had atomic bombs, we wondered: When will the Russians get it? By September 1952, I'd begun to doubt if Truman understood the significance of Operation Ivy. We were sure our Russian friends were trying to get a fusion device with huge yields out of less active material. But Truman was interested only in us not blowing it up on election day. At a meeting he requested in Chicago with Gordon Dean and the rest of us, he tried hard to get us to postpone it, insisting that exploding it on election day would get him called a warmonger and hurt Stevenson. (The press was picking Stevenson to win at the time.) On 22 October, Truman even sent air force Assistant Secretary Eugene Zuckert to us with a last-minute appeal for delay.

Faced with weather and other unfathomable ways a test can be derailed, it's foolish to risk it for such a flimsy reason, I said bluntly, adding, "Who are we to judge political angles in this thing?" And Dean told us to stay on course. (In his memoir, Truman has Dean fearing that an election-day test "might be judged a political maneuver," and him saying "Political considerations should not be tolerated in the Nation's atomic program"—when, in fact, the whole program already was soaked in "political considerations.")

So, on 31 October, west longitude time 2:15 P.M. 1 November in the States, the Mike shot was detonated—but it wasn't a hydrogen bomb, as the media reported. It was a two-story-high, sixty-two-ton monster. It contained deuterium (an isotope of hydrogen) and liquid hydrogen encased in a uranium U-238 shell. Its spark plug was weapons-grade U-235 in the center with tritium and lithium-deuteride. Tremendous temperatures are required to release the high level of energy we knew was in a hydrogen atom, so we put a seventy-kiloton atomic bomb on top of the structure to initiate the action. When it blew up, all the neutrons came down, struck the core, and caused what we called a radiation implosion. All the atoms, even the U-238, fissioned.

We wanted to prove we could make a fusion weapon with a radiation implosion, and prove what the lithium exchange ratio is—to prove that if a lithium atom (we thought it had to be lithium 6 but it turned out any lithium isotope will do) is hit by two MEVs (million electron volt) neutrons, it gives rise to tritium, helium, and a fourteen-million-volt (MEV) neutron. That set Teller's "Alarm Clock" going and, timing being critical to getting energy release, we wanted to confirm the timing. We'd proved most of Ed Teller's theory at Sandstone, but the Mike shot proved it all. Called a "wet bomb" since it used liquid hydrogen to create the thermonuclear reaction, it dug a 280-foot channel off Eniwetok and blew away most of the island.

There was hot debate over whether Mike produced a yield of 6 or maybe 12 megatons. Someone put 10.4 megatons in the history books but the yield can't be known for sure, because the fourteen-MEV neutrons fissioned the uranium 238 shell in addition to the U-235. In any case, "Mike" showed that we could initiate neutrons fast enough to create megaton yields. That's all it proved. It was far too big for a bomber, of course, and with all its predictable operating impossibilities, we never built a liquid-type hydrogen bomb. We went to the solid type. Since we never wanted to do only one test at a time at that remote proving ground, and fission being at the higher end of the element scale, fusion at the lower end, we also blew up a five-hundred-kiloton fission device, proving that with design improvements, we could get a much better fission result with less plutonium.

In my diary I wrote, the day after "Mike" blew, "It won't be long now before our Air Force friends deluge the press with cries that we now have the ultimate weapon and the only hope for us, the uneducated masses, is to give them what will make them supreme. They will demand an immediate operational capability with a 'wet' fusion weapon." And I wondered for a week, have the American people decided to go the way of socialism or did they elect Ike president? Then on 9 November, I learned, "Ike in a landslide!" And already the rumor mill was saying he'd replace Gordon Dean, as AEC chairman, with Lewis Strauss, "The most arrogant man I know," I wrote. "Along with Borden and the Air Force, he will do his best to do in Oppenheimer."

Our first thermonuclear reaction without fissionable material occurred at 2025 hours on 6 November at the Los Alamos test site. That and Ivy announced, in effect, that we were heading in a new, megaton-yield direction in nuclear weapons. Still, the air force kept on pushing for a "wet" fusion weapon, and spent millions on it. The fight was bitter in the scientific community, many of them acting as if, since they knew the technology, they were masters at devising nuclear strategy, too. They bought the air force "big bomb" idea. Industry also was on the air force side, one motive for both parties being to keep the so-called "aerospace" industry large and profitable.

Before Operation Ivy, Truman had asked AEC's view on MacArthur's request to hit China and North Korea with atomic bombs. I'd sat in on Omar Bradley's Joint Chiefs meetings on the subject. AEC had lots of test vehicles but a stockpile of only seven weapons because of the pace of production but also because, in 1952, we still had to change the initiators every 140 days. I was told to develop an answer. Truman took the air force out of it by saying the bombs were not to be stored in Japan. He also wanted assurance of accurate delivery. Radford helped me dig into it. We'd modified six Skyraider dive bombers by then to carry Mk-7s, but the only carrier available and equipped to handle them was the *Midway*. And what would be the military targets of our seven-bomb salvo, Yalu River bridges? Not the front lines, surely. We'd kill our people, too. Then there was that lurking truism.

The purpose of nuclear weapons is to deter war. If used, they've failed to do that and we've lost the war—even if we win it in the usual sense. After reading my report, maybe Bradley, probably Truman, said no to using atomic bombs in Korea.

When Ike came in, a rumor ran around the world, saying, "This general is liable to drop atomic bombs on Korea." Allan Dulles's CIA folks yakked that up in bars all along the Pacific rim—just a ruse to scare North Korea into being reasonable in Panmunjom. Ike was against using the weapon there. I'm glad he was. It would've been dumb to do. Ike was a bright man in many respects, sharper than most people gave that modest man credit for being.

After we detonated our first fusion device, Truman asked me to come visit him in Independence, Missouri. He said he wanted to hear more about the impact of our A-bombs on Japan. During that pleasant and, for him I hope, helpful chat, I asked why he'd given the French one of our older carriers with a full complement of aircraft and military advisors to train the French crews. French President Charles de Gaulle wanted it for his war with Ho Chi Minh's guerrillas in Indochina, Truman said, but he didn't want to sign the NATO treaty. Without France, NATO'd be a mess, he added, so he'd swapped de Gaulle the carrier for his signature.

As I left, he asked what I wanted to do next. Since 1944, I've had shore duty, mostly, I said, and I'd like to command a ship. I didn't say my chance to make admiral was zilch without that nor that I couldn't stand ex-Rear Adm. Lewis Strauss and would want out of the AEC if he came in. And surprise, surprise, in January 1953, as we firmed up plans for more nuclear-bomb tests, I was told I'd spend June in school, studying the doctrine on a ship captain's duties, then in July (my service record made it May) take command of the escort carrier *Point Cruz*, stationed in Japan. Truman had delivered, but looking back at where I'd been and what I'd done since 1944, I knew that if I hadn't been in league with Deak Parsons, Lauritsen, "Dusty" Rhodes, Schoeffel, Ofstie, Stump and the like, I'd have been bounced into oblivion like John Crommelin.

I also knew that a badly done job on the *Point Cruz* could end forever my shot at a flag. So I read every book I could find on ship handling, rules of the road at sea, even the General Signal–ship communications book. Meantime, in March, the Russian tyrant, Joe Stalin, died and Georgi Malenkov took over, with V. M. Molotov replacing Vishinsky as Soviet foreign minister. Suddenly, by design, probably, the Russians suspended their bluster in Europe and began urging the UN to "seek peace" in Korea. The UN wasn't the problem, of course, but if Moscow was saying in China and

North Korea, privately, what it was telling us, publicly, I told Lili that a truce agreement might finally be signed in Korea. The crew of my about-to-be ship would be delighted, I was sure. I left my apartment on 15 May, my AEC tour over, my efficiency report a nice one, thank you, written by an army general, head of its Military Applications Division.

Five days later, Ike, ignoring Radford's "Revolt of the Admirals" role, nominated him to succeed Bradley as Joint Chiefs chairman, surprising us, disappointing the air force. In June, another friend, Adm. Robert B. "Mick" Carney, "Bull" Halsey's chief of staff in World War II, became CNO, and I left on 30 June for the *Point Cruz*. Commissioned in 1945, she was named, as were most escort carriers then, after a marine corps battle. Hers had been on "Cactus" where marines had been hurt badly, crossing a Matakama river sandbar. She had a one-thousand-man crew, twelve-thousand-ton displacement, was 553 feet long by 75 wide, and carried up to thirty-four fighter-size aircraft. With a deep, thirty-nine-foot draft, geared-turbine single-screw propulsion, she'd be hard to maneuver, but success as her captain could mean a flag for me. With children's school costs draining my bank balance, rear admiral was a pay scale sorely needed.

When I took command at Yokosuka on 8 July, my XO was Dick Friede; my operations officer, Lou Hardy from my old VB-106 outfit; Butch O'Neil was air boss; our very able navigator was Bob Boe; and "Jonesy," a young Seventh Day Adventist, was my mess steward, an excellent cook, I learned, but appalled at all the books I'd brought aboard. Our ship's chaplain was Reserve Lt.(jg) Edward O'Neill Riley, a special person like my bluejacket mentor, Father John Brady. To officers and sailors alike, his rank was unimportant. He was "our chaplain," in charge of morale, emotional security, inspiration. He really made that ship. I was just his backup.

We were to sail on 12 July for Hakodate, a port on the south side of Hokkaido, the largest island in northern Japan. In my aviator years, a nightmare repeatedly haunted me. In it, I was in a captain's cabin when a voice at the door said, "Captain, the ship is in all respects ready for getting under way." Ignorant of what to do, I barked, "Well, get her under way." I always woke up then. Now, nightmare was reality. Half-reading a physics book, I could hear bustle up top saying the time was near. A captain stays in his cabin until his XO comes there to say, "Sir, the ship is in all respects ready to get under way." When Friede did, I rose, shaky, knowing crews watch a new skipper's every move. Up top, I saw a flight-deck antenna unhoused. It would have hit the dock on our way out. I barked an order, then gave the deck crew my opinion of their foul-up. Their stunned faces said they were impressed.

On the bridge, always quiet when a ship is getting under way, I told the watch, "Look, I don't know all there is to know about this business. So anyone, I mean anyone, who thinks I'm doing something wrong or not doing what should be done, tell me. Tell me, right away."

Eyebrows went up, faces asking, *What kind of skipper is this?* but after that, no crewman was afraid to say he thought I was making a mistake. It saved me a few times and prevented a tragedy two months later in the Yellow Sea. The night was pitch black, making a vessel's red warning light easy to see, but I hadn't. We'd have crushed a fishing boat if a young sailor hadn't rushed up to warn me because he remembered what I'd said my first day on the bridge.

Underwater steel mesh nets, to keep enemy submarines out of Tokyo Bay, opened only a few times each day. At one of those, 1300 hours on the 12th, we left Yokosuka, fighting a rough tidal flow. We inched past mines and a herd of ships without hitting any. That lifted a ton of worry off me. Not yet expert at ship-handling, I'd get lots of practice, both going north and on the swing in to anchor behind Hakodate's breakwater. The *Point Cruz* took ages to stop and had weak backing power. That, with her deep draft and single screw, made her hard to maneuver. We finished Individual Ship Exercises, as specified, on 15 July, as the Chinese hit us in force along the thirty-eighth parallel. "It's hard to figure," my diary says. "They talk truce one day, fight a day later. This is a strange war, nothing like what our JCS masterminds have in their war plans."

15

The George Cruz Ascom Story

At sea on the 16th, we ran a week of ASW (antisubmarine warfare) drills against our own boats. Our "hunter-killer" group, VS-23, flew the Grumman AF Guardian. It had 315-mph speed, a fifteen-hundred-mile range, and was configured two ways to work as a team. The AF-2W "hunter" had a big radar dome hung under its belly, run by two men in its four-man crew. The AF-2S "killer" carried one 2,000-pound torpedo, two 2,000-pound bombs, or two 1,600-pound depth charges. We worked the drills day and night with fair success, then returned to base on the 23rd, me so tired my bones ached. A captain, I learned, is not allowed to rest. We smelled Hakodate before we saw it, an offshore breeze bringing us the odor of a thousand squid hung out to dry. Our sidle up to a tanker for fuel was slow because of my ship's weak backing power. Then we nudged into port. I could handle the ship rather well by then, not great at it but not a "hack," either.

That first day, I played the city's three-hole golf course six times, and the mayor invited Bob Goldthwaite, the admiral on my ship, and some of us to dinner. It began as a kind of Japanese royal banquet and ended wild. A Mama San met us to see to our "hot battsu" bath. Then in robes and slippers, we were served dinner on tiny individual tables by delicate geishas, with lots of hot sake first, then a parade of fish dishes, a main course of rice, clear soup, vegetables, more fish, and seaweed. The geishas entertained as we ate, drank sake, then a tasty Japanese beer, then more sake. By 2300 hours, we were as witty as Bob Hope, the geishas easily the world's most beautiful women, Goldthwaite so jubilant that Mama San advised him to take another "hot battsu" before we left. Our other five days in port, I rested or read one of the books I'd brought aboard.

A truce was signed on 27 July, bringing a fragile peace to Korea. But we'd lost the war, lost it by never giving MacArthur a certain-sure objective. Not deciding what we wanted to achieve, we were bound to fail. And Arleigh Burke, on our truce talks team for a time, said, "Our State Department undercut us in 1951 by accepting that year's battlefront as the cease-fire line. So the enemy had no reason to agree to anything else. And they seemed to know before we did what State Department orders to us were going to be each day." That farce cost us. Of thirty-four thousand Americans killed in the war, two-thirds died during the two years of negotiation.

Next day, we left for Yokosuka with typhoon Lola south of us. Two days later, we were refueling destroyers when one came too near our starboard quarter. If a small ship does that, our deep draft will suck it into us. The collision damage was minor, not as harsh as my lecture to that skipper, a young commander. Next up was John Bulkeley's ship. He'd lived a bit of history when his PT boat had extricated MacArthur and his family from Corregidor in 1942. I'd met him when the AEC sent now-Commander Bulkeley to Los Alamos to count our atomic material, see if it matched the stockpile the Military Liaison Committee had ordered. Before he inventoried what he knew would be bad news, he always enjoyed asking, "Are you war-ready?" During a refuel, we often offered a destroyer's crew ice cream. In Bulkeley's case, I sent a message over the high line, "Our ice cream is war-reserve qualified. Are you war-ready?" My XO thought I was nuts but it broke John up.

On 2 August, we learned that Soviet fighters had shot down a B-50 bomber off Vladivostok over international waters, but they ignored our protest. A military counter-hit would stop that stuff, I argued, but we didn't think Washington had the guts for it. We got hit that night, too, our first crash. With the ship riding a moderate swell, wind over the deck at thirty-five knots, the marine pilot, coming in, bounced into the No. 3 barrier wire, then hit the island, making a mess but no injuries. And on the 5th, I saw my worst-ever aviation show: two crashes, two flat tires, two hours taken to bring seventeen piston-engine marine F4U Corsairs aboard. (We didn't carrier-deploy lots of jet-powered combat aircraft 'til Ike got into our act in 1954.) If you don't shape up, I told my guys, we'll soon be out of airplanes.

Orion's belt was on the horizon at dawn that day. It'd be up high enough to navigate by, going home for Christmas. We left Yokosuka early, on 8 August, a Saturday, hoping to reach Sasebo, on Kyushu island in southern Japan, before Mamie, a typhoon born after Lola died. Flight operations I ordered, taking on nine more F4Us, went much better than the show on the 5th. In Sasebo on the 12th, after lunch with "Jocko" Clark, now Seventh Fleet commander, I met Task Force 95's and my boss, a British admiral on the carrier *Ark Royal*. I'd be Task Force 95.1, my marine air group—"The Polka Dots," they called themselves—flying patrols from Seoul's Han River to the Yalu and to Bo Hai, off China, mostly for Gen. Randolph McCall Pate's First Marine Division north of Seoul. Our F4Us, the last piston-engine fighter built by any service, had 470-mph speed, 1,120-mile range, six 50-caliber machine guns or four 20-mm cannon and two 1,000-pound bombs. We were in our Yellow Sea operating area by the 13th, with a new typhoon, Nina, headed our way. "The Yellow Sea," I noted, "is a 'Cul de Sac' for typhoons."

We and the *Ark Royal* were to alternate, three weeks on station, three off, under the best air-control concept I'd ever seen. Its Tactical Air Center was south of Seoul and had in it a navy commander, air force colonel, and a marine in direct contact with all frontline forces, including our First Marines north of Seoul. Every aircraft, both from our carriers in the Yellow Sea and East China Sea, and from the air force bases in Japan, had to call into the Center to get its mission orders. This efficient setup was air force Gen. Pete Quesada's idea, but some air force brass didn't like "the navy" telling their aircraft what to do. So much for the 1947 "Unification of the Armed Forces." As Jim Forrestal wrote, then, "The spark of Unification must be fanned into flame by the thoughts and actions of generals, admirals, ensigns, lieutenants, soldiers, sailors, airmen and civilians. We must all learn we are working together for a common cause." My "Polka Dots" understood. To tell me they did, they gave me one of their patches, five black polka dots on a silver box, and "promoted" me to "Colonel, USN."

With renewed fighting a constant threat along the thirty-eighth parallel, the "truce" issue now was prisoner exchange, a hassle as weird as the war had been. Usually, after a war, POWs (prisoners of war) may return to wherever they want. Not now. The "Commies" wanted all ours sent forcibly to China, which, ethically, we could not allow. Many of our POWs actually were Nationalist Chinese, some still wearing Chiang Kai-shek's "Sunflower" tattoo. We also had South Koreans who'd been conscripted into the Red army and others who, whatever their origin, simply hated communists. Shoving them into China would mean to prison or execution.

The truce called for an Indian Army to be in the so-called "demilitarized zone" (DMZ) between us and the enemy to enforce the cease-fire and repatriate prisoners. Hardly a DMZ, it bristled with weapons like an angry porcupine, and a Neutral Nations committee, set up to run the prisoner exchange, was hardly "neutral." Chaired by an Indian General Thimaya, on it were a Belgian, replaced by a Swiss as we arrived; a Swede; and a Czech and a Pole both spouting Moscow's party line, of course, insisting they question every prisoner, forcibly if necessary, to find out where he was from and what he'd done. The Swiss and Swede voted no to forced interviews, so the Indian general's vote would decide. He asked Indian Prime Minister Jawaharlal Nehru, "How should I vote?" and Nehru wired back, "Vote yes."

South Korean President Syngman Rhee, a Nehru-hater, erupted. If the Committee, which he thought a political phony anyway, tried that, he'd order ROK (Republic of Korea) soldiers to seize the Indian camp. And, he snarled, he'd jail any Indian who set foot in South Korea. Lt. Gen. Maxwell D. Taylor, who'd succeeded Ridgway at Eighth Army and was our top man in the POW snarl, gave him a quick answer. He had General Pate's First Marines move up to the DMZ line to fend off the ROKs, then devised a way around Rhee. When the five-thousand-man Indian army arrives at Inchon in transport ships, he said, we'll helicopter them from the *Point Cruz* to their camp, now called Freedom Village, in the DMZ, a forty-five-mile flight. He'd get me every available helicopter in the area for the mission. He named it Operation Platform.

I'd be needed there by 30 August, he said, when the first Indian troop ships should arrive. A thorough leader, he said he'd come aboard then, to hear if I needed anything more from him. Three days later, on the 16th, an F4U crashed off our bow on takeoff, caused by engine failure. We swung sharply to port to miss him but the crash broke the pilot's back. Then the helicopter sent to pick him up crashed, why, we couldn't be sure. But it did remind me of why career military folks hate war. We've lost too many friends we had to order into battle, written too many letters telling families that a son or a husband isn't coming home. Maybe that is why I read "think" books that week: Aristotle's *Alexander of Macedon*, H. Van Straelen's *Through Eastern Eyes*.

Nina was a howler by then, winds up to 173 knots. In the Yellow Sea, fabled for bad typhoons, this was one of the worst. That sea is fairly shallow, so waves made by high winds can be huge. We spent a week buttoned up, being pitched and rolled. Knowing that typhoons seldom go as far north as Bo Hai, I did take us up to there briefly, on an ECM (electronic countermeasures) run to see what frequencies the Commies were using to talk to each other. But on the 19th, Nina veered off, into China, so we went to Sasebo to rest a week before Operation Platform. The highlight there for me was going upriver one day with Jocko Clark and a guide, fishing for ayu, a small troutlike fish that doesn't smell like a fish. Our bait was a female ayu on a long leader with trailing hooks. A male ayu, going upriver, goes for the female and gets caught on a hook; funny way to fish, maybe, but the natives say a very effective way to catch ayu, which taste delicious. I had mine for breakfast, next day, cooked superbly by "Jonesy."

On 26 August, we headed for Inchon. Admiral Olsen, who'd replaced the Brit as head of Task Force Ninety-five, and four of his staff were with us. He didn't say why but I thought it was just to get in on the media attention we'd surely receive at Inchon, a guess I kept to myself. By then, our schedule said, after Operation Platform ended, we were to be in Hong Kong in October, a lovely time of year there; be six weeks off Tokyo Bay; then leave for home on 2 December.

A biting cold front swept in from Siberia on the 27th, making temperatures nosedive and the Yellow Sea white with foam. Going up the tortuous thirty-three-mile-long channel by a flock of little islands to Inchon's inner harbor was as bad as I'd been told it would be. Any charts of it usually were inaccurate. The channel was fed by a rushing stream—the only kind they have in Korea, I think—and the thirty-four-foot tidal swing was one of the world's highest. Consequently, sandbars built and shifted in a hurry. Merchant ships had to enter the inner harbor at high tide. Then, as the tide ebbed, floodgates closed, locking them in until the tide came back. Too big to get in there, we had to anchor the *Point Cruz* outside, in the river. From there, I put our marine F4U air group ashore at an air base near Seoul to make room for all kinds of helicopters: marine Sikorskys, army Chinooks, British "birds" off the *Ark Royal*, and a half-dozen other kinds of "whirlybird."

Then I went to the UN base near Munsan Ni, where the truce talks were held, to scout Freedom Village. I rode an army helicopter up but didn't fly it because I'm a lousy helicopter pilot. (They say the trick to flying a "helo" is to be able to "scratch your head and rub your stomach at the same time." But thirty hours' instruction after World War II and four hours' credit flying solo had taught me I just could not do that.) During my tour of the DMZ, my escorts told me detail on the "truce violation" lies being thrown at us by M. G. Sanjo Lee, the Red's top negotiator at the time. On the 29th, I had dinner with Maxwell Taylor and his chief of staff, Paul D. Harkins (who, in 1964, would be the first general whose career was wrecked by the political vicissitudes of the Vietnam war.) We also visited Queen Min's castle, a classic oriental palace with colorful, precisely manicured gardens, where we learned the history, the how and why, of the Min and Dai families being Korea's dynasty.

On the 30th, I flew back to Freedom Village to greet Ens. Roland C. Busch from Cape Girardeau, Missouri, our first naval aviator POW released. In a VF squadron on the *Valley Forge*, he'd been flying cover for a downed pilot when his Corsair was shot down at Hap-su in May 1952. As senior naval officer there, I got to pin new wings on him and notify him officially that he'd been promoted while in prison. The navy wings, looking out of place on his Chinese-coolie clothes, put me near to crying. I noted in my diary, "I'll never forget seeing these prisoners released. I wish all Americans could see what freedom means."

We *Point Cruz* folks were tense and a little excited on 1 September 1953, the day Operation Platform started. First aboard were the 458 men we were to airlift that day, led by Brig. Gen. Gurk-Bakh Singh, a burly six-foot, six-inch-tall Sikh, and Maj. Gen. S. P. Thorat, a slender, much shorter Hindu. Father Riley told me of another difference: a Sikh drinks but won't smoke and a Hindu smokes but doesn't drink. In fact, Father Riley became our unofficial envoy, chatting up the generals and their troops, learning what their religions let them eat, what not. Most were short men who'd not seen a helicopter, before, let alone ridden one off a carrier. Bitter ethnic rivals back home (still are), the Hindus and Sikhs got along well enough aboard ship since they were in the same outfit, and had to obey orders and work together. We'd stocked food for them, no beef, of course, but they brought their own, even their goats, with them.

So now, in addition to running the largest helicopter airlift ever up to that time, my guys got very interested in this strange new culture parading by them. So did Sen. William F. Knowland, a California Republican who'd come aboard, part of a "fact-finding" mission he was on for some Senate committee. And, of course, we had the press swarming all over, assuring we'd hit the papers in a big way back in the States. In my notes I wrote, "If I say anything tactless or am mis-quoted, I'm sure I'll hear about it but at least we should have some good pictures for the family scrapbook."

A second Indian transport with nine hundred troops arrived that evening, and our airlift of them began on the 2nd. Every day that our helicopters were doing the forty-five-mile run to Freedom Village, we kept the deck clear of everything else; and they flew dawn to dusk, a continual stream, each one taking from eight up to twenty or so fully equipped troops to the Village, and returning right away for more. In the midst of it all, Jocko Clark stopped by. Acting lately as if I were his long-lost nephew, he wanted my advice, this time, on a book he was writing on his experiences in World War II. After reading it I suggested, bluntly since he was a friend, that he give a few other people some credit too. "You make it sound as if you won the war in the Pacific all by yourself," I said.

My diary entry for 4 September is all over the lot: "General Jonathan Wainwright, the 1942 hero of Bataan, Corregidor, died today. How many remember those grim, desperate days? Asia is still aflame 11 years later, looming ever larger in American affairs. Our preoccupation with the Plains of Europe has hurt us badly in Asia. Is there a solution to the French-Indochina war or for Chinese Nationalists? Konrad Adenauer, 'Der Alte,' was elected West Germany's President. All NATO seems stronger for it. But I told my guys we have to stay sharp. Korea is just a cease-fire—which is when Communists like to attack."

On 5 September, we airlifted a third transport's 971 Indian troops to the DMZ and were under way to Sasebo for supplies, then out to our operating area where, on the 9th, we moved up off Shantung Peninsula to do some ECM on the Chinese. They tracked us too, but didn't react. What, I asked myself, do they thinking we're doing, anyway? By the 14th, we were back in the world's worst port, Inchon, to airlift 978 Indian troops to the DMZ, taking up what was left on the 15th.

When they'd first come aboard, they'd criticized us severely for not making any headway in the truce talks on the prisoner-repatriation business. Next day, I followed them up to Munsan Ni for their first session with North Korean dictator Kim Il Sung. It all seemed "sweetness and light," as they say. Gurk-Bakh even tossed down a few vodkas there and left with, he said, Kim's agreeing that the North Korean POWs we held could be released to go wherever they wanted. He's easy to talk to, Singh said. Later I learned, when he returned to Munsan Ni next day, "to work out the details," Kim had glared at him and asked, "What agreement?" That night at dinner on the ship, I needled Indian Generals Thorat and Singh with a "Now you know what we've had to deal with" sermon.

Korea, "Land of the Morning Calm," was hardly that in late 1953. Nor was Washington, in a nasty debate over what to do to American POWs who, under torture or fear of it, had given up military secrets or said things publicly damaging to our country. Eventually, they amended the Uniform Code of Military Justice so that a POW, in certain situations, may give more than just name, rank, and serial number without necessarily facing prison or dishonorable discharge at home. But the top Turkish POW who transited my ship told me they'd handled it differently. Only one Turk, under torture, had babbled "Red" propaganda to the newsreels. Back in prison, the Turks killed him. No more Turks talked to the Commies at all about anything.

As we absorbed that story, we slipped easily out of Inchon on 17 September because I now knew how to use the anchor to swing my single-screw ship around against tide and wind. We were supposed to operate in the Yellow Sea, but an Indian ship brought five hundred more troops to Inchon, raising to fifty-eight hundred the number at the DMZ. So we had to go back on 25 September to wrap up Platform. What had begun as exciting was now an exasperating mission, and I felt sorry for the Indians. They'd be stuck in the DMZ all winter, probably, with those Siberian cold fronts slashing at their tents like icicles. By then, the whole drill was the weirdest POW story in history. The first of our POWs the Committee questioned refused to say where he wanted to go. They didn't interview any more, finally, in January 1954, releasing all POWs to go anywhere they wanted. I think it was Aristotle who said, "A peace is more difficult to organize than victory but, if it isn't done, the fruits of victory in a war will be lost."

All that was in the future when we left Inchon on 1 October as my crew turned on me. The World Series was coming up, my Yanks against the Brooklyn Dodgers, the "Bums" heavily favored. My crew backed the "Bums," of course, knowing I was a Yankee fan. (The Yankees won to my delight, my crew's dismay.) Because of the Platform interruption, we went directly to Hong Kong, arriving at 0900 hours on 7 October. Its harbor is beautiful, the city fabulous in ways my words simply can't describe. But it was a tense test, edging in to anchor off Kowloon, with Chinese junks like water bugs seeing how close they could come to having my ship cut off the dragon's tail supposedly dragging behind their boats.

I didn't let the Chinese pilot do much, handling most of that myself. At anchor, we were greeted by harbor officials and a horde of chattering Chinese offering deals to let them set up shop on our decks. The most intriguing was a cute gal named "Garbage Mary" who said her girls would paint my ship's exterior if I supplied the paint and gave her our garbage. I agreed, I said, if she promised not to say which, if any, restaurant got our garbage. Her little "painter-sans" did a superb job and my guys just loved watching them at work. During our week there, I spent all my money, loaded my quarters with loot, played golf at the posh Fan Ling course on the border, was mesmerized by Repulse Bay, and saw Macao (worst "Sin City in the East," I was told). I met some fascinating folks, especially missionaries Father Riley invited aboard to see our baby, now a big problem for me. Under way, as we passed Admiral Goldthwaite's flagship, he signaled, "Your ship had never looked better." Thank you, "Garbage Mary."

We'd been launched into the baby business in late July by a message from General Pate. A footnote in it said a medical corpsman—marine or army, he didn't say—had found on 28 July a two-day-old, part-Caucasian baby in a trash can by an ASCOM (Army Storage Command) sick-bay tent at the DMZ line. I heard nothing more until, off Inchon in early August, Father Riley collared me. He'd gone ashore to see a friend, Sister Philomena, a strong-willed Irish nurse who ran Inchon's Star of the Sea orphanage. Staffed by French nuns, some Korean nurse's aides, it was full of babies in cribs made from apple crates. So, with my okay, he'd taken along medicine, powdered milk, and such from our ship's stores. And come back with a problem.

Sister Philomena had shown him a blue-eyed baby boy an army orderly had delivered to her orphanage. It had to be the one Pate's message mentioned. The infant was emaciated, its abdomen swollen, burns on its back hinting it had been born in a Korean hut, the kind heated by shoving hot coals under the stone floor. Always short of food and clothing, the sister had to send her charges out on their own when they became teenagers. If she did that with this child, she said, he'd be a Korean pariah. Certainly, no Korean couple would adopt him. Even now, her Koreans saw to him only if ordered and if she watched to see that they did. "He's an American, Father," she'd said. "Can't you do something?"

Now, he asked me. "Bring him here," I said. "We'll keep him 'til he's healthy and I find out how to get him adopted in America. Won't be easy. Regulations forbid having civilians on a navy ship in a combat zone." When he frowned, I added, smiling, "But this time, we'll follow Pete Mitscher's rule. It says, 'A good leader knows when to intelligently disregard a regulation.'"

Quickly, our carpenters built a nursery in sick bay, with a playpen, a crib, and a baby buggy from a bomb cart. Our laundry guys made diapers from sheets and next day, as Father Riley arrived in a motor launch with the baby, my entire crew was topside, grinning and cheering. By then, the baby was George (for the guy who'd found him) Cruz Ascom (Army Storage Command), and his nursery was awash in homemade toys. Our ship's doctor, flight surgeon Bill Tooley, a bachelor and pediatrician—who hated children—showed our galley how to make baby formula. (In my experience, all flight surgeons on carriers are either a pediatrician or an obstetrician.)

Within days, George was chubby, rosy-cheeked, and active like all healthy infants. We put a big black pharmacist's mate and a corpsman in charge, both fathers, and tough enough to do what I called "crowd control." They, in turn, made anyone changing George's diaper take a test to prove he could do it "correctly." And so many guys wanted to visit George, I had to institute "baby-san call" to get any ship's work done. Every afternoon after his nap, our public-address horn would blare, "Attention, all hands! Baby-san on the hangar deck from now until 1600 hours." Petty Officer William J. Powers, in charge of the hangar deck, told a journalist, later, "That baby had a thousand uncles. All we wanted was to go home and along comes this little boy to hit us right in the heart. It was as though he was what we'd been fighting for."

Absolutely. When I took command, my ship's crew had been overseas five months. They were tired by then, with petty arguments and fighting every day. They blamed me for taking on Operation Platform, delaying their return home. But once Father Riley got George aboard, morale soared. Father Riley and George pulled those one thousand guys together as if by magic. Now, the only arguments were over whose turn it was to change diapers and was the baby formula warm enough. And my wish to keep George hidden ended quickly. The crew began to hang his fresh-washed diapers on our signal-flag yardarm to dry. Passing ships would signal, "What's the white flag mean?" and my guys answered, "Baby-san on board!" Two weeks of that got nearly every American sailor in the Far East aware that we had a baby aboard ship.

So was Lt. Hugh Keenan, a Spokane, Washington, surgeon serving on a hospital ship off Inchon. Even before Father Riley, he had gone to the orphanage to help out, and Sister Philomena had handed him George, saying, "Here, you feed him."

Keenan and his wife had two very young girls and had just lost what was to be their first boy, either by miscarriage or a fire in their home. (I heard both explanations.) Four more trips to the orphanage, a letter to his wife, a longer one back from her, added up to their deciding to adopt George—in spite of a navy rule banning adoptions by personnel overseas. Dr. Keenan didn't contact me, but Mrs. Keenan wrote to Father Riley to tell him why she wanted George. In spite of her letter and their decision, I still had to get, in the meantime, guaranteed support for the boy in the States. Without it, I couldn't get anything. So I sent Father Flanagan at Boys Town a letter. His quick answer back said, sure, he'd sponsor the child.

Back at Inchon on 19 October to be in a marine-landing drill, my boys noted how wrong I'd been in saying, as we left for Hong Kong, "Well, we've seen the last of this depressing place." Leaving Inchon, we hit sand in mid-channel where charts said we had 40.5 feet of water. It fouled up our starboard main engine condenser. Then, barely back on station, we had to go to Sasebo for work on our boilers. By 8 November, we were on our last tour in the Yellow Sea, off Tokcholk To, Korea, for a marine landing exercise in wintery winds, then back in Inchon on the 14th. We were to go home on 2 December and I still hadn't solved our "baby-san" problem.

The crew wanted to take him back to the States with us, so I went to see the American consul in Pusan. George would need a Korean passport, he said. After being stiff-armed a few times by South Korean bureaucrats in Seoul, I called in Father Riley. Handing him the bottle of medicinal whiskey all good carrier skippers keep in their safe, I said, "Padre, take this to Seoul in my helicopter,[1] and don't come back 'til you have a Korean passport for George."

It being my duty as captain, I signed and gave him a form, certifying that three-month-old "George Cruz Ascom" was American. In CBS-TV's movie, *1,000 Men and a Baby*,[2] Father Riley wins the passport in a poker game, beating a South Korean's full house with four treys. Maybe. I know only that he was gone three days, including a Sunday. (Later, when a BuPers jerk demanded to know how I could deny my crew church services, I growled back that I'd done the service, which, I was told, got a "big laugh" out of the BuPers chief, Admiral Holloway.) When Father Riley returned, he didn't bring back my scotch but he did have a long, thick Korean passport.

Army detectives had flimsy clues, by then, that George's father might be an army soldier or his mother an army nurse, but they'd run out of leads. I was glad they'd stopped hunting. What facts we did know convinced me George's best shot at a good life was with the Keenans. Still, to cover the possibility that he was part Korean and needing a visa to enter the United States, I called a State Department contact I had in Seoul. "Gee whiz, skipper," he said, "with the backlog we've got, it may be three years before I can get a visa for that child."

"Three years!" I snorted, "Will he have to live on my carrier that long?" Then in mid-November, I was invited to dinner at our Korean ambassador's mansion in Seoul, to celebrate the armistice and receive a medal so important I forgot within a day what it was. The banquet hall was full of stars, generals and admirals, Maxwell Taylor, Jocko Clark, plus new Vice President Richard Nixon and Ellis Briggs, a high-level Foreign Service officer and Nixon's escort on their junket. At a lull in the drinking, old Adm. Harold M. "Beauty" Martin, asked the crowd, "Have you heard about Chick Hayward's baby on the *Point Cruz?*"

Every flag officer in Korea knew by then about "Chick's baby," but I had to recite the story anyway, and did, including how tough it was going to be to get George a visa. At that, Nixon turned to Briggs and asked, "Can't you do something about that?"

The visa was waiting when I got back to the ship that night. We went to Sasebo on the 14th for supplies, then to Tokyo to stay until we slipped the chain to head home on 2 December. Adm. Bob Briscoe, COMNAV-Japan, our top admiral there and my hunting companion in the past, met us in Tokyo. His first words were, "Chick, I understand you have a baby aboard."

"Yes," I smiled and took him to see George, telling him the whole story as we went. He suggested I tell "Mick" Carney about George. "Wouldn't hurt," he said, "to have your friend, the CNO, on our side if you run into any more problems."

We didn't have shipboard phones, back then, only teletype machines. Forgetting it was near-midnight and a different day at the Pentagon, I sent Carney the message and promptly back came a wire, "Baby? What baby? Get it off your ship immediately!"

Well, of course, within an hour, my crew knew what that order had said. Since day one, they'd said they'd decide the adoption issue. Maybe Father Riley sold them on Mrs. Keenan. In any case, by the 20th when we reached Yokosuka, they'd picked her for George. That night, a chief petty officer brought me a thick envelope, and said, "Skipper, we got it all settled. Here's $1,500 we've collected. We can send George and Father Riley home on a Pan American airplane."

That was a lot of money, what with most of my thousand-man crew being enlisted men and a dollar able to buy about ten or more times what it can today. Still, I told him, "Chief, we don't have a visa to let him into Japan. He'll have to go home on our ship. And if the navy wants to fire me for it, they'll fire me."

Briscoe, hearing my angry tale, agreed Carney had made an awful mistake. So we re-sent the message. Now, his answer was to applaud us and order, "Bring the baby home."

He hadn't seen our first message. A commander on midnight watch in the flag plot, tops in class on navy regulations, probably, was who'd "zapped" us, then filed the cables away. Now, trying to help, Carney had two nurses sent out to us. The men were furious. "Skipper," they snarled, "We don't need any nurses to take care of our baby. Get them off our ship, please, Sir!"

My having to shoo them away made my crew happy, the nurses sore since, they said, the CNO himself had sent them. However, Carney's "bring him home" did inspire a new plan. Our destination was San Diego, but Mrs. Keenan, with two small girls in grade school, was unable to leave Spokane. So we decided to send George directly there on a transport, which, with nurses aboard, was better for him. Thus, before we ourselves sailed, George Cruz Ascom, IB/fc (Infant Baby, first class) was piped over the side, a rite reserved for VIPs, with his escort, Father Riley. All hands were on deck to watch. In Seattle, the commandant of that area's Naval District met the ship on Carney's order, and escorted George and Father Riley to Mrs. Keenan in Spokane.

Meantime, at Yokosuka, we off-loaded our Marine F4Us and took on VS-23 with its TBF torpedo-bombers and twenty-five-thousand-pound gross-weight Grumman AF-2W and -2S aircraft, plus destroyer screens—the usual hunter-killer setup—for ASW exercises to be done as we sailed for Hawaii. By 6 December, we were five days out, the Able and Baker phases done and Charlie up next, to be run six hundred miles off Hawaii with eight submarines to find and "sink." I know someone scored the thing but no one told us the result. (In early 1954, the navy got its first Grumman S-2 "Tracker" aircraft, which put both "hunter" and "killer" capabilities in one aircraft. It carried APS-38 search radar, sonobuoys, searchlight, a magnetic-anomaly detector, and 4,800 pounds of homing torpedoes or depth charges or, after it was developed in the late 1950s, an air-to-underwater missile. The aircraft had a 255-mph top speed, 1,150-mile range and, like the AF-2W, used a crew of four including two radar operators.)

Worse than not getting a report card, however, was newspapers reporting how the Korean "no-win war" had clobbered reenlistment rates. In my diary, I groused, "Instead of screaming to the Press that it's Congress and the Public who make it hard to keep career people, the Navy ought to put some intelligent thought into what it could do, itself, to help solve the problem."

On 16 December as a band played "Aloha," we pulled away from the dock under Honolulu's Clock Tower, heading for the "barn" in San Diego. The sea was soft, the breeze warm as we plowed east. Homeward bound after months at sea is the best trip a navy ship makes. George was with his mother now. The paperwork I'd done would prevent anyone else claiming the boy later, but my crew was who really saved George's life. I was just their pliable tool. On the 21st, we tied up to the North Island pier, with wives, mothers, girlfriends, families, the press, and a band there to greet us. Of course, we told them all about "our baby." Predictably, it was reported the next day in the *San Diego Union* and *Los Angeles Times* newspapers. *Time* and *Newsweek* put the story on their magazines' Christmas-issue front covers.

And I ran off to Pensacola for Christmas with Lili and the children, a little sad that Mommie Floss was gone but glad to be "home" up on the Sound. Father Riley left the navy a short time later. For a few years, many of us *Point Cruz* guys tried to track him down. The closest we came was hearing that he'd gone to one of Mexico's impoverished mountain provinces or "maybe Central America" to do missionary work. Unable to find him, at least to correspond, was a frustration the crew of the *Point Cruz* would bear for years.

By mid-January, off our big gray ship at North Island, I flew up to Spokane to give Mrs. Keenan George's $1,500. Hollywood wanted to do a movie on him but Mrs. Keenan had said no. His new parents would tell him how he'd become an American and they were not interested in Hollywood's money. They'd named him Daniel, after Dr. Keenan's father, Edward, after Father Edward O'Neill Riley, she said. She didn't want to accept the $1,500, either, but I told her it was from the crew. I was just the delivery boy with no way to give it back. Use it for his education, I suggested. She was a good, caring lady, convincing me the crew'd made a right choice.

Back in San Diego, I was told to put the *Point Cruz* in the Long Beach Navy Yard by 7 February for boiler repair and turbine work. To my amazed amusement, one of the yard's security guards was Bob Meusel, a New York Yankee left fielder in the 1920s—when his brother, Emil, played outfield for the New York Giants. The yard released our ship in mid-March, a mess as usual, the only way navy yards, unlike private ones, seem able to release ships. On the 24th, flight operations got three planes creamed. This newly manned VS-23 group is just not ready to deploy, I wrote, but the long trip to the Orient will give it lots of time to train.

On 31 March, I was told Fuller Brush, not "the Fuller Brush man" but his real name, would relieve me. But when? That same day in a grand bit of posturing, Russia said it wanted to join NATO. As background, I noted, "The H-bomb has everyone by the neck, hysterical, the Air Force now fighting on three fronts: the Army, Navy and Marine Corps. We have night trials next week and a battle problem and I'm tired of it all, now. More boiler trouble, aircraft repairs, have delayed to 27 April our departure for Hawaii. The Geneva meeting began today, already looks inane. It's called a 'nuclear disarmament' conference but its aim is just to slow the pace of nuclear-arms expansion. Surely its fatal flaw, true of most such treaties, is it won't have police power, no effective sanctions against a treaty violation."

In late April, I got a thank-you letter from Syngman Rhee for my article, "In Defense of Syngman Rhee." Written while he was mucking up the truce talks, all my essay said was that he had reasons not to trust us. We'd reneged on every promise to Korea since the 1908 Root-Takahira treaty by Secretary of State Elihu Root and Japan's Ambassador Takahira, the first time the executive branch signed a binding pact without asking the Senate to ratify it. It said Japan's purview in Manchuria and Korea was "paramount." That canceled our treaty-promise of 1876 to defend Korea's independence. As a result, in 1910, Japan murdered Queen Min and seized Korea. After World War I, buying Woodrow Wilson's promise of a people's right to self-determination, Rhee led a rally in Seoul, demanding "a free Korea." Japanese troops crushed that and he fled to Shanghai. He returned after World War II to learn we'd given half his country to Kim Il Sung.

I'd offered my essay to Jocko Clark for his book. He said no. So I sent it to The Naval Institute's *Proceedings* magazine. On its advisory board by then was Vice Adm. C. Turner Joy, who'd once headed our UN truce-talks team. He, too, vetoed it, letting his bias against Syngman warp editorial judgment, I think. How the Koreans got it, I don't know nor saw a copy, but a Korean naval officer told me they'd published it. Also in April, General Dynamics asked me, again, to be their Atlas program manager, but I'd been told that on 15 June, after fifteen days with Lili, I was to take command of NOL (Naval Ordnance Laboratory) at White Oak, Maryland, near Washington, D.C., a job with all sorts of intriguing attractions.

In May, the French in Dien Bien Phu gave up to Ho Chi Minh and began to evacuate Vietnam by way of Hanoi. Ho promptly began building in Laos the "Ho Chi Minh trail," mostly a paved highway, in fact, to outflank South Vietnam. Our State Department urged Ike to send "a strong force" to fill the "power vacuum" but army Chief of Staff Ridgway talked Ike out of it. "South Vietnam has no infrastructure," he said. "Its transportation assets are totally inadequate for a modern, mechanized combat force, and its government is very unstable. If we go in, we must go in to win. We can do that but it will cost us more casualties than Korea did." Too bad, sad, really, our next president didn't listen, too. For saying to Jack Kennedy what Ridgway had to Ike, George Decker lasted as army chief of staff just one two-year tour. Long forgotten by then was one irony in all the tragedy: Ho had been our ally in World War II.

Finally, on 18 May, one day before Armed Forces Day, Fuller Brush relieved me and my ship's crew gave me what a newspaper called "a rousing send-off." I was a bit teary-eyed. That was and would remain the finest crew I'd command on a ship, ever.

16

On Making Admiral

Nine days later, AEC's Commissioners voted four to one to cancel Dr. J. Robert Oppenheimer's security clearance, a decision Eisenhower endorsed. The "rip-off" (my label, not Ike's, certainly) had begun when Oppie chaired AEC's GAC (General Advisory Committee) while head of Princeton's Institute for Advanced Scientific Study from 1947 to 1952. The accusations behind it were from William Borden, Congress' Atomic Energy committee staff director when Oppie was at Los Alamos. That told me Oppie's enemies just wanted to get rid of him.

But it killed Deak Parsons. Very fond of Oppie, when he heard about the mess on 4 December 1953, Deak, who rarely lost his composure over anything, was horrified, his wife, Martha, told me. Chief of BuOrd at the time, he said, "This is the worst mistake the United States could make." That night, a pain in his arm told him he might have suffered a coronary. Martha took him to Bethesda Naval Hospital next day for an electrocardiogram. He died as it was being given. He was fifty-two. He'd have been the strongest witness in the world for Oppenheimer.

North Carolina University president Gordon Gray chaired the three-person panel that ran the inquiry. After their two-to-one vote against Oppie, the AEC said it had lifted his security clearance because of "his earlier left-wing associations and opposition to . . . development of the hydrogen bomb." Phooey. When I was in the AEC, Gray, Bill Borden, and Morgan, like Ed Teller and the air force, all wanted to make "big-bang" hydrogen bombs. Like Lauritsen, Parsons, Bradbury, lots of us, Oppie just wanted to build small-yield, tactical weapons first. As for the "left-wing" business, Oppie like a lot of us had favored the Soviet-backed Loyalists in the Spanish Civil War. We didn't want the Hitler-backed rebels to win. And his brother, Frank, a professed communist? Groves and Vannevar Bush had known all that in 1943. A rumor Ike denied said they jumped on Oppie because they feared Senator Joe McCarthy would beat them to it. McCarthy's voodoo hunt for "Communists" anywhere he sniffed a headline was infamous by then, so "Why," I asked, "doesn't Ike just disown McCarthy?"

The reason Lewis Strauss dumped Oppie, I believe, was the result of how Oppie'd crushed Strauss' very large ego earlier. I was a "horse-holder" at the event, a GAC meeting on Norway's request for isotopes for medical research. Strauss had said no. Such a gift will jeopardize our military security, he said. Oppie, usually kind and soft-spoken, could be arrogant at times, especially on a science subject. In a nasty lecture, he called Strauss' claim "Ridiculous!" None of that was in the hearings, however, and I wasn't asked to testify at the "hanging."[1]

He got an apology, sort of, when the AEC, endorsed by President Jack Kennedy, gave Oppie its gold Fermi Medal and a $50,000 honorarium on 2 December 1963. By then, a study by Dr. Harry Smythe of Princeton had proven that the only thing Oppie was guilty of was disagreeing with Strauss. Yet still today, the charge echoes in the media, and Russia's Foreign Intelligence Service, once their KGB, replies, as it always has, "Reports which appear regularly in the press are untrue. Dr. Oppenheimer never gave information to Soviet intelligence—absolutely never."

The "Oppenheimer affair" was gnarled history when my family and I arrived at the Naval Ordnance Laboratory in June 1954. I was the first aviation officer to command it. Until then, an Ordnance postgraduate had filled the slot. Lili and I, the three children still at home—Vicky, Jennifer, and John, Jr.—lived in posh (for us) quarters facing NOL's main gate. I could walk to work, play all day with NOL's toys spread over 875 acres—and avoid the Pentagon. A $50 million capital investment, NOL had a $34 million annual budget and thirty-three hundred employees, many of them very able scientists or engineers, involved in all phases of ordnance, mines, bombs, aeroballistics, proximity fuses—everything.

NOL had built, in 1949, the first hypersonic (Mach 10) wind tunnel in the United States. Data derived from it was being used to help develop all the services' first fifteen-hundred- and six-thousand-mile-range ballistic missiles. NOL also had a "shock" tunnel and an enclosed one-thousand-foot precision missile-firing range. (I tested my shotgun there, for the navy, of course, to record the pellet dispersion so I'd know where to aim to hit birds.) For hypersonic flight, all three kinds of tunnels/ranges were needed because hypersonic flow fields are so complex that a model simulation of a full-scale vehicle is almost impossible in a single facility.

NOL's technical director when I arrived was Ralph Bennett, but I'd hardly sat down when he left to go work for General Electric. Greg Hartman, our yield-measurement guy when I was at Bikini, had our Explosives Division. He was the obvious successor to Bennett but it's a Public Law 313 job with lots of Civil Service paperwork and reviews. So I wore two hats until March 1955, when Hartman became technical director. Don Marlow headed Engineering, Paul Fye the Research and Chemistry Departments, Dave Bleil his Physics Branch, James Ablard the Explosives Research. A Mr. Hightower ran the T&E (Test and Evaluation) Division and our test ranges at Pax River and Lauderdale. Herman Kurzweg, a German physicist, was Aeroballistics director and wind-tunnel boss. In 1938 at Peenemünde, he'd done the aeroballistic design work on the German V-2 rocket, using a Volkswagen racing down the autobahn as his "wind tunnel." He was Peenemünde's deputy director by 1943, when our army hired Wernher von Braun. Herman, who had no use for von Braun, asked the navy to hire him, which it did in 1946.

Even for the nine months when I was my own technical director, I let my civilian experts run the show and, unlike at Inyokern, this time I got no growls from upstairs. BuOrd had gone more quickly than other Bureaus to the OSRD system of having scientists run its labs. And after the Naval Aircraft Factory, Inyokern, and Sandia, I knew by now both the problems and potential of that approach. I did hear one jarring note from home in mid-July. The Hyer Towing Company was near broke. I felt sorry for Daddy Pete, but after Mommie Floss died, he'd just walked away from the business. "Men, alone," I noted, "don't seem to do as well as widows, alone."

One of our top priorities was to produce the Mark 52 mine, the first in a sophisticated, esoteric new family, a bitch to defeat, much harder than sweeping up simple pressure mines the air force had dropped in Japan's Inland Sea. And Japan never had been able to defeat even those. The Mk-52 had a pressure sensor, an acoustics sensor, and a total-field magnetometer to detect a ship's magnetic field. Since a ship can't be demagnetized, it can't defeat the mine. Each sensor can work separately or interactively, and the Mk-52 can be set to count both elapsed time and ships. Thus, an Mk-52 can let the first ship pass and detonate at the second or third. An enemy can't know what to expect next. The field also can be set to shut itself down months, even a year after it was planted. In short, not one mine but the mine*field* is the weapon, and a highly cost-effective one. Yet even in the air force, though B-29s did a superb job of shutting down the Inland Sea with the old-technology pressure mines, their Ellis Johnson had a tough time selling General Curtis LeMay on building Mk-52s. And in the navy, which clings to the ancient view that only ships can lay down a minefield, use of the air-dropped mine is our most neglected battle tactic.

I wasn't the only one with ailments. With Oppie gone, the air force was demanding anew an atomic monopoly. My log says, "It wanted the second atomic weapons laboratory. We beat that down with Livermore. They'll not win this one, either." Then on 8 December, Daddy Pete died of pneumonia after only four days illness, a blessing, really. He'd not be there to watch his company fail. Lili was estate executor but I doubted there'd be much to inherit. The *Forrestal* carrier was launched on 11 December, and Harry Sears made admiral, an early selectee. "The odds I'll make it are poor," I wrote.

My year-end summary on the Mk-52 says, "It's giving me 'the haunts.'" I didn't achieve much, having Hightower and Fisher, heads of our T&E group, meet with Greg Hartman and all my other technical people. T&E had to certify the Mk-52 for production but didn't want to. Like many in the R&D business, it was a case of the best being enemy of the better, this time T&E wanting better sensors. I told them, "Morale in all of NOL will nose-dive if you give it a 'down-check.' So if you do, your technical reasons better be sound, provable ones. It's already the world's best mine but it does our navy no good stuck in this laboratory."

In March 1955, the Mk-52 finally went into production at the Naval Gun Factory. Then we added a solion device that, under outside pressure, would change the mine's chemical structure and set it off. Thus, a field of Mk-52s could make submarine and ship operations impossible in much of the world. Yet the navy refused to buy very many. To most sailors, a mine, for all its potential, just wasn't sexy. They even made Panama City our countermeasures test site, saying people who make mines should not help build defenses against them. Crazy.

In April, my diary reports, "Charles Bergin (BuOrd director of research) gives me fits, can't make up his mind on anything. Lili is in the hospital, hepatitis. Home is empty without her. P. D. Stroop (Chief of BuOrd) says I'll get a 'Big Command' next year, but I doubt they'll ever select me for Admiral. Still, NOL is good training for the civilian technical world and they have me traveling a lot in addition to going to BuOrd or the Pentagon to argue with people on various technical subjects."

A typical duel was with Adm. Don Felt, OP-03 (ACNO for R&D) at his big Pentagon seminar on AI (Aircraft Identification) radar and night fighters. When the F4H Phantom came in, in 1958, as a high-altitude (thirty-five thousand feet) interceptor with a pilot and a radar operator, we had made huge gains in radar technology.[2] But in 1955, we were just getting into it. At his meeting, Felt, an aggressive guy anyway, stood on stage and gave a windy speech, mostly on how easy radar night fighting really was, and asking at the end if everyone agreed. I raised my hand and said, loudly, "No. The range, radar frequency, the whole system isn't like you say it is."

He abruptly dismissed his shaken audience and snapped, "Young man, (I was forty-six, then) come to my office." He was a short guy who made up for it by acting tough. I knew that if I didn't stand up to him, he'd drive me into a corner. Anyway, I told him he'd ignored what a combat pilot has to do, and, if AI R&D doesn't focus on that, I said, its work will be irrelevant. Basically, I said, whether it's an aircraft or whatever, first, of course, the pilot must find and identify it; then, and the hardest task usually, classify it. Is it friend, foe, or a UFO from outer space? Then he has to decide, if it is an enemy, "Do I kill it or let it go? And if I decide not to kill it, what do I do, stay with it or leave the scene? And why, either way?"

He finally bought my sermon, common sense, really, and, I learned, like King, he wanted not "Yes, sir," but frank, honest opinion. He was on my side from then on in my career, though by then for every Felt, I could count twenty who were not on my side. Then in August 1955, Eisenhower gave my morale a boost. He reached past ninety-two more-senior officers to name Arleigh Burke as CNO (Chief of Naval Operations), confirmed by the Senate on 17 August. It pleased me, one reason being that I was still prouder of the Distinguished Flying Cross Burke got me, after Kavieng in World War II, than of any other medals I'd received. And Lili came home, her four-month hepatitis battle won.

In August, my diary announced, "Shelley just gave birth to an 8-lb. boy so I am now a Grandpa. Lili is with her in Atlanta but will be home by 1 September, and I leave for Europe on the 3rd." The European junket was principally to the air show in Farnborough, England. It was better than the Paris Air Show, which occurs in alternate years, since more companies show up. But, I said, Paris' soon will be much larger, one of my most accurate and least exciting forecasts. But the trip's best fun was gassing up at Shannon Airport, Ireland, to fly home, nonstop, for the last phase of NOL's atomic depth-charge program.

Besides mines, the depth charge was NOL's top-priority project in 1955. Designated the Mark 57, it was a twenty-kiloton atomic bomb, actually, with a three-mile kill radius. We worked with Los Alamos to make it since they had the active material, and my knowing their players was why I was at NOL. For a test of it, called Operation Wigwam, the bomb, code-named "Betty," was lowered from a barge down to two thousand feet below the ocean's surface. On a wire down to the bomb, we attached structural devices, "squaws," every seventy feet to measure the pressure all the way to the surface when the bomb exploded. It was detonated 187 miles southwest of Coronado, California. Capt. Ed Hooper from the AEC's Division of Military Applications, a friend of mine from my time there and a good friend of Deak Parsons, ran the test and would go over to BuOrd after it was done.

When we detonated the bomb, under a cloak of secrecy, its energy rose like an inverted cone all the way to the surface—and stopped. Only a very small bubble surfaced. One thing we'd wanted to know was, will the radioactive force move up through water into the atmosphere? It didn't. That said that surfaced is the only safe place for a submarine when an atomic depth charge explodes. Still, even without an above-water burst, the water was as "hot" as it had been at Bikini. Unlike Bikini, the radioactivity didn't disperse. The irradiated salt just sank to the ocean floor. Being reasonably sure of that result was why we put the bomb down two thousand feet. But its radioactivity would infect marine life down there for years. Salt water is deadly when it comes to that.

At NOL in 1955, as the Russians exploded an H-bomb device, the whole mine program was under assault. NOL's Jim Dare was working on an atomic mine designed so that it could be laid down efficiently from the air in low-level high-speed drops. A problem he had was that the navy didn't have an aircraft able to lay down a good minefield. On 16 November, as I argued the case for mines, I got orders to take command of the *Franklin Delano Roosevelt* in Bremerton, Washington, on 1 February 1956. Our shipyard there was refitting it to give it a reinforced, angled flight deck, stronger elevators, a new electronics suit, and steam catapults to handle the new seventy-thousand-pound A3D and A4D attack bombers joining the fleet. After commissioning, she was to sail on 15 May by way of Cape Horn to home port in Mayport Naval Base at Jacksonville, Florida—"Jax," the natives called it. A small air station before, its port facilities had just been expanded to take ships the size of the sixty-two-thousand-ton *Roosevelt*. She'd be the first big carrier to base there.

Willie Wilbourne, a non-aviator classmate of mine, was to replace me at NOL. I hated to leave. It was the best navy job I'd had, in a science-and-engineering sense. In January, Admiral Burke called me in to explain, "A new Board begins work in July and you need this ship so they can select you." Besides, he added, "I sure as hell can't pick a Republican to command the *FDR* and you're the only Democrat I know around here who has the time to do it."

His trying to make me a rear admiral was nice, but I was happier that the climax of my career would be command of a big ship; then, passed over, as I thought I would be, I'd retire. Ending up as a navy captain is like shooting par golf, a rare, laudable achievement; and commanding a big ship was the highlight of any naval officer's career. So I wanted my family there for the commissioning. Especially Vicky in high school and Jennifer were very upset at being pulled away from their friends, but not nine-year-old John, Jr., who hated school anyway, sad to say, like his father at that age. But they all finally agreed, reluctantly, to be there.

I sold our sedan, bought a station wagon, and, with twenty days to reach Bremerton, we took off. Ice storms and icy rain chased us through Tennessee to Arkansas as ever-pessimistic Jenny said, in the midst of it, out in the woods, wouldn't it be awful if we had a flat tire. We reached Dallas ahead of an ice storm, nearly were hit by a sandstorm near Los Angeles, had a great time at Disneyland, and met a snowstorm at Portland. But as we rolled into Bremerton, the sun came out. We checked into an apartment and put Vicky in a Seattle school, the other two in a Catholic school in Bremerton.

Before we'd left NOL, Secretary of the Navy Robert Anderson had said I could invite along on our cruise anyone I wanted—male, of course—and Arleigh Burke had said the same. So from Bremerton, I began enlisting. Henry Jackson, Washington's U.S. Senator, could not make the trip but did agree to be at the commissioning. I enrolled Tom Clagett, Dean of the University of Washington; Vlad Reichel; Ernest Lawrence; and Norris Bradbury, all enthused as children over taking a "pleasure cruise" on a big aircraft carrier, but most said they'd have to join us along the way, in San Francisco, San Diego, Panama, or wherever.

But at Bremerton long before that would happen, my XO, Jack Howland, and operations guy, Whitney Wright, said the *Roosevelt* was a mess and would be hard to clean up. That put the commissioning date back to April 6th. Then happiness came to town, the *Point Cruz*, headed for the junkyard. Harbor pilot Pappy Beacham invited me to go with him to bring her in. It was a fun day with half my old crew still there, all wanting to join us. I told them to come aboard while I asked Adm. Mel Pride, now commander of Naval Air Forces, Pacific, and pal from my *Langley* days, to transfer them. When they came over, with them came everything of value not welded in place on our old ship. And two of my best people at getting done all the detail work needed to launch the *FDR* became Chief King, who'd been *Point Cruz* chief quartermaster; and Chief Starrett, who'd run the flight deck.

In mid-March, after dock trials of the *FDR* went well, running first two, then three, shafts up to forty rpm, we left dockside to go out into Puget Sound, a tricky maneuver since Bremerton's channel requires an extremely tight turn, tough for a 980-foot-long ship to do without running aground. After a lot of twisting, just as we got straightened and reached the critical turn point for the channel run, the engine room reported gear trouble and said it had to shut down all four shafts. Immediately, I let go the anchors. Later, I told Pappy Beacham I'd had aircraft-engine failures before but never all four engines at once. It would delay our departure date yet again.

On 3 April, I moved aboard ship, and Admiral Burke said he and Admiral Libby from OP-06 (Plans and Operations) would be out on the 20th to see how we were doing. Coincidentally—or was it?—at our conference that day with BuShips, they said they'd give us anything we wanted. The inclining experiment also went well. For that, the ship's bow was inclined to a certain metacentric height in order to measure the "riding moment," how long the ship takes to return to horizontal. That's done, of course, to see how stable she is with a lot of topside weight on her.

Then came 6 April, the commissioning, a grand gala for me. Admiral Stump, now CINCPAC, was the speaker. And Senator Henry Jackson, whom I'd first met when I was in the AEC, joined me on the *Roosevelt* where a wise-guy reporter took our picture. Its caption in the next day's paper called us "The Nuclear Noodles." And, unannounced to surprise me, my mother was there. Inside a parcel she handed me was a ship's life preserver with "U.S.S. Theodore Roosevelt" stenciled on it. Whitney Wright had helped her dream that up, she said, because they were good Republicans who couldn't stand FDR. And Ernest Lawrence, who would sail with me all the way to Mayport, appeared to enjoy the party immensely.

Party over, we went back to "drudge" work. Constant gear problems during April pushed our scheduled departure back to June and "upstairs" decided, when we did leave, our dependents would be sent to Mayport on a special train dubbed, aptly, "The Diaper Express." On 12 April, the InSurv (Inspection and Survey) Board, headed by Burton Davis, a stranger to me, arrived to inspect and accept or reject the ship. I quickly learned he was both thorough, and quick to raise hard-nosed hell for poor workmanship. For instance, finding a very unauthorized side container welded to the bridge wing, he refused to leave until someone cut it off with a blowtorch.

His first day there, I offered him lunch in my cabin. No thank you, he said. All he wanted was a "buttermilk sandwich" at the Bremerton Elks Club—where I learned that the "sandwich" was a stiff belt of bourbon with water. This will finish the InSurv, I thought, but I was wrong. Back on the ship, he eagle-eyed everywhere until 1800 hours, when he returned to the Elks Club. Next day, as he finished up, he told the yard he'd not accept her until it fixed the gear problem. He reminded them that our sailing date was June.

After that, everything imaginable went wrong. The yard used the wrong-sized bolts to secure the double-A firebrick on the boilers, for instance, and in a sea trial one day, we had ten steering breaks and seven burned-out condenser-pump bearings. But in fog and bad weather, we got her up to thirty-six knots, very good for that old lady. But the dark cloud over me was the General Board, about to meet to select the next round of promotions. I'd learned that fifty-five aviators were ahead of me in its promotion zone. "I'll stay, anyway," my log says, "as long as I command this ship, but it's funny, 31 years, apprentice seaman to Captain, and this is the climax. I wonder if my classmates are going through the same trauma."

On the 24th, a Wednesday, P. D. Stroop stopped by, mainly to tell me, if I was selected, I'd replace "Red" Raborn as the escort carrier *Bennington*'s skipper and Red would relieve him. Then the 1 May InSurv revisit brought more trouble. Oil was found in the feedwater and, on the 7th, while we were going out for full-power runs, something tore through a reduction gear and, with tugboat help, we limped home. On the 9th, General Electric came in to fix our gear ailments and I wrote, "It's not the personnel failure I'd thought, but the flexible couplings between turbines and reduction gears, a peculiar design with the thrust bearings way back on the shaft," which shows, if nothing else, how carefully I'd gone over that whole ship.

On 15 May, I took my family on the ferry to Victoria, Canada, to spend the weekend in a quaint boarding house run by a former RAF aviator. The children loved it. But at work, besides the gear ailment, now we had catapult troubles, and I was appalled, listening to NAFE (Naval Aircraft Factory Experimental) people, to hear how little they knew about catapults. On 4 June, with my family on its way to Florida, the *FDR*'s crew and I spent a week in the cold north Pacific, doing speed trials, refitting, and all the other things done on a ship that's been in the yard a long time—and on the 8th, ran into sand in Puget Sound. "We have," I snarled, "miles to go in crew training." Finally, on the 14th, we left for Mayport. My watch people were novices so I'd be on the bridge a lot, under way. The beautiful blue Pacific's deep swells can be dangerous, what with the traffic it carries.

On the 16th, we were off San Francisco, the day clear. It was for me a thrill, taking my ship all the way into the navy yard. Going by Oakland Bay Bridge, when she began to set down on its foundations, I give her a power-shot to swing her smartly clear, power we hadn't had in the *Point Cruz*. Norris Bradbury and Tom Clagett, my guests, joined Ernest Lawrence on board, as did ten guys from Washington, D.C., as navy secretary guests, such invitations a common practice then—and eagerly accepted if a ship was going to Hawaii, which we weren't. Vlad Reichel and Mac Stone, a Duquesne Instruments executive, friend of Adm. Burke's, would join us in Rio de Janeiro.

The next two days, I felt terrible, hoping it was only transient, but knowing that a rough life and the stress of war can take a toll, both then and later too. Still, 'Frisco was a favorite port. I'd seen it first in 1927, I told Ernest, as a midshipman on the *Nevada,* the ship we pulverized at Bikini. Leaving 'Frisco on the 20th, we'd had no fog coming or going, which folklore says is a good omen for a cruise. In San Diego on the 21st to load fuel, at a big dinner party that night at Trader Vic's, Admiral Schoeffel, my ally from his BuOrd days, said Felix Stump was chairman of and Mel Pride was on the selection board. A break for me? My sister Eleanor and her husband, Dan Pollock, rode a boat out to chat as we rolled at anchor in the Pacific's big swells.

On 23 June, going south at twenty knots, the sea glassy, the air getting hot, my captain's inspection forced me to order that ship cleanup would be first priority all the way to Panama. I was not about to turn a dirty ship over to Mayport. And my troubles "Before the Mast" were still chronic, but I vowed I'd get rid of the bums quickly. My crew had been drafted from other ships, and those captains had dumped on me as many misfits as they could. And others in the crew had signed on just to get a transfer to the East Coast, not caring how they got there.

We were hit with showers and wind on the 24th. Our position at 0800 hours was 17 degrees, 18 minutes north latitude; 102 degrees, 13 minutes west longitude, the intertropical front line. We were in the doldrums on the 25th, in squalls along the intertropical front, and my log entry adds, "What is it makes my blood run so fast? I should learn to sit and contemplate." In Balboa, Panama, on the 27th, its hottest time of year, we'd stay only two days, discouraging visitors, which was fine with me since the ship was still a mess. I had dinner at El Panama with Ferris White, an American Can executive, one of the two navy secretary guests who'd come aboard there. (I avoided the other, a "VIP," he told me, from the Washington, D.C., Metropolitan Club and all that, a "more money than brains" kind of guy.)

The review air force Lt. Gen. Harrison held for us on the 30th at Albrook Field was impressive as was that evening's reception, but the State Department canceled our stop in Peru, wisely, I thought. Somebody had tried to overthrow Peru's dictator, and State didn't want it to look as if we were taking sides. So we went directly to Chile, leaving Balboa at 1300 on 3 July, the temperature falling to the low sixties as we went south. It'd be much lower when we reached the Cape, midwinter in the southern hemisphere. But I was smugly content being captain of this lovely ship with no admirals or anyone else around telling me what to do.

And I learned the precept to be followed by my selection board, to begin its work in July, was remarkably straightforward: Pick the people clearly best fitted to be admiral. Still, it came down to whether my class was in the zone and, if so, how many aviators they would pick. As if to match my frets, reports said Cape Horn's weather was gale-force, blowing snow. I wanted to stay on the ship in Valparaiso, Chile. It's an exposed anchorage and, in winter, fronts can go through there in a hurry. But I hadn't factored in the Chileans. While we were at anchor on the 9th, our Ambassador Lyon, a Foreign Service type, came aboard. Our chat was enlightening, and I told my XO we need more career people like Lyon and far less politicians holding ambassadorships.

Next day, I met Chili's El Presidente Arias, who took me to lunch with his top military officers. I was impressed, but our big carrier had impressed them too. After lunch, they escorted me to their Bernardo O'Higgins statue to lay a wreath at its base. Purportedly Irish-American, he was the soldier-statesman who, in 1817 with Jose San Martin, freed Chili from a military junta put in by the Spanish after they themselves had been beaten by Napoléon Bonaparte. (San Martin also tried to "liberate" Peru, the only result being a dispute that's gone on ever since, occasionally with bullets, over where the Peru-Chili-Bolivia border is.) Chili'd bought one of our *Brooklyn*-class cruisers and, named the *O'Higgins,* it was their navy's flagship.

Invited on board, I met her captain and his wife, noting in my log that "she is a beauty as are all the gals down here." Then I erased that, recalling how mad Lili had been, at home babysitting, when Lt. Hayward on his first South American visit wrote her about how pretty the girls were. From the ship, they took us to Vina Delmar for a late-running cocktail party and I noted, "If I keep this up, I'll be dead by Friday." On "Black Friday" the 13th, we headed south for the Cape at 0800. A storm hit us on the 14th and for two long days we rolled and groaned in a gale, giant seas forcing me to slow down. By the 16th, we were thirty-two hours behind schedule.

Off the Straits of Magellan that day, the barometer falling, the temperature high only 41 degrees, the sea was a long, slow swell, indicating a storm out ahead of us somewhere. I was back up to twenty knots to make up lost time and also to be off Cape Horn on 17 July at 0940, sunrise there. The sun sets soon after that, only a small daylight window when I could give the crew a look at Cape Horn. And on the 17th off Cape Horn as scheduled, the squalls eased off to let my crew come topside. After I photographed the Cape, I went below to update my diary.

I had finished writing, "My left eye has bothered me for a week and seems to get no better," when my orderly handed me a dispatch. It said I was one of those the selection board had recommended for promotion to rear admiral. "What a thrill," I jotted down, "after 31 years!" In an outfit where captain is tops for most people, I'd gone from apprentice seaman to admiral. We were at exactly 58 degrees, 23 minutes south latitude, due south of Cape Horn, when I got the word. I knew Lili and my mother would be happy but that evening I received the most touching tribute of my career. The chief petty officers invited me to dinner at their mess where they gave me an admiral's hat they must have bought in Bremerton. "How did you know so early?" I asked. "Oh, hell," one said, "we always knew you were going to make admiral."

I did not let them see the grousing and bitching in my diary almost every day that year, over "not being selected." If I had, they'd probably have carted me off to a padded cell. But it wasn't so pleasant two days later, going into the Straits of Magellan at Drake's Passage, where eighty-five-knot winds hit us and did so steadily for a week, forcing us to tie down our aircraft tightly and hold the ship to barely the twenty knots needed to keep the deck dry against an across-the-bow sea. We also ran into a six-foot-deep ice field and I thought the ship was going to sink. It was dark twenty-four hours a day, and the ice told me to head east "Right Now!"

On the 21st, at the tip of South America's eastern coast, we learned that Antarctic gales had pushed a two-hundred-mile-wide ice field up against the Falkland Islands. As we swung east by Port Stanley to miss it, I asked, why would anyone want to fight over that place? (The Falklands had been British for some 250 years, but off and on since 1831, sometimes with a flex of military muscle, Argentina had claimed that the Falklands belonged to them.) By the 22nd, we were north of there, the weather warm. And my left eye was still acting up. That old injury from my landing on the *Saratoga* in 1935 was my weakest link, the first to hurt when I got tired.

On the 24th, we flowed into Rio de Janeiro, the Paris of South America, and I began the courtesy-call bit. The first and happiest was on our ambassador to Argentina, Ellis Briggs, my pal in the "George Cruz" drill, a superb man, now often-praised ambassador. Then I got word that a retired Rear Adm. J. B. Sykes wanted to visit. Yes, Vlad and Ernest Lawrence said, the same miserable Sykes who'd given me my only "Unsat" fitness report ever. Reading his note, I said, "I'll fix that character," but Ernest and Vlad said, "Like hell you will. You will put out the red rug, shoot the guns for him and all the rest."

So I broke out the guard and band and gave him a welcome-aboard as if he'd made rear admiral before he retired, not in fact on condition that he retire. I'm glad I did. There's no profit in nursing a hatred. He was there visiting his daughter who'd married a Brazilian, a fine young man Sykes brought with him. As he left, he said of my promotion, "I always knew you'd be a success." Quickly, Vlad put a hand on my arm to make sure I didn't do anything rash.

On the 28th, we headed north, and half the days until 8 August when we docked in Mayport, I sat in my dark cabin, nursing my eye. The doc, not an oculist, said I had iritis. All our families including mine met us in Mayport. So did a Jacksonville welcoming committee, since a big carrier basing there was a first for them. Two days later, I went to the naval air station hospital, where the doc said my eye seemed to be mending. I thought he was just trying to cheer me up, though the eye did feel better than it had earlier. Still, the doctors might tell me I'd have to take a medical discharge. Or I might have to give up flying. Or they might just give me a shot or something to cure the problem; but I knew, no matter what, I'd be deeply disappointed if it meant I had to give up my ship. Three days later, on Monday the 13th, Dr. Lawrence, a very caring, careful man, had an operation done on my eye. For two weeks, I lay in the hospital, very discouraged by the whole business.

By 27 August, I was back on the *FDR*, and by the 29th we were at Gitmo for training, twenty-five hundred men in the ship's company and twenty-two hundred in the air wing learning to work as a team, basically. We'd be at it until 21 October when we were due back in Jacksonville. Flight operations began at 0630 and we flew all day. A typical night operation was a radar approach into "alpha," Guantánamo. After the long trip from Bremerton and a mere two-week leave before Gitmo, the grind inspired lots of guys to suddenly visit sick bay. I did give them a weekend, twice, in Port-au-Prince. The other weekends were at Gitmo where one Sunday night what began as a beach beer party ended up a drunken brawl, just like the "old navy" used to be in my bluejacket days.

On our trip, I'd called the *FDR* "my lovely lady." As we worked, I began to back away from that nickname. In spite of her upgrades, she was still basically a World War II ship with all kinds of engineering troubles and equipment breakdowns—obsolete, really, unlike the *Forrestal* being built, our first carrier designed from the keel up to handle jet-powered aircraft. Our aircraft were piston-driven old ladies, too. Even so, with the Suez Canal hassle turning more critical by the hour, we'd probably have been deployed to the Med if we hadn't been in training. Egyptian President Gamal Abdel Nasser had seized it on 26 July 1956. Behind him was Moscow, selling weapons and fighter aircraft to Egypt, Syria, and Iraq. Nasser's piracy had provoked the Israelis to attack across the Sinai desert. As the Israelis reached the Red Sea, British and French warships off Port Said landed paratroops on Egypt's coast without either one telling us what they were going to do.

In a meeting with Ike and Secretary of State John Foster Dulles, Admiral Burke had argued, "Nasser has to be shown thievery is unprofitable and violation of international commitments will not be tolerated." (Nasser earlier had agreed to let an international commission control the Canal.) Burke wanted Ike's okay to give logistics support to the British and French. Dulles had argued that the best way to restore order was to make them withdraw. Ike did what Dulles wanted—later saying it was his dumbest foreign policy decision as president—and Britain and France angrily pulled out all their forces. Back in his office that day, 30 October, Arleigh had wired his Sixth Fleet commander: "Situation tense. Prepare for imminent hostilities" and "Cat" Brown wired back: "I am prepared for imminent hostilities but whose side are we on?" Arleigh's answer: "Be prepared to fight anybody."

Meanwhile from our Gitmo training site, I flew to Jax twice to have my eye examined, knowing, after I cheated reading the eye chart, that it was dying. On 2 November when it broke down again, I was off my ship, relieved by Tommy Hopkins. When it failed a third time, they told me the August operation had saved my left eye's optic nerve, but the eyeball was dead. They suggested I consider a cornea transplant. (Five years later, they said operating might make the right eye fail, too. So since 1956, I've been like Popeye, a one-eyed sailor.)

In Europe that fall, Poland and Hungary revolted against Soviet domination. Their under-equipped forces were crushed and thousands of hapless, unarmed Poles and Hungarians killed. Then, in what Ike called "pure, barbaric vengeance," the Soviets, violating a pledge of safe conduct to the West, executed Imre Nagy, leader of the Hungarian uprising. Meantime, the Haywards were at ease in Admiral Ward's house on Duval Street in Atlantic Beach, east of Jax, with the children in school, Mother visiting, and me going to the hospital for eye treatments.

In December, I was told to be in the Pentagon in January 1957, where they were putting me in the Strategic Plans business. That wily Burke, I growled. I'd wanted to keep the *Roosevelt* since she was to be two years in the Med, one of my favorite spots in the world. Still, the nature of my first assignment as an admiral hinted that they wanted me for future use. Otherwise, they'd have put me in a lab. At least in Plans, I thought, "You can say what you think."

17

The Pentagon: A Place to Pass By

On 7 January 1957, I reported to Adm. Ruthvan Libby, OP-06 (Assistant CNO for Plans and Operations) in the Pentagon, a place I'd tried to avoid ever since navy headquarters moved to there in 1947. Libby's Strategic Plans deputy, -06B, and my boss was Adm. Charles A. Buchanan.[1] After renting for $175 a month a house in Arlington, Virginia, I spent the month learning to cope with "the world's largest office building." And agreed with the adage, "Things happen very fast in the Pentagon but it takes a long time to get something done."

Its "Puzzle Palace" nickname fit at least one activity, Joint Strategic Planning. The first "plans" I read were just pleas for a piece of Ike's defense budget, which had air force first among equals, navy second, army a weak third. The planner types avoided the hard question: would we "nuke Russia" if it attacked Europe? French leader Charles de Gaulle thought not. That's why he wanted his own atomic capability. And the Pacific was secondary to them even though our latest war had been in Asia. Nor was I popular, arguing that our island nation needed a strong navy, a carrier task force being the best and often the only practical way to handle a crisis overseas. Our own Old Fuds were busy refighting the battle of Midway. But our CNO wasn't. Arleigh Burke had a very perceptive feel for how the so-called "technology revolution" could alter navy plans, doctrine, and force structure.

OP-06B's job was, in part, to reflect that in studies of every situation where navy forces might be used, for whatever. Head of Strategic Plans from 1951 to 1954, Burke had said of it, "Planning for a conflict short of general war is far more difficult than for all-out war, a rather clear-cut proposition. It is much harder to visualize situations complicated by unpredictable political and economic factors where there's hardly a clear-cut, easy solution. It's extremely difficult, for example, to think through a future Korea or Indochina, Formosa or Middle East."

In February, I took leave of that enigma to go help Lili move to Virginia. While in Jax, I was coaxed into buying a house in Atlantic Beach for $16,000 with a $3,000 deposit I also had to borrow, putting us once more deep in debt, me wondering why Lili puts up with it and how she keeps us afloat. (In 1998, the IRS claimed that beachfront house was worth $615,000.) Later, in the Pentagon, Admiral Stump told me that Captain Sykes had cinched my being put on their admiral's list. "If he'd given you a good fitness report," Felix said, "we wouldn't have selected you."

My first assignment was to write plans for the Middle East. In a 1953 reorganization, the authority to specify theater commanders was taken from the Joint Chiefs and given to the secretary of defense. CincusNavEur (Commander in Chief, U.S. Naval Forces, Europe), Adm. Holloway, was specified commander for the Middle East, tied to NATO South in Naples, but his only command, really, was the Sixth Fleet. In May, while diddling with the plans for the Middle East, I was named navy member of the tri-service JSPC (Joint Strategic Planning Committee). Like the other services' Plans folks, -06B wrote position papers for my boss or Libby to show the other service deputy chiefs for operations (OpDeps) at their weekly meeting in "the Tank," as they called it. The JSPC debated the papers before, in our case, my boss gave ours to the OpDeps. Once a paper cleared the OpDeps, it went to the Joint Chiefs who at the time met only twice a week. So "positions" piled up unresolved. A more inefficient red-tape process I could not imagine.

I took some time off in mid-1957 when Ed Teller and I were summoned to an abandoned auto-repair shop in D.C. by the "Company," what some called the Central Intelligence Agency (CIA). (Less enamored "spook" employees called it a "Glue Factory.") There we studied pictures of Soviet atomic-production plants taken by a Lockheed U-2 reconnaissance plane they'd had overflying Russia for some time.[2] The photos could help reveal how Russia was building nuclear weapons. Also, radiating U-238 to make plutonium emits a gas that, in the air, gives rise to an isotope called krypton 87. When the U-2 told us how much was there, we knew how much plutonium was being produced. (They capture the gas now, so we can't do that any more.)

By June, I'd lost interest in the rain dance. I told my chief that nobody ever read all the reams of paper I was grinding out, and bet him a martini lunch that I could prove it. He asked how. "On page eight of a ten-page paper I'm doing, I'll say, 'The solution to all our Middle East problems is easy. Just trade Israel the Gaza Strip for Miami.'"

If we don't hear from anyone within a week, I told him he'd buy lunch. After six days' silence, late on deadline day, Admiral Felt called a roomful of us to his office where he began reading my paper, droning out the fatal sentence as if by rote. As he did, laughter roared but it was obvious he hadn't, until then, read my paper. Immediately, he ordered everyone else out, but I jumped him before he could me. You just proved, I said, writing these papers is wasted effort. He said I didn't understand the bureaucratic process, but he did calm down, even chuckled. Actually, I thought it was a pretty good swap. 'Course, I doubt if the Cubans would have.

Our main JSPC drill was the SIOP (Strategic Integrated Objectives Plan). It was supposed to plot ways to gain our Joint Strategic Objectives. But no one gave us those. So we debated tactics, the military "limited war" role implied in George Kennan's "Containment of Communism" plan, and what Soviet act would justify Eisenhower's promise of nuclear war. The "I" in SIOP implied that we agreed. Actually, we'd just agreed not to disagree in public.[3] In sum, a SIOP echoed budget ambitions, but rarely noted real-world politics and military or technology trends. My diary notes, "We're just shovelling feathers in the JSPC."

On 30 June, I sneaked out to Burning Tree Country Club in Maryland to play golf with Gene Sarazen, inventor of the sand wedge, and by then in golf's Hall of Fame.[4] Ike was there already, practicing. "I'm glad to see," he said, "someone in the Pentagon has some sense."

I kept my mouth shut. Used to being boss, he could be mean to whoever in the outfit made him mad. Anyway, I'd soured on his presidency. Unlike Secretaries of Defense George Marshall in 1951 and Robert Lovett in 1952, Secretary Charlie Wilson ran the store, and left up to Eisenhower the setting of strategic objectives, military plans, and policy. But Ike had become pretty vague in that department, great at holding elaborate National Security Council meetings, slow at deciding.

Besides, I had my own beef by then. I'd been selected a year ago and the Senate had yet to vote on it. A rear admiral's pay doesn't start until they do. (My diary notes, "If my debts don't put me in jail, I'll be lucky.")[5] The whole routine was callous. There was a limit on how many generals and admirals we could have, so new ones weren't let in until we fell below the cap. Even then, we'd be just "lower half" rear admirals, same as a one-star brigadier general, though all rear admirals wore two stars. But if a "lower-half" isn't promoted to "upper half," equal to a major general, within five years, he or she must retire.

Also in July I was told I'd replace Adm. Felt as OP-03, Assistant CNO for R&D (Research and Development) in October. Given the R&D itch Burke and new navy Secretary Tom Gates had, I got into it right away with several trips overseas. One reason was that OP-03 supervised our navy role in a MWD (Mutual Weapons Development) program to codevelop military hardware with other NATO nations. And Adm. Gerald Wright, then CINCLANT, wanted me to go for various reasons. So did Arleigh, mainly to tell and sell our flag officers overseas on our R&D program. So in my last months in -06 and all my ensuing years in R&D, I was both a doer and seller of the doing.

The record on my first junket, to London, Paris, Brussels, Oslo, and Bonn said, "Norway's R&D effort excellent; French electronic work ahead of ours; the ASRE (Aviation and Space Research) Lab at Farnborough, England, one of the best, fine technical people." I added, like a news analyst, "Russia signed up Syria on the 26th. If it goes communist, we can lose the Mid-east. Indonesia is falling apart and I give India only 5 years. Religions cause so much trouble: Sikhs vs. Hindus, Muslim-Arabs vs. Israelis, North vs. South in Ireland. It won't end in my lifetime. As Father Farley told me years ago, 'Never confuse the Lord with his followers.'"

I was back in Europe on 1 September for the Farnborough Air Show, then to Paris to pick up Ed Teller and fly him, on the 6th, to Athens to meet Vlad Reichel; Herb York from the Lawrence Livermore Laboratory; Sandia's Bob Pool; and Ernest Lawrence, "a neat guy," to me, far more politically savvy than Ed Teller. From there, we all went COD (Carrier On-board Delivery) in an AD-5 Skyraider modified to carry a few passengers, out to the *Coral Sea*. I had Teller in the copilot's seat as we whistled aboard. Next, we helicoptered to Sixth Fleet commander Cat Brown's flagship, the cruiser *Salem*. Admiral Burke had devised this tour and had me lead it because I knew all those people. Its purpose was my missionary work with the fleet admirals in reverse—to show the technical people how a task force operates.

After dinner and a soft, moonlit night on the Grecian Sea, Cat told my friends eloquently in an all-day session how hard it would be for Soviet atomic bombers to destroy a task force. Afterward, Teller said he'd already sent *Newsweek* magazine an essay on how helpless surface ships are against submarines. Arrogance, I muttered, experts in one field posing as expert on everything. But Cat just laughed, "knighted" Teller "the new Don Quixote," and to end the meeting, quoted Edmund Burke: "He who calls in the aid of an equal understanding, doubles his own; and he who profits by a superior understanding raises his own to that level."

Then, to educate Edward, we put him in a submarine with the *Salem* as its prey. In the first drill, with the boat abaft of her, the ship at thirty knots simply outran the fourteen-knot boat. (Some of today's nuclear-powered submarines do thirty knots submerged, but they didn't exist in 1957.) Then our skipper shut down two of his four screws so his ship sounded on sonar like a freighter. The boat, hunting for the *Salem*, ignored us and we ran right over it. The *Salem*'s sharp skipper was Capt. Frankie Williamson, one of three bluejackets—him, me, and Al Cunningham—that Coolidge gave his "Persistence" plaque to in 1925. Williamson had failed the academy entrance exams the year I passed, but repeated and did pass in 1927. A down-check at Norfolk had kept him out of flight school, but of six Williamson boys in the navy, he was the first to become an officer.

Back in Washington on 1 October to become, officially, assistant CNO for R&D, I told Arleigh Burke about Frank's success, adding, "I'm sure there are a lot of smart enlisted men in the navy, right now, not all under twenty, who ought to be officers."

"You're always saying that, Chick," he said, "Why don't you do something about it?"

So I started the Naval Science and Education program. Charlie Martell, who'd made admiral the year before I did, Joe Jaap, and Rivets Rivero were all in the personnel business at the time, so getting my package approved was easy. Under the rules, a sailor as old as twenty-seven could join. The swap was, if he agreed to go to OCS (Officer Candidate School), we paid his way through college. The first of many fine people who signed on was Seaman First Class Jeremy "Mike" Boorda. We sent him to the University of Rhode Island for an oceanography degree and in 1994 he was the first bluejacket, non–Naval Academy graduate to be CNO.[6] (That Science program no longer exists. To encourage all the young talent in the navy, I think it should be revived.)

That fall, 1957, I also was mired in endless meetings with the navy bureaus and other agencies, trying to shape an R&D budget for Congress. And Arleigh told me to get to know all about all our programs so he didn't have to ask twenty different people every month what they were doing in R&D. In budgeting, all the combat specialists, pilots, sailors, missileers, and so on, all had their pet rice bowls, of course, competing for funds. Then, after resolving those priorities, we had to sell our package to the defense secretary echelon and White House OMB (Office of Management and Budget). And in January each year when the defense budget request hits Congress, the battles start anew.

First, through the House and Senate Armed Services (now National Security in the Senate) Committees, Congress authorizes what's to be funded. Then its two Appropriations Committees vote how much to spend, largely a repetition of the arguments to get authorization in the first place. This two-step drill is supposed to be completed by 1 July, back then the start of our fiscal year. It almost never was, which meant an "Apportionment" fight with

OMB. In that, if funds are appropriated by, say, November, OMB may say, "Admiral, you can't spend all that in what's left of this fiscal year, so we'll give you just 40 percent of it." Then we start "Obligations," issuing and continuing contracts; and "Expenditures," payouts.

And since the Constitution says a Congress may not authorize an appropriation to be spent beyond its own two-year life, yet building a ship, for instance, takes about four years, all the same debates have to be repeated no less often than every two years. And of course, each year, new programs are added to the list. In sum, during a calendar year, we were spending old money held over, new fiscal-year funds just appropriated and allocated, and preparing a budget submission for the fiscal year starting some sixteen to eighteen months in the future. In my R&D business, the easiest cash flow to track was basic research since the wise way to run it was as a yearly level-of-effort. For the rest, my tour in the R&D business taught me a lot about how to juggle money.

As I waded into that swamp in October 1957, it also put me at the door of the Vanguard satellite launch, an orphan to the ballistic missile program that itself had a hectic history. The ICBM (Intercontinental Ballistic Missile) work had nearly stopped in 1949 after a three-year, $2.3-million study concluded that an inertial guidance system such as Vlad Reichel and Stark Draper were studying, to put a missile accurately on a target five thousand miles away, was ten years in the future; we didn't know how to prevent the bomb's burning up at atmospheric reentry; we didn't have a booster with ICBM range. If a 200-pound missile can hurl a one-pound bomb five thousand miles, they said, we'd need a 1,200,000-pound missile to throw a 6,000-pound Mk-IV nuclear warhead that far.

The technology dam broke in 1952 when we blew our hydrogen device. From that in mid-1953, Defense Secretary Wilson ordered a review of guided missile projects. The ICBM portion was done by a science group chaired by my old "airsick copilot," Johnny von Neumann. It reported in February 1954, that off Teller's "Alarm Clock," they'd soon be able to put a megaton-yield nuclear warhead in a missile nose cone; and problems in guidance, booster power, and a way to prevent the warhead from burning up at reentry, all would be solved within a few years. Even I sort of got into that act. When I was in JSPC, I wrote an essay, rebutting the air force demand for first call on atomic material. "Demand and technical gains will supply all the active material needed for as many atomic bombs as we want," I wrote. "The problem is missile size and weight but, as it gets lighter, the guidance more accurate, they will become harder to stop. Even clandestine delivery will be possible some day." My colleagues deemed my forecast "irrelevant to the SIOP."

Off the von Neumann study, Ike set up a Strategic Missiles team, headed by James Killian, president of the Massachusetts Institute of Technology (MIT), to assess the merits of possible missile developments, worldwide. My pal, then-Cdr. Pete Aurand, funneled navy missile data to them. So, of course, its report in 1954 included a call for a fifteen-hundred-mile-range FBM (Fleet Ballistic Missile). So did a NOBS (Naval Operating Base Study) that Ed Teller had helped do. But to fire the cruise missile it wanted, the boat had to surface, its most vulnerable location. A submerged ballistic missile was far better.

Then in mid-1955, Killian, now Ike's science advisor, urged two competing ICBM programs be run concurrently to insure against failure and obtain the weapon as early as possible. By 1956, the air force was working on Atlas and Titan ICBMs and the army's Jupiter had flown thirty-three hundred miles by then. But as an answer to Russia's twenty-five thousand combat aircraft and 170 army divisions, Ike's first priority was to deploy in Europe an IRBM (Intermediate Range Ballistic Missile.) The air force Thor was to deploy by 1957, so Wilson put it at the head of the parade. For insurance, he told us to team with army on a Jupiter IRBM backup. That's why Arleigh visited army's Redstone Arsenal in September 1955, right after he became CNO.

Also, to end a BuAer-BuOrd squawk over whether a missile is an "air vehicle" or a "gun shell," Arleigh set up a SPO (Special Projects Office), a setup unique in the navy back then, to be our half of the team. And we were told to contract with Chrysler, the Jupiter contractor, for a liquid-fueled missile; even had an auxiliary ship picked out for it when I reached OP-03. By then, IRBM/ICBM R&D outlays, $14 million in 1954, twice the total for all of 1946 to 1953, soared to $1.365 billion by 1957. Then came the Gaither Report from a group set up by Ike in April 1957. Led initially by H. Rowan Gaither, Jr., of the Ford Foundation, it included Robert Lovett, 1949–1952 U.S. High Commissioner for Germany John J. McCloy, and MIT's Jerome Wiesner.

Their 1958 report said, in brief: the Soviets with a third our gross national product were spending as much on military programs as we were; had active material for fifteen hundred nuclear weapons and forty-five hundred jet bombers; had three hundred submarines able to reach our shores, surface, and launch long-range cruise missiles like our Navajo or supersonic Regulus II; had a strong air-defense system; had been making seven-hundred-mile-range ballistic missiles for a year; could launch, possibly by 1959, an attack on us by one-hundred-megaton-warhead ICBMs; and the report said also that our people and Strategic Air Command (SAC) bombers "are defenseless."

Meantime, in October 1957, I had appealed Wilson's army-navy team decision to Arleigh Burke. Jupiter, like Thor and Atlas, used a liquid oxygen (LOX) fuel. Cryogenics in a submarine, I said, was begging for disaster, and a bad idea on ships too. Mixing LOX with hydrogen got a specific impulse (burn rate) of four-hundred-plus pounds per second, almost an explosion. Besides, giants like Atlas take hours to fuel. We need a slow-burn solid fuel, I said, and NRL had been doing research on solids since 1948, slowly increasing the grain (cylinder) size they could cast without a crack in it. "NRL," I said, "now casts grains with a 240-pounds-per-second specific impulse, enough for a missile range of twelve hundred miles." When I then said that my science friends, von Neumann and Lauritsen, had tried and failed to get the army to use solid fuel, Burke asked, "What do you want to do?"

"We can't use Chrysler, and the air force missile guy, Gen. Benny Schriever, has all the airframe people chasing Thor except Lockheed. It's been dropped from the competition."

"Come with me," he said and, down the hall, we barged into Navy Secretary Tom Gates' office. There I repeated my tale and Gates asked, "Why don't we call Courtlandt Gross," and he promptly called Lockheed's chief executive to ask if he was interested in doing the airframe; then Dan Kimball, Aerojet General's chief executive, on the solid propellant. They were in his office next day, an all-night plane ride since we didn't have jet transports then. We gave each a $100,000 check to start what was revamped into our FY 1958 budget as the Polaris FBM program. We ordered the nuclear boats from Hyman Rickover, of course, and he told us, "Don't put a submariner in charge. They don't have any imagination."

So Arleigh picked an aviator, Rear Adm. William Francis "Red" Raborn, class of 1928. Red and his technical director, Capt. Levering Smith, lined up more than two thousand contractors and set up a routine called PERT (Program Evaluation and Review Technique). For it, each month, piece-part makers reported to them any problems meeting delivery dates to subsystem builders, who did the same relative to subcontractors and the Lockheed prime. And Red visited them all regularly to urge the workers on, "Twisting the tiger's tail," he called it. The payoff of it all was the first Polaris fired under operating conditions, from the *George Washington* submerged in July 1960, three years ahead of schedule. Noted ex-Defense Comptroller Wilfred McNeil, wryly, "Polaris went into production faster than the army's new M-14 rifle."

But for seventeen months after that, Defense Secretary Robert McNamara wouldn't decide how many sixteen-missile FBM boats Red should buy. So I took Red to Sen. Henry "Scoop" Jackson, and he agreed to get Red money to build forty-one boats at a one-a-month rate. McNamara was sore, of course, and tried but failed to make the navy buy them from its already-approved budget. Nor did I apologize for end-running him. Jackson's decision stabilized the program with Red able to save millions by ordering costly, long-lead-time items in quantity. I did spend $58 million to adapt our first nuclear-powered ship, the cruiser *Long Beach*, for Polaris, but we didn't put it on her. We should have. Why have a missile stuck in a fixed easy-to-target spot on land when it can be mobile at sea? Arleigh did get Italy to put tubes in some of their ships, but at home, since we had the mobile weapon, anyway, we didn't press the point. I did learn that one way to get Arleigh mad was to call the *Long Beach*-Polaris project "Burke's Folly."

Meantime, in August 1957, Moscow announced a test "to its full range" of an ICBM, and Atlas wasn't due to try that until 1958. It showed that Russia was way ahead of us in developing powerful boosters. Then four days after I joined OP-03, on 4 October 1957, they launched Sputnik, Russian for "traveling companion," the first man-made satellite to orbit the Earth. A month later, they put up a bigger one, a half-ton capsule with a dog in it. That shook the country but Ike saying it wasn't a threat to our national security amazed me. Where was his science guy? To escape the Earth's gravity and go into orbit, a vehicle must go twenty-five thousand feet per second. For a suborbital ICBM/IRBM trajectory, just lower the propellant's specific impulse.

After Sputnik II, White House Press Secretary Jim Hagerty issued a press release saying we'd launch Vanguard on 2 December. Who told him so was my mystery of the day. I promptly told Secretary Gates, "Look, that's a test shot. We don't even have the satellite, yet. What it will carry is a transponder ball so army radar stations downrange can track its course and velocity. And the one rocket stage we had to build ourselves is untested, and we've never had all five rocket stages together until now. It could easily blow up in front of God and everyone."

And on 6 December 1957 at its Cape Canaveral launch pad, the Vanguard rose maybe six feet, then burst into ball of flame and ink-black smoke. As it crumpled, "The nose cone toppled over," said one reporter watching on television, "looking remarkably like a dunce cap."

From day one, it had been hampered by dumb decisions. When Ike announced in July 1955 that our contribution to the July 1957 to December 1958 International Geophysical Year would be a satellite in orbit, Moscow just hours later said they'd do that too. Then Dr. Clifford Furnas, involved in picking a Vanguard rocket, vetoed an offer to make it a joint army-navy effort. "The engineering obstacles to achieving a launch would be less difficult to overcome," he said, "than the obstacles of interservice rivalry." Phooey. All the odious "interservice rivalry" I'd seen or been in had been in Washington, scrapping for a slice of a budget pie. In R&D, "interservice rivalry" usually had been very helpful. It was a reason Polaris had beat its schedule, and the army said Thor competition was why Jupiter had deployed a year early.

Anyway, under fine program director John Hagen of NRL, Vanguard was a pretty solid program in 1957. For its second stage, which NRL itself had to make, it created UDMH (unsymmetrical dimethylhydrazine), hellish to work with but the first storable liquid fuel. Vanguard 1, a navigation aid launched in April 1958, was still orbiting decades later. The only other two Vanguards put up, in 1959, proved that a satellite could gather weather data, measure the Earth's magnetic fields and the Van Allen radiation belts, and track micrometeoroids. About all Vanguard didn't do was what we never were asked to do: beat the Russians.

Donald Quarles, defense deputy secretary from 1957 until he died in 1958, prevented us doing that by ruling that we couldn't use any IRBMs. If we'd had a Thor, Hagen could have had a satellite up in 1956. Indeed, after a Jupiter test in 1956 showed it could, too, Quarles made the army sign an affidavit, promising not to do that. Vanguard's failure brought out the dragons, of course. In January 1958, Sen. Lyndon Johnson's Space Committee began hearings on, "Why the United States is behind in the missile-Space race."

One clue to the mood swing: in 1956, Congress had said a navy request for $20 million to start Vanguard was "a frightening waste of the taxpayer's money on a scientific game." In 1958, Congress railed at the Pentagon about the $110 million we'd spent on the program by then being "callous pennypinching stupidity." Nor were they impressed by the army's using a Jupiter to orbit the Explorer I satellite on 31 January 1958. The "Great American Public" was appalled, angry. That galvanized a typical political reaction, a rash of Congressional hearings, raving critics ignorant in the science but anxious to be on the evening TV news. I especially resented snide remarks by those in the Pentagon who'd put Vanguard in a box in the first place. Predictably, their testimony to the Johnson Committee, which took all of February, evaded that issue.

When the Committee finally did get to this just-caught rear admiral, it was some comfort to see Henry Jackson, Stu Symington, Margaret Chase Smith, and other longtime friends on the panel. But I was sure I'd get whacked. Rickover, an expert at this sort of thing, advised me, "Never agree to testify except under oath and never, ever give a Congressional committee a prepared, written statement."

I did it anyway, on 3 March 1958. Committee Chairman Lyndon Johnson began the grilling with a sarcastic, "What's the matter with you people?"

The Vanguard blowup in December didn't really hurt the program, I answered, "Since, Mr. Chairman, we didn't have a satellite on it."

"You mean to tell me the president lied to the American people?"

"No, not at all. He was just misinformed."

From then on, the questions fired at me implied that I'd caused all Vanguard's problems. I fired bullets right back at them and a blast at Quarles and others who'd denied us use of proven IRBM rockets long before I was OP-03. After that, Johnson adjourned the Committee, sending me back to the Pentagon in his big black Cadillac. There, Neil McElroy, defense secretary since 9 October 1957, bluntly summoned me, Gates, and Burke. As we went to see him, Arleigh said, "Boy, if you say anything, I'll punch you."

And McElroy really let me have it about loyalty and the rest of the normal "hoomalimali." He also said he'd nominated me for a third star and was very disappointed by my actions. That's the first I'd heard of that, but I didn't deserve criticism. I didn't make policy. My job was to follow it. But because I was under oath, I had to tell Congress what I honestly believed. Ours is the only democracy, by the way, where the military testifies under oath to the legislature on the executive's policies. Congress is wrong to make an officer do that—"legalized insubordination," Ike called it. Anyway, two weeks after that session, I was invited to a Pentagon meeting where Gen. Benny Schriever, the top air force missile-development guy, was to tell us about their new Minuteman ICBM. I said as I entered the room, "The only minute man I ever knew was the guy who double-parked outside a whore house and left his engine running."

That got a big laugh except from the air force generals who frowned and growled, "Who's this navy character?" What Benny said was, encouraged by the success of Polaris, Minuteman would use solid fuel. Turned out, a Polaris man had supplied most of what Schriever used to justify his decision. Later, the army said that after Jupiter, it would develop a solid-fuel Pershing, so-named, said a colonel, because "The air force won't touch it with a ten-foot pole if we name it Pershing."

Like air-cooled versus liquid-cooled aircraft engines in World War II, the navy went solid fuel, army–air force cryogenic, proving anew that competition in R&D yields big rewards. Admiral Burke often backed two guys chasing the same goal in differing ways and, until one got there, let both think he liked both. The tough question, of course, is when to cut it off. A new idea will have everybody saying it won't work. That's what we were told about solid propellants and for years they were right, but we kept at it and it paid off.

That got lost in Congress' media-fed reaction to Sputnik. There, the consensus was, "The Russians beat us because we weren't organized properly and had all this interservice bickering." "Reorganization" became the biggest game in town. By late 1958, a truckload of it was in place. Oddly, ours, Congress' Defense Reorganization Act of 1958, enacted in August, affected most not R&D, but the command structure. Its keys: (1) The president and the secretary of defense, aided by the Joint Chiefs, "will establish and decide the forces assigned to unified and specified combat commands," their commanders to report directly to the defense secretary; (2) a "truly unified" six-hundred-person Joint Staff, set up along conventional staff lines—J-1 for Personnel; J-2, Intelligence; J-3 Operations; and so on—to support the Joint Chiefs whose strategic planning and other JCS work "shall take precedence over" their separate service roles.

It was Ike's heavy hand at work, really. He wanted a single military Chief, the JCS chairman, with a single military staff like the German General Staff the army seemed to admire. Fact is, in World War II, it didn't run the show. Hitler did. The German General Staff just deployed forces. Besides, we also beat the Japanese, who had two staffs; and the Italians, who had three. Burke opposed it since it made the military departments mere suppliers of people and hardware to the combat forces, and put a CNO outside the chain of command. Being the tough-minded leader he was, he didn't buy at all the idea of agreeing to a plan and then making someone else responsible for carrying it out. The idea of having a CNO had first come up in 1915. Over nasty opposition in the navy secretariat, it was created but that CNO was not given any money or tools, except his own prestige, to do the job. The 1958 act took us back to 1915, the CNO once more becoming just chief of naval staff in Washington, D.C. Burke was our last real CNO.

What the act did in R&D was more encouraging. It raised the assistant secretary for R&D to defense director of research and engineering (DDR&E), later naming my Lawrence Livermore Lab director, Herb York, as the first DDR&E. In Pentagon jargon, an assistant secretary advises the boss. A director can issue orders on behalf of the boss. More importantly, the DDR&E would receive all appropriated R&D funds and distribute them to the services. My diary notes, "This won't be the last reorganization, I'm sure." It wasn't.

Then in mid-August, my left eye flared up again and I had to spend a month in Bethesda Naval Hospital. The doctors wanted to retire me on a medical disability but Gates, bless him, said no. "We need Chick here," he said, "and he'll not be flying airplanes anyway," a promise I ignored. After that, Burke said he'd not give me sea duty because of my eye. Truth is, he just wanted me around. He thought I knew where a lot of bodies were buried. Maybe I did.[7]

All was not quiet elsewhere in the world. In the spring of 1958, Sen. Henry Jackson lured me away from the turmoil, briefly, to be a member of his wedding when he married Helen Harding from Albuquerque, daughter of an air force junior officer. In August 1958, after the Lebanese government requested it, Adm. Jim Holloway himself came down from London to lead a smoothly run amphibious landing of a force to protect Lebanon's mixed Christian-Muslim population from Syrian incursion. And in September, Taiwan, Quemoy, and Matsu were under Chinese Communist fire. Ike sent in the Seventh Fleet to cool off the conflict, helped by Taiwanese pilots using our new Sidewinder missiles to clobber Communist fighters.

Meantime, at home, cascading down from the 1958 Act came the Franke Board, chaired by navy Under Secretary William B. Franke. Its package, proposed in May 1959, included folding the navy assistant secretary for air's role, among other tasks, into a new assistant secretary for R&D office. On the military side, they created a chief of naval material. What it did was merge BuAer and BuOrd into a single Bureau of Naval Weapons, headed at the outset by Rear Adm. P. D. Stroop. Before, BuAer and BuOrd had been private fiefdoms reporting to the navy secretariat who, being appointees who usually serve only two to three years, could be outwaited if a bureau got orders it didn't like. And the CNO, who knew best what the fleet needed, had no influence on the bureaus except through the secretary, since Congress funded them directly and money rules in that business. Now the money would flow through the CNO to Material to the Bureau of Weapons.

The Franke Board also divided OP-03 in two to create a chief of research and a DCNO (deputy CNO) for development. I had no problem there. The level-of-effort plan says basic research must be free to roam. Only when a piece of it seems ready for big-bucks development is it time for a DCNO to step in as, for instance, when Deak Parsons saw how radar could evolve from navy research on radio waves, and solid propellants from NRL's chemistry research.

Even before Franke gave his plan to the secretary of defense, in April Burke made me a vice admiral and DCNO development with Charlie Martell, my deputy in what we named OP-91. We set the approximately three-hundred-person shop up by weapon-system groups: ASW (Anti-Submarine Warfare); Naval Warfare; Supporting Warfare; Advanced Technology; Guided Missiles; and Nuclear Energy offices; and, of course, a Plans, Programs, and Budgets office. The Nuclear Energy Office, formerly the

CNO's Atomic Energy Division, was to be run as it had been there, by Dick Ashworth, the Nagasaki weaponeer. We also controlled a big pile of money, now not only R&D but T&E (Test and Evaluation). In short, we were boss of the OPTEVFOR (Operational Test and Evaluation Force) that the navy had set up late in World War II.

Meantime, in late 1958, NACA (the National Advisory Committee on Aeronautics) became NASA (National Aeronautics and Space Administration). Dr. T. Keith Glennan, president of Case Institute of Technology, was named NASA's first administrator, Hugh L. Dryden his deputy and a man I knew well. I'd done early test work on Dryden's first guided missile, called the Bat, when I was at the Naval Aircraft Factory. I'd fired it off an XBT-2, prototype of the Douglas SBD Dauntless dive-bomber. The main effect of the NACA-to-NASA change was to slow down the pace of work while people got themselves reorganized.

And the "space cadets" got carried away. They put their launch site at Canaveral where they can't do a polar orbit. If they tried and had to abort, they'd have large North or South American cities in the flight path. One result was that the only test firing under operating conditions has been of our navy Polaris. The space cadets also built or expanded labs everywhere, some simply because of who chaired the Congressional space committees. Today, they have the same problem the navy has: too many labs, upkeep costs taking maybe 40 percent of what should be spent on R&D. And the so-called "space race" made them neglect the preeminent role NACA had given us in aeronautics, though in 1959 they did complete creation of NACA's variable-geometry wing design. Stuck out like a conventional aircraft's for subsonic flight, folded back toward the fuselage for supersonic flight, that wing design ended up on the F-111 attack-bomber and B-1 bomber.

Of course, the House, in its wisdom, duplicated the Johnson Committee with a House space committee, chaired by Louisiana's Overton Brooks (the reason a NASA facility was built there). Soon, they had science and space subcommittees breeding like gerbils. I had to testify to all of them, of course, in the create-NASA debate and on the navy's R&D programs. Finally, exasperated, I told a Senate space subcommittee, "You have too many space panels, usually two or three to a subject. It's the most complicated way I can think of to do business."

Still, it was throwing boatloads of money at us. When I'd testified before a House subcommittee in December 1957, the entire Pentagon R&D budget for FY 1959 was to be only $525 million. So little to cover too many programs, I'd thought then. A year later, the $42.412 billion defense contracting authority for FY 1961 included $4 billion for RDT&E.

But concocting budgets was just scorekeeping on projects begun or expanded while I was in R&D business. One big program was ASW (anti-submarine warfare), "a top priority," Admiral Burke said. It was, too, for the British, and they, Burke and our CINCLANT headquarters, wanted us to set up an oceanographic laboratory. So in August 1957, I was told to fund it and pick a site. Dealing mostly with civilians for that nine-nation setup, I picked La Spezia, my maternal grandfather's home, since it had a fine harbor and a big Italian Navy base. We hired Penn State University to run it, but the deal said eventually NATO would fund it and name its director, with his deputy always to be an American. Over time, it has done excellent work in both the Mediterranean and the Arctic.

We needed it because, like air wars, the basics of ASW are these: find a target; fix its location; identify it to, as Vice Adm. "Jimmy" Thach[8] said, "be sure it's a burglar and not just a stray cat"; and report that to a sub-killer. But the ocean is opaque, sonar our only "radar." To function well, a boat skipper needs to know ocean currents and ocean-floor topography, know where underwater storms are and how water-temperature variants are affecting sound propagation. So does the aircraft, ship, or attack submarine that's chasing him.

By 1959, I was spending $132 million a year on ASW projects, millions more on R&D that might be useful in ASW. One reason was Hyman Rickover's nuclear submarines. A rule as old as a caveman's first spear is, while making a new weapon, work on ways to defeat it. The other reason was Russia. Hitler had begun World War II with fifty-seven U-boats and nearly isolated us. By 1958, Russia had 475 submarines its top admiral said would be their primary force "for control of the seas and enemy supply routes to its forces ashore." Not surprisingly, Russia's largest contribution in the International Geophysical Year was not Sputnik but in oceanography.

An ASW aircraft I ordered in 1958 was the P-3 Orion. Derived from Lockheed's four-turboprop Electra commercial transport, it was deployed starting in 1962. Its ten crewmen included five to work its APS-80 radar and ASQ-10 radome, air-dropped sonars, and a searchlight under the wing; and to handle the seventy-three hundred pounds of bombs, depth charges, and atomic or standard mines in its bomb bay, and twelve thousand pounds of ordnance or guided missiles (including the 75-mile-range Harpoon developed a decade later) under its wings. At its patrol altitude, fifteen hundred feet at up to 240 miles an hour, it can stay three hours on station at the far end of its 1,550-mile range.

18

From R&D to Cuba

The P-3 was and is an excellent ASW machine. My problem was being told to compete it among possible suppliers against only a performance specification, a purchasing idea popular then in the Pentagon because, they said, it let companies be "cost-effectively creative." Maybe, but how could I disqualify anybody on only that? I had to fall back on lowest bid, knowing, when we did, I'd best keep some cash tucked away to buy later the product quality we really wanted in the first place.

Over time, by using only performance specs, the services also lost an ability to write engineering specs. One example of their value was the mine cases I bought for the Ordnance Lab. A mine case has to be a very precise item. So before I went out, I asked our old Gun Factory folks, who knew the engineering, what it should cost. About $1,900, they said. But a Richmond, Virginia, boilermaker bid $600. They were solvent, not in default to the government, and, of course, powerful Sen. Harry Byrd of Virginia was in the act, so I had to give them the contract. Two years later, we had no mine casings. So I had to rebid the job. A capable contractor, AMF, bid just under $1,900 apiece—making me pay two companies.

Another myth popular at the time, still is, was that second-sourcing can beat the price down. In 1940 at the Naval Aircraft Factory, I second-sourced a lot, to increase production and broaden our supplier base. But it always cost me money. And can cost more. In 1958, I second-sourced the Mark 46 torpedo and the new guy put the veteran out of business. Only one guy makes it, now, and we'd need $200 million up front to get another one. But today's contracting red tape makes it so hard to get anything done. We did Polaris, Sidewinder, and the *Nautilus* each in less than five years. Now it takes ten—took twenty on the B-1 bomber—to get something from design to hardware.

In mid-1958, fighting the low-bid syndrome, I went to Europe again, mainly on MWD (Mutual Weapons Development) business. One project we ran was developing the British 1127 engine now in their Harrier VTOL (Vertical Take-off and Landing) fighter. Another was the Atlantique ASW aircraft. In Paris, Adm. Rene Bloch, whom I'd met earlier, introduced me to his uncle, industrialist Marcel Dassault, an alias he'd used to protect his family when he was in the French Resistance in World War II and kept after the war. Rene was in charge of the Atlantique, to be built by Marcel and have British Rolls Royce engines, and I ran the team making its all-U.S. instruments. They'd never have gotten the plane if I hadn't led that group and knocked a few heads.

On that trip, I also turned over to West Germany the Rhine River Patrol. Our navy had been doing it in specially designed boats since 1945. Mine was its last, a colorful trip from Koblenz to Sheerstein and Wiesbaden. Then in Munich, I flew a German light aircraft with pusher engines I could rotate down to about a 45-degree angle, like putting flaps down on a standard aircraft, and almost jump into the air. I took off from a street, flew all over the city, and landed on a street, "A very interesting idea," my diary says, "for short take-off and landing."

Back home, we sped up the McDonnell F-4 Phantom fighter development begun in 1955. To be armed with air-to-air missiles and use an already developed General Electric J-79 turbojet engine, it had to fly Mach 2-plus without burning up, yet be made of aluminum since the best heat-shield answer, titanium, cost a king's ransom. And it had to land on a carrier deck inside 750 feet. Since it was to be an all-weather fighter, we spent $100 million in R&D on its AI (aircraft identification) and fire-control radars. And, certain that the pilot couldn't fly Mach 2 and operate the radar too, we put a second man in it, renaming it the F4H Phantom II. (A navy aircraft directory says that was decided in 1955 when I was still at NOL. If so, why did a top navy R&D guy and a lot of fighter pilots bitch at me for making that decision?)

In 1958, that F4H Phantom II won out as our primary air-supremacy aircraft over the F8U Crusader, a good supersonic fighter but a dog to land. Deployment began in 1961, but it was rough getting there. It always flew in a buffered zone, so the pilot had to be careful. If he exceeds his "Q," one-half the air density times velocity squared, and easy to surpass at low altitude, the plane will rip apart as a couple did, trying to set speed records. We put slats in them to solve that problem. The operational F4H surpassed fourteen hundred mph at thirty thousand feet, climbed on afterburner at sixteen hundred mph, and became the most popular aircraft of that era, with five thousand built for countries all over the world.

I also ordered, in 1958 from Pratt & Whitney, a Mach-3 engine. They built the J-58, its thirty-five thousand pounds of thrust enough to drive a ship. We didn't have an aircraft for it, but I'd learned that engine work has to start first to have a smoothly run program. Navy never used it, but Lockheed's Kelly Johnson put it in the air force SR-71 Blackbird he started in 1960 and delivered in 1964. At its eighty- to eighty-five-thousand-foot cruise altitude, it can sustain Mach 3.2, thirty-six hundred feet per second, which just outruns twelve-hundred-feet-per-second rockets. (A Soviet MIG-25 fighter can do Mach 3.2 in spurts but then has to replace its engines.) The SR-71's titanium skin resists 300-degree temperatures and from eighty thousand feet it can photograph a continuous one-hundred-mile-wide strip of Earth with a resolution so sharp that street signs can be read. They gave me a commendation for the J-58 and a ride in the airplane.[1]

But in navy ships, I was snared in arguments from the time I entered that R&D arena in 1957 until after I left in 1962. Money, not technology, was the problem. In 1957, Congress whacked our shipbuilding budget. Needing 43 new ships a year to maintain the existing 864-ship fleet, we got funds for only 20 in spite of, as my diary notes, "Nothing but trouble in Laos and Indochina. . . . Indonesia is in turmoil as is India. . . . If Communists retain Syria, Arleigh thinks eventually we'll lose the Middle East. It's the most serious situation since Korea, he says."

My strongest allies, promoting nuclear engines, were Hyman Rickover and Deputy CNO for Air Bob Pirie. Everyone in the neighborhood beat on Rickover but Congress listened to him. By 1958, we had the nuclear cruiser *Long Beach* under way and Congress had authorized building the nuclear carrier *Enterprise* with eight DW-2 reactors in it. But after eight more two-DW-2 cruisers were built, they went to gas turbines since General Electric's LM-2500 engine, for one, meant no more getting up steam, just push a button to start the engine. It's also better for smaller ships, the destroyers. A DW-2 reactor is pretty big.

And the *Enterprise* was our only carrier victory. The *Forrestal* and its sister ships, the *Saratoga, Ranger,* and *Independence,* launched between 1955 and 1959; and four more so-called "supercarriers" commissioned between 1961 and 1964, all were fossil-fueled.[2] My argument was that the *Enterprise* had a nearly unlimited range; and carried more aircraft, double the jet fuel, and more weapons. A conventional carrier had enough bombs and bullets to fight for three and a half days. The *Enterprise* would have seven-plus days worth of ammunition. "We should build fighting ships," I argued, "not floating hotels."

But even Admiral Burke; then George Anderson, CNO, 1961–1963; and David MacDonald, CNO, 1963–1967, argued against me. The *Enterprise* will cost $380 million, they said. But a *Forrestal*-class ship costs only $280 million plus some $80 million more than a nuclear carrier to operate "over an expected 20-year useful life." (The *Enterprise* is still in service, today, thirty-seven years after it was commissioned.) And they believed that the only way to get the new carriers they wanted was to "buy cheap." But I stuck to the precepts from Coolidge's "Persistence" plaque; wrote several strong letters to the secretary of the navy; and testified to Congress on the subject. Anderson even raged at me once, in 1961, over a Norfolk newspaper article that was simply unfair. The reporter had asked me about the *Enterprise* and I told him four times, "See CINCLANT." But he dug out my well-documented views on nuclear ships and "quoted" them as if he'd interviewed me.

Meantime, Hyman Rickover had a war on, too. Adm. Grenfell, our senior submariner, and the BuShips chief, Adm. Al Mumma, wanted conventional submarines, hated Rickover, and fought his nuclear-boat program constantly. But he had the navy pretty much in a box because he wore two hats: head of the AEC's Reactor Division and BuShips assistant chief for nuclear propulsion. So he wrote AEC letters to the navy, then answered them for the navy, agreeing with himself, of course. Finally, he came to me with a way to end that feud and, more important, to give us a far better product.

"I need $20 million," he said, "to build a convection-cooled reactor at ARCO in Idaho."

The critical speed of a submarine is its tactical speed, how fast it can go and still listen. In the first nuclear boats, that was twelve to thirteen knots, the speed of a conventional submarine. The more power they pull from the reactor, the more water they have to pump to cool it. Above fifteen knots, the pump's noise drowned out the sub's sonar, while a sub hunter could track it easily. If the *Nautilus,* for instance, went all out, we could hear it half an ocean away. That's what was behind making a convection-cooled reactor. So I robbed the till, helped by "middle-management" pals in the budget shop, and Rick built his reactor. Some physicists said it wouldn't work but it did. Today, *Trident*-class boats, run by a convection-cooled A1-G reactor, can sonar-search while doing twenty-three knots. At a higher speed, the sonar is blanked by the flownoise of its own big boat moving in the water. A conventional submarine still may be the toughest to find since it can lie dormant. That makes a hunter go from passive to active sonar and if he does, everyone knows exactly where *he* is.

Meantime, as Jack Kennedy became president, Cuba crept onto center stage. For me, it had done so the first week in June 1957, when we had Ike, John Foster Dulles, CIA Director Allen Dulles, and some other VIPs on the *Saratoga* off Florida. I was the navy R&D-type on the trip, and flew home with the Dulles boys, who argued about Fulgencio Batista and Fidel Castro. Allen's hard-nosed point was, "The trouble with the State Department is you're so busy pulling down someone, you're never ready to work with his replacement. Don't even know what that next government will be."

That fall, Castro rebels seized a busload of our sailors returning to Gitmo from leave inland, why, the rebels didn't say. At a National Security Council meeting, Admiral Burke asked Ike's okay to send marines out to rescue our sailors. John Foster Dulles, for one, objected, claiming Castro was a reformer and a potential ally, and Arleigh snapped back, "He's not a reformer. He's a communist son-of-a-bitch and we've got to stop him."

Ike denied Burke's request but, without comment, our sailors were released anyway. By 1958, ex-President Truman, national newspapers, and academe were praising Castro as "the best Cuban liberator since Jose Marti." And on 1 January 1959 when Batista fled the country, Castro took over, vowing, now avidly communist, to export his "success" to his neighbors, and military arms began to flow in from Russia. By 1960, when Guatemala and Venezuela asked us to help them fend off Castro and our ships stepped up Caribbean patrols, Ike decided Castro had to go and told the CIA to organize a team of CIA-equipped Cuban exiles to invade Cuba.

Then in November after Kennedy's election, Admiral Burke took me with him to the White House for a briefing to Kennedy on the CIA mission. The reason I was Burke's horse-holder was that back then, the CNO shop had a designated flag-rank duty officer each week, and I had the duty that day. A CIA guy did the briefing. Kennedy acted uninterested, half asleep, said nothing at the end, and I wondered, what the hell goes on here? Driving back to the Pentagon, Arleigh asked me what I thought of it. "Don't know," I said and he said, "Well, let me give you some good advice. Don't believe anything you hear in a briefing."

That month, I gave a speech on "Science and Education" to some science enthusiasts at the Mayflower Hotel in Washington, D.C. Many critics blamed our education system for Russia after Sputnik being, they claimed, "so far ahead of us." But, I said, basic-research scientists worldwide regularly and openly publish the results of their work. Leads go to whoever moves fastest and wisest to turn that information into better things of one kind or another than people already have. From the audience, later, some Augustinians (the educational order for Villanova University) said I'd inspired them and they asked me to head a committee to raise the $3.5 million they needed for a new science building. "Old Joe" Kennedy, Jack's father, they said, would know whom to see for the money. Later, in Joe's Boston office, I learned that his idea of a good lunch was martinis. Lots of martinis. Nothing but martinis. Off his list, I got $100,000 from Cardinal Cushing, and a Jewish American gave me a check for $850,000, but Cardinal Spellman didn't give me the time of day. He just didn't like Old Joe. Later, at the groundbreaking, the Augustinians gave me my only ever honorary high school diploma—and a martini shaker.

Then on 17 March 1961, Jack Kennedy okayed the CIA landing of Brigade 2506, fifteen hundred anti-Castro Cuban exiles, on the south shore of Cuba at 2 A.M. on 17 April. But, he said, its top priority was that he must be able later "to plausibly deny" that we were involved. Burke promptly sent a carrier task force to Cuba, telling its skipper, Rear Adm. Johnny Clark, to do as the B-26s had done, paint out his aircrafts' insignia "to prevent their being identified," a silly gesture, he knew, but he'd be able to tell Kennedy he was "sensitive to the president's priorities." Arleigh ordered Clark to be ready to provide air cover if Kennedy agreed to it. On station, Clark wired Burke, "What are my orders?" It was a dumb question. Arleigh fired back, "Your relief will have his orders!" He'd told me often, "If you don't push responsibility down in peacetime, how are you going to find out how your fleet commanders will behave in war?"

CIA planners advised by marine Col. Jack Hawkins, on loan at their request from corps Commandant David Shoup, wanted to land at Trinidad, eighty miles away. Most of its eighteen thousand residents were anti-Castro, and anti-Castro guerrillas were operating in the Escambray mountains there. But Secretary of State Dean Rusk vetoed it, saying it would look like an invasion. How, we wondered, can an invasion not look like an invasion? The site the White House accepted, Bahia de Cochinos, the Bay of Pigs, also was chosen because of the "plausibly deny" business. But its landfall, though isolated, was mostly a swamp. Then the White House cut in half, to eight, the number of B-26 bombers, ten 500-pound bombs in each, that they'd let the CIA use, and based not near Cuba but in Nicaragua, a seven-hour flight. So for their first attack on Castro's three air force airfields, prior to the landing, they could use only four B-26s since they were to attack again at dawn. Then, after an ineffective first attack, the White House canceled the second one.

Next day, after the landing, Castro's aircraft sank two of Brigade 2506's supply ships, and the others fled. Still, the Brigade, under Jose San Roman, fought for three days, inflicting three thousand casualties, before they ran out of ammunition. Yet the White House told our carrier aircraft to stay at least twenty miles from Cuba, and refused a plea by the *Essex*, one of our carriers, to rescue what was left of 2506. San Roman's last message to her was, "Why are you doing this to us?"

Admiral Burke said of it, "It might have worked: if Castro's aircraft had been destroyed; if ammunition had been off-loaded right away; if the guerrillas had been marines; if the supply ships had not been lost. Three times I was told no when I asked Kennedy to let us go in." Next time, he noted, "Be sure experts handle it; make sure they understand it must succeed." But he said, "The biggest problem was trying to run a detailed operation from Washington." A later Board of Inquiry report, a creative revision of history, didn't mention that at all; but much later, Bill Chapman, the *Essex* air boss then, noted, "The slippery slope of Washington micromanagment that led to Vietnam had its beginning at the Bay of Pigs."

On 21 April, Joe Kennedy called, giving me hell for letting it happen. I told him we'd been ordered to stay out of it. "Damn it!" he said, "You're not supposed to let my boy make stupid mistakes like that!" President Kennedy was leery of the Joint Chiefs after that, and virtually did away with using the National Security Council.

Then I went to Paris in a Kennedy entourage to see Charles de Gaulle. A hero paratrooper in World War II, Jim Gavin, was our ambassador, subsidized by Kennedy since Jim didn't have the money and JFK wanted him there. And Jackie Kennedy just mesmerized de Gaulle. For all the attention Jack got, he might as well have been on the moon. My reason for going was that the French CNO Calbanniet and Rene Bloch, in the French navy often an R&D program director, wanted details on our tactical atomic weapons and Polaris. We were helping Britain integrate Polaris into its fleet, and I assumed that we were going to give the French Polaris, too. I was ready to do a briefing and discuss weapons with them, but it all fell apart at the last minute—why, I don't know, nor who scuttled it. Gavin didn't know, either.

But it made de Gaulle furious. Our snub forced the cash-strapped French to build, over time, their own nuclear capability, both land-based and a submarine nuclear weapon. (More far-sighted than we, by 1985, France had 85 percent of its electric power coming from nuclear power plants, ending its reliance on Arab oil.) Eventually, de Gaulle just pulled out of the military side of NATO, forcing its headquarters to move to Belgium—but he stayed in NATO's political side, and in the economic part to help sell French military equipment to the rest of NATO. The whole Mutual Weapons Development program died, finally, its goal—arming all of NATO with the same weapons—lost in bitches over what was a "fair share" of the contracts each participating country's companies should receive.

Much earlier, at our very first navy budget meeting with McNamara, I'd gotten in trouble. Burke, Vice CNO Jim Russell, Curtis LeMay, and Defense Deputy Secretary Roswell Gilpatric—we're all there, and I start off by telling a joke I'd just heard. "You guys know the definition of a complete failure? It's a pregnant prostitute driving an Edsel with a Vote for Nixon sticker on it."

Burke kicked my shin and later told me my little joke probably cost our budget $100 million. I'd forgotten that McNamara had been president of Ford Motor Company when it introduced the Edsel because Mr. McNamara's statistical analysis, for which he was renowned, had said there was room in the auto market for another medium-price car. It had been a flop, of course, proving that Mr. McNamara's charts can be wrong, sometimes. I soon realized that he had no sense of humor and, clearly, his view of us was that he was "Management" and we were "Labor."

At the time, I was running a Missileer project. It included a 140-mile-range supersonic Eagle missile, built by Grumman, with a big lock-on radar; and a Douglas F6D aircraft powered by a subsonic TF-30 engine, our first forward-fan jet. (Up to then, jet engines had all been aft-fan.) Jim Russell had snarled at my using a subsonic airplane. We're putting the performance in the missile, I'd told him, and Bob Pirie, DCNO for air, had backed me up, saying, "There's no need to have aircraft boring a lot of supersonic holes in the sky."

Well, of course, the new "Whiz Kids," as the media were calling them at the time, had to cancel something, so McNamara canceled Missileer. (Then he canceled the air force Skybolt, a concept like Missileer's but with a longer range. That inflamed British Defence Minister Denis Healey, who'd spent millions modifying British bombers for it—and delighted Moscow, I'm sure.) So I changed the name to Phoenix, as in "It rose from the ashes of the Eagle," and continued with the program. Today, Phoenix technology is good enough to launch it from an F-14 fighter.

The TF-30 story didn't end there, either. When he came in, McNamara set up a new Systems Analysis shop, run by Alan Enthoven, a "Whiz Kid." Its job was to do "cost-effectiveness" studies. From one of those came the TFX (Tactical Fighter eXperimental), eventually designated the F-111B, a high-priced attack bomber. It was to be a supersonic, on-the-deck fighter for both the air force and the navy so we'd "stop wasting money duplicating each other's efforts." Its tactical requirement was inane. Vulnerability to a Mach 2 missile isn't any less if the TFX top speed is only that. Moreover, at that speed, to avoid meltdown, it had to be made of very expensive titanium.

Worse, yet, when McNamara heard that we'd already spent $85 million on the TF-30 engine, he and Enthoven chose to put it in their F-111. A forward-fanned subsonic engine, no afterburner, in a supersonic plane? Later, they tried it in the F-14. It stalled out there, too, under supersonic stress. But we couldn't convince McNamara nor any of his people. Only after it had failed (as we knew it would) the carrier-suitability tests—and McNamara had left the Defense Department to run the World Bank—were we able to bow out of Mr. McNamara's multimillion-dollar military "Edsel."

As we started down that rocky road, Arleigh Burke retired, effective 1 August 1961. He'd turned down Kennedy's offer to extend him for one more two-year tour. He'd been CNO a record six years, he said, and if he stayed, a lot of very bright people the navy needed might get discouraged and retire. I told him I was writing a book on the navy's best leaders since John Paul Jones. My two picks for the twentieth century, I said, were Ernest King and Arleigh Burke. He just growled at me, said he didn't want the publicity.[3]

Then in the spring of 1962 my ego got rubbed pretty good. First, French Ambassador Alphand, on behalf of de Gaulle, gave me their Commander of the Legion of Honor of France for our Atlantique work. Then Leslie Groves got me the American Ordnance Association's (now Defense Preparedness Association) Blandy Medal, and the Navy League gave me its first William S. Parsons Science award. I also became the first military man awarded ONR's (Office of Naval Research) Robert Dexter Conrad science Award, named for Bob Conrad, Oxford student and naval officer who helped start ONR. A law, changed later, said I couldn't receive its $10,000 honorarium, so they made my medal solid gold.[4]

My diary notes, "Nice kudos but silence from the CNO is deafening," no surprise, really. Though he'd married one of Lili's cousins, making us sort of kinfolk, George Anderson, a desk jockey most of his career, acted as if I were a threat. I wasn't, of course. No way would the "Establishment" let me become CNO. I was a maverick.

However, in February 1962, JFK asked me to be CIA deputy director. He wanted a technical guy there, he said, because science had become so important on the world stage. I'd be promoted to four stars, he said. "Can't do it," I said, "I'm taking a demotion to rear admiral so I can command our first nuclear-powered task force, Carrier Division Two."[5]

"You don't want it? You'll take a demotion instead of this?" Kennedy was incredulous.

But Washington just wasn't my bag. "Yes," I said, "I want out of this town. So does Lili."

I'd been told they'd not let me go until I found my OP-91 relief. I got Red Raborn. He wasn't crazy at all about it but did take the job. My reverting to two stars put me on the way to being the most demoted naval officer since John Paul Jones. That aside, I felt the navy people ashore and those at sea were drifting away from a feel for each other's problems. Besides, I'd done some of the heavy pushing for nuclear-powered ships so, fair's fair, I ought to go see how well they really worked in a task force at sea.

We can't modernize all ships all at once, so some in a task force will be all modern, some nuclear-powered but with older equipment, and some dogs in the pack will be vintage World War II. In my Carrier Division Two, the *Enterprise* and three of my cruisers were nuclear-powered. The others were not. And equipment was a mixed bag. The *Enterprise, Long Beach,* and two other ships were the first to have the new computer-driven NTDS (Navy Tactical Data System) for command-control. (Today, one hand can hold the computer power my two five-foot-high computers had in 1962.) The other ships didn't have NTDS. And the *Enterprise* had the first big fixed-array radar, the SBS-33, giving me but only me long-range vision. Of course, we had a small morale problem, too. The *Enterprise* could circle the globe nonstop. That's why most sailors didn't like nuclear ships. They might not get any shore leave in foreign ports. It added up to a tough test. If a commander doesn't know the differing capabilities of his ships, he easily can mess up a mission.

A sea change rose up to drown that thought. My log for 30 March sums up it: "Nice trip to Dallas. Dresser Industries has made me an excellent offer. I'll take it if they can wait until October, a big decision reinforced by finding Lili in tears on my return over John, the deputy-CIA offer, everything—except she's not crazy about living in Dallas. And I've failed completely with John. He's basically a good lad, much smarter in many ways than I was at his age. Lili sort of shudders at my leaving her home with him. It's difficult to watch your son not do what he must do to go forward. I know how my father felt 40 years ago."

But on 9 April, with my flag flying for the first time at sea, we put out on the *Enterprise* to rehearse for JFK's visit. It went lousy. Notes my diary, "I am unimpressed with the ability of my people. We even lost an F-4H in the AirOps drill. We sail for the Med on the 19th, ships from Newport and Mayport to join up at sea. Are we ready?"

The all-day JFK visit was 15 April. In his entourage were McNamara; Gilpatric; Fred Korth; House Armed Services Committee boss Carl Vinson; John Stennis, his Senate counterpart; and ambassadors from Africa and Central and South America, the one from Chad wearing a leopard skin. We were told to pair them up in private cabins, for toilet facilities and such. Equal Rights disciple Martin Luther King was front-page news, that day, when I asked JFK whom I should put with the guy from Chad. "Why," he said, "Senator Stennis from Mississippi, of course."

We were to fire our Talos, Terrier, and Tartar surface-to-air guided missiles, which had replaced antiaircraft guns as carrier defenses. Trouble was, not one "T" missile worked in spite of them all having been in R&D since I was at Inyokern. After that fiasco, I took JFK on a tour. In my quarters, he looked at the plush furniture, boxes of Cuban cigars, a bar well stocked with scotch and brandy, and said, smiling, "*Now* I know why you took a demotion to get command of this ship."

Then on the 19th as we headed east, I was made to look a fool by a newspaperman. Right after the Kennedy business, he'd come aboard on some pretense, then eavesdropped on my chat with my chief of staff about the "T" failures and what a mess they'd made of our show. He reported it as if he'd interviewed me. Anderson was livid, of course, but at least CINCLANT Adm. Robert Dennison defended me. My diary says, "I am pretty dumb in this press business."

And as we neared the Med, the *Enterprise* developed turbine problems. She had to turn and head back into the yard to get a fix. The good part was, I had to transfer my flag over to the *Independence* where my flagship's skipper was Pete Aurand. My log entry reads, "Pete just may be the best carrier skipper alive. He really has the *Independence* in tip-top condition. If he'd been at the JFK show, it would have been different."

We were in the western Med on 3 May when I wrote, "We blew up our third bomb out in the Pacific and the so-called 'neutrals' howled. JFK's decision sure was right. We'd said we wouldn't test any nuclear weapons if the Russians didn't and the Russians did, so Senator Russell demanded and JFK went back to testing. The situation in Laos is hell in a handbasket."

Most of early May we spent in Cannes, "The film festival in full swing," my diary says, "and I played golf in Monte Carlo." Then on 17 May, the day the White House gave up on Laos but put troops in Thailand, we ran an exercise on how we'd help NATO repel a Soviet attack. It seemed a waste to me. Earlier, Navy Secretary Tom Gates wanted me to tie our mobile carriers into an air force SIOP. They weren't, at that point, nor did I think they should be. Polaris, I said, will play our role in nuclear deterrence, so a task force's goal should be to survive a first strike with enough weapons and aircraft for a decisive counterstrike. I'd lost the argument.

Then on 25 May, Kennedy excited the nation by proposing a program to "put a man on the moon . . . within the decade." Emotionally, the idea upstaged the Russians who already, twice, had put a man in orbit. It also shelved a NASA plan, revived in 1997 as an international program, to put up a big space station, then go to the moon. We were having a big debate then over who, the navy or air force, should be lead agency in military uses of space. With three-fourths of the Earth's surface water, we said space surveillance was far more important to us than to anyone else. We lost to the air force Skylab—which McNamara canceled in 1965. But I did write Kennedy a long letter with examples to back my opening, "Space is not a program. Space is a place, a place where mankind can learn to do better many things we do on

Earth." That did inspire one person, whoever put my letter on display in the Kennedy library.

For four months after I sent that letter, most of our few days a month at sea were spent working on proficiency, chasing Russian submarines and trawlers that were spying on us. They went to sea more each month. We even saw a surfaced *Foxtrot,* first one seen in the Med, on 16 June. There was no way to tell, of course, if they were in a stepped-up training mode or just giving us a little "in your face" bluster. Our own Sixth Fleet exercises did not impress me. The Sixth Fleet had lost ground a lot since Cat Brown left. The most glaring need was in communications, which, I wrote, "are horrible and, the way BuShips does business, won't get any better."

Because of our few days a week at sea, the cruise was relaxing, giving me lots of time to read, to study. That changed when we met Lili and two of the children, Jenny and John, on 20 June in Taranto, a port on the heel of the Italian boot. From there, the wife of the Italian naval attaché to Iran took us to Truli village, a vestige of early Greek settlers. Then we went to Corfu to be met by the mayor, a woman; spent the day on the beach; then to the Club Med where our ship's band entertained us. We were in Cannes by 2 July, and my diary says, "I'm sure glad I took this job just for the fact that Lili could come over and have plenty of fun for a change."

On the 3rd, Lili and I went to Paris to party for a weekend in Vlad Reichel's favorite town. On the plane, we saw Louis de Polignac. We'd met him in Monaco when we'd gone with Paul and Ginny Gallico to visit Princess Grace Kelly. Louis, who owned Lanvin Cosmetics and a champagne business and had tons of money, was riding in tourist because, he said, all the interesting people travel tourist. Back at work by the 10th, I took John, Jr., along when we went to sea for three days, me doing some flying and deciding to wait until my return to Norfolk to put in for retirement. Then Lili and I went to Paris, again, dancing at Maxim's, the works; then back in Cannes with Prince Ranier and Princess Grace aboard for lunch, bringing the Monaco symphony orchestra along to play us a concert. My diary astutely notes, "She's a doll."

On 29 July, we ran Exercise Full Swing, a drill to support ground troops in Western Europe, using a low-altitude approach to targets 265 miles away in France. We used thirty-three of our eighty-four aircraft, specifically, eight ADs, sixteen A4D Skyhawks, and nine A3Ds, which delivered on target thirty tons of conventional ordnance. We kept our air-cover fighters aboard ship because they couldn't go top speed, 265 mph, at low altitude. We just proved that if we wanted air cover, we needed a high-speed-at-low-altitude, three-hundred-mile range fighter. My record notes, "We learned that in War

II, one reason we put rockets on the fighters. Obviously, we've unlearned it since then!"

Then after a night at the Eden Roc with Rene and Yolande Bloch who were down from Paris, on 3 and 4 August, from Naples, Lili and I went to Venice where the Italian Navy gave us a barge to go all over—the Murano Glass Works, Portofino, the grotto with the Christ under the sea. On the 18th, she and Jenny left for home. On the 19th, I was relieved from the task force command and we sailed for home with Secretary Korth's okay, in writing to avoid squawks later, to take John, Jr., with me on our ship, the *Independence*. Off Gibraltar as we moved at a twenty-five-knot SOA (speed of advance), my log says, "go rhumb line to the Azores, then great circle route to Norfolk. Due in on the 27th, my air group will fly off on the 26th. My decision day approaches. Wish it was settled. Uncertainty is not my cup of tea."

During the cruise, I had one nasty hit, a visit by Adm. Page Smith, ComNavusEur, a crony of CNO George Anderson. On 2 August, he came aboard just to growl at me that, "The navy's senior officers have lost confidence" in me. At sea, I wrote, "We are frauds. The War Plan 217-61 Anderson signed admits we can't do what we say. And hanging on is my January testimony when I told Congress Navy R&D is mis-managed. Fleet requests go up one way, but funds flow down to Bureaus with no tie-in to the fleet. General Electric uses R&D to obtain market share, diversify. Navy R&D should be profitable to the Navy mission. But too many senior Navy people can't hack it technically. [Rickover had said that, too, but only to me, not Congress.] That was clear on Vanguard, the Thresher submarine sinking.[6] Yet, arguing what I believed was what Burke wanted and he gave me great fitness reports. Too often, I guess, I challenged an officer's competence to testify on a technical issue. The 'Establishment' hated that. So why not just leave quietly, remembering the successes, the flying, the fine people I've worked with?"

In Norfolk on the 27th, as I left on ten days leave, "Miles" Beakley, an admiral now and CINCLANT Dennison's vice chief, told me to see him when I return. "We have a problem," he said, nothing more. Neither Anderson nor Korth had any news about my future. Henry Jackson said not to quit. My diary says, "Confused, confused, I need a psychiatrist, I am sure."

Back in Norfolk on 3 September, I learned Beakley's "problem" was that U-2 reconnaissance flights over Cuba had found eight newly built SAM (surface-to-air missile) sites, and inland from Mariel, west of Havana, they'd seen construction started on a dozen launch pads for Soviet IRBM-range missiles. And our ASW patrols had seen Russian freighters bound for Cuba, some large-hatch so they could put IRBMs and nuclear warheads in the hold, he said, and some with Ilyushin IL-28 Beagle nuclear-capable bombers crated on deck. He told me to take a P-2 patrol plane out to look at the Russian ships, then find out where we'd stored our low-drag bombs from World War II, the one bomb we had that can be hung on the outside of our fighter-bombers.

We may need them, he said, because the Joint Chiefs had given President Kennedy three options. One was a naval blockade to stop any more shipments. A second was a fighter-bomber "surgical strike," they called it, to take out the missile sites and bombers already there, the ones we knew about. But the Joint Chiefs said, we may not have seen them all; so the only way to remove the threat for sure, the third option, they said, was to invade Cuba. "If Kennedy orders a blockade," Beakley said, "Corky Ward's Task Force 136 will do it. He'll have the carrier *Independence*, with its airborne early warning capability, as his flagship, and put ships out on a five-hundred-mile arc northeast of Cuba. And the *Essex*, due there for ASW exercises anyway, will join him and look for Soviet submarines that might deploy to the area in large numbers."

In Beakley's plans, I'd command Task Force 135, the *Enterprise, Saratoga,* nineteen destroyers, and some submarines for ASW work. If Kennedy ordered the surgical strike, my aircraft were to take out the missile sites. If we invaded, I was to cover an amphibious landing at Mariel by marines and two army mechanized divisions from Ft. Hood, Texas. The army divisions already were in south Florida, he said, and a marine division from California was aboard ships off Jamaica. And, he added, fighters would be available from MacDill Air Force Base, Florida, already on DefCon-3. (In those days, DefCon [for Defense Condition] -5 or even -4 meant, roughly, relax; DefCon-3 was "get ready to fight;" -2 meant deploy; and DefCon-1 was "attack!") And I was told not to talk to his staff, to Dennison's, to my own people, Lili, anybody until after I left Norfolk.

Congress was beating war drums too. In August, Republican Senator Kenneth Keating had raged at Kennedy for "doing nothing about the Soviet SAM sites and combat troops" in Cuba. Senator Strom Thurmond said we should invade, and a House-Senate resolution, passed on 7 September, authorized Kennedy to invade Cuba. After a U-2 spotted Soviet MIG-21 "Fishbed" fighters in Cuba that same day, an F4H Phantom II squadron and Grumman Tracer planes, to give the F4s AEW (airborne early warning) command-control, were moved to Key West, ninety miles from Cuba—where the B-26s should have been during the Bay of Pigs foul-up.

And as soon as my flagship, the *Enterprise,* left the navy yard on 3 September, I changed her aircraft mix, off-loading our A3D nuclear bombers because they took up a lot of room and Beakley's "surgical strike" plan ruled out Mk-VI atomic bombs. It would be like swatting flies with a sledgehammer. We put on sixty A4 Skyhawk jets, dubbed Ed Heinemann's "hot rod." With 670-mph speed at sea level, they had two 20-mm forward-firing guns, carried nine thousand pounds of bombs or rockets externally, and were well adapted for either of my two missions. For more muscle on the amphibious-landing support mission, I added a squadron of AD Skyraiders, our last piston-engine aircraft, with four 20-mm cannon and up to four tons of rockets, bombs, and/or napalm. We also had a few F4H Phantom IIs for air cover and to defend Guantánamo, our secondary mission.

As we did the switch, I scoured the East Coast for our low-drag bombs. Turned out, most were in magazines at Guantánamo. But not until 12 October, the day Wally Schirra completed six orbits in a NASA space capsule, were we ordered to get under way, "immediately!" Our cover story was that we'd left "to avoid a hurricane," and, at sea, I finally could tell my staff the truth. By then, I knew who was what. Dennison was to fend off CNO Anderson, McNamara, and Kennedy, while Beakley ran the show (and earned a Distinguished Service Medal for it.) But deploying the ships was not difficult. An *Independence* fleet already was there on an ASW exercise under Adm. Red Stroh, as was the *Saratoga*. He took her when Ward took the *Independence*. And I learned that the "Cuban Missile Crisis" was news to Red. He knew nothing about any of it.

On 14 October with us two days at sea, a U-2 confirmed that Cuba now had sixteen IRBM launch sites, twenty-four mobile launchers for Soviet one-thousand-mile SS-4 missiles, and forty-two IL-28 bombers. That put everyone on DefCon-3, on Strategic Air Command worldwide, and our six Polaris FBMs sent to preset launch sites at sea. Then Beakley told me that Kennedy wanted to know if I had any requests. "Well," I said, "I assume we're not going to do a Pearl Harbor, not attack without telling the whole world, first. So, ask him to announce in the evening what we're going to do. That will permit a dawn attack, next day, the best time to gain tactical surprise."

"Fine," he said, "fine."

Then he ordered me not to communicate with him or anyone else higher up except by our KW-7 encryption machine. Then on 22 October, the KW-7 told us the president was going to tell the nation about the Cuban crisis at 7:30 P.M. that evening. So I loaded all my airplanes with bombs, fuel, intelligence data, photos of missile sites and the terrain around Mariel and Gitmo, everything. Then I began to get silly messages. The first one, unsigned, asked how many fighters I was going to put over Camagüey, a large inlet 300 miles northwest of Guantánamo, 250 miles east of Mariel. My answer was, "Enough."

And Beakley clattered on the KW-7, "Damn it, I'm not asking. Kennedy wants to know!" Here we go, I muttered, more White House micromanagement like what ruined the Bay of Pigs. I gave him a number and promptly forgot it. Then I pointed us toward Cuba at an unhurried twenty-five-knot pace, noting in my log, "In 43 minutes, Kennedy will speak to the American people. Each turn of the screw brings us closer to our launch point," adding, irrelevantly, "China and India are in a full-fledged border war." That evening, Kennedy announced that we were imposing on Cuba a "quarantine," a euphemism thought up by a State Department lawyer who'd said, "If you call it a blockade, it could be construed as an act of war, unwarranted, armed aggression." Lawyers.

19

From Cuba to the Naval War College

When the "quarantine" began, at 10:00 A.M. on 24 October, reports said twenty-five Soviet ships, five of them large-hatch types, were headed for Corky Ward's picket line with orders not to stop. An ancillary mission of ours was to be in position for my submarines to attack them, and stay on top of any Soviet submarines moving into the area. So we did shift around some, to Jamaica and Puerto Rico. But I wasn't given a tie-in to Corky's blockade. I had aircraft that could radar-scan out to three hundred miles but, when the air force sent U-2s over Cuba, it didn't tell us. Maybe the White House knew where and when. I don't think Beakley knew or he'd have told me. When a Cuban SA-2 SAM shot down a U-2 on 27 October, killing the pilot, we could have prevented it had we known about the flight. I'd seen that happen often when I was in R&D, intelligence boys with such a phobia about "security" that their information never gets to the guy who could use it and doesn't care where it comes from. But they don't tell him.

Meantime, though it was quiet where we were, Anderson later revealed he'd had a fight with McNamara during the crisis, in the Operations center with George Anderson's staff watching. McNamara had started yammering at him to move one of my ships "fifty miles closer," and then, maybe fifteen minutes later, "farther back. George said the *Enterprise* would have needed eight hours to do some of those "right now!" moves. Finally, exasperated, he'd yelled, "Look, I run the navy, you don't, and I am damn sure that task force commander knows better than you where his ships ought to be."

McNamara had stomped out of the room. We were delighted to hear George had stood up for us but dismayed in May 1963, when he agreed to be ambassador to Portugal. It put him in a closet, unable to rebut the McNamara-Rusk crowd's euphoric praise of Kennedy for "facing down Nikita Khrushchev, yet avoiding nuclear war." Fact is, Nikita got more than he gave.

To get him to take his weapons home, Kennedy pulled our Jupiter IRBMs out of Turkey, as my diary entry for the 28th predicted he would, one of the few times my crystal ball worked. Worse, Kennedy agreed to "protect the territorial integrity" of Castro's Cuba. My diary says, "Our Press will call it a Kennedy victory, but it wasn't. We've given Castro a long life when we should have shot him down. From now on, he will be a threat to us and our neighbors."

Interestingly, Kennedy and Khrushchev cut their deal on the 28th, one day after a powerful Senate leader, Richard Russell, said publicly that we must invade Cuba. Our invasion plan called for our army–marine corps team not only to seize the missiles but to take over Cuba, and it gave them just ten days to do that. Most people think the blockade turned back the Russians. I don't. Moscow knew we were set to invade. Unlike the blockade, an invasion would not have been a direct confrontation with Russia. And though the White House feared he would, it was very unlikely that Nikita would react to an invasion by sending the Red Army into West Berlin or invading Turkey. Either move would have mobilized NATO and our then-superior numbers of nuclear weapons. Only a Kremlin nutcase would have traded Moscow for Washington over a ragtag island in the Caribbean. We should have taken Cuba.

By the end of October, we were well south of the islands, staying mobile because of the many submarine contacts that our destroyers were reporting—which I thought were not valid submarine contacts, just a lot of underwater noise. Still, I told my staff, we could be down here until December, at least. On 3 November, north of Jamaica off Montego Bay, we tracked a U-2 at sixty-eight thousand feet as he made his west-to-east run, seeing if Khrushchev was doing what he said he would, and guarded five more flights on the 4th. And on the 15th, my diary notes, "Marvelous birthday party put on for me by the officers and crew. People make this Navy go!"

On the 18th, I groused, "We have been at sea 40 days, now, working day and night, still no word and they wonder why we can't keep people. Why join the Navy when you can go to the Air Force or Army and be home with your family every night? They could at least let us go into port but the hardest thing for the crew to take is the uncertainty. This is particularly so since the rest of the Defense establishment has now gone to DefCon-5 while we're still on a 2-hour alert."

However, next day, I was told we'd be relieved on 6 December by the *Saratoga*'s Carrier Division Six. Then on the 25th, after Kennedy had called off the blockade because Khrushchev said he'd remove his IL-28s "within a month," we moved over to protect U-2 flights over western Cuba. There, that afternoon, we lost Lt. Sutro in an F4H when his catapult went haywire. Its link to the plane, once called the "car," now braked hydraulically, carried away out of control at 450 knots, which happens every so often, and he ran off the angled deck. My log notes, "It's sure death to go over the side that way. This is the second pilot we've lost on this mission and a miserable Thanksgiving present for his wife and family. His father, a strong Navy fan, is a well-known San Francisco stockbroker. Once more, I have to write a sad letter."

Two days later, as we were guarding two U-2 flights over northern Cuba, we had another fatal accident when a young seaman named Evans ran into a cable on the elevator well while operating a "mule" truck. And as he had been all month, Smedburg in BuPers still was saying he had no job for me, as yet, "So," I wrote in my log, "I will go back to the Med in February, to Task Force 60 or a Battle Group, nice jobs, a happy way to leave the Navy."

On 4 December, I was the first U.S. official to call in Kingston on Jamaica's new governor general, Sir Clarence Campbell, a coal-black native, very friendly toward us. Then we headed for Norfolk, and Red Stroh said Frank O'Beirne, COMAIRLANT (Commander, Naval Air, Atlantic) was sore at me because I'd ignored him in the chain of command. I told Red and confirmed in my diary, "O'Beirne is odd-ball, first class. I reported to CINCLANTFleet. In that set-up, O'Beirne's just logistics support, had nothing to do with operations. I need only one boss at a time. If O'Beirne picks a fight, I'm sure Beakley, my boss, will saw him off."

When we arrived back in Norfolk on the 6th, we'd been gone forty-nine days. I did my debriefs to everyone, then flew an F4H Phantom II to Mayport to spend a quiet holiday with Lili, Jenny, and John, Jr., in Atlantic Beach. Back in Norfolk by early January, we put to sea twice, briefly. Both were my first look at a pair of new aircraft. One was the twin-jet, high subsonic Grumman A2F, later designated the A-6 Intruder. An answer to navy close air support problems in Korea, it could fly on the deck, under enemy radar scans, in all weather. Its side-by-side cockpit seats held the pilot and a navigator-bombardier who used data from DIANE (Digital Integrated Attack & Navigation Equipment) to find and hit small targets with up to eighteen thousand pounds of ordnance.

The other, a twin-turboprop Grumman W2F, was designated E-2 Hawkeye later, in its final preproduction phase. Work on it had begun in 1957 as I became assistant CNO for R&D. It was to be the key to our new Navy Tactical Data System. It was to relay back to a task force commander real-time information on the over-the-horizon location, numbers, and headings of all ships and aircraft, friendly and enemy, in his assigned operating area. The W2F had a big disc, a long-range search radar, atop its fuselage to detect targets; and digital computers, not only to record that data but also to pick automatically by location, armament, speed, and range the best airborne intercept to send at the target. Basically, it was an airborne air traffic control machine.

Except for that show, I had as easy a ship captain's job as the navy offers because we stayed in port all month. I did go to Washington, hunting orders and to give Fred Korth another sermon on the merit of nuclear-powered carriers. All I got out of that were orders to head for the Med on 6 February to relieve Johnny Hyland as commander, Task Force Sixty. And on 29 January 1963, I took the *Enterprise* to sea with a mob of visitors aboard, led by Sen. Barry Goldwater and Adm. Robert Dennison, "the hardest person to talk to on any subject," my diary says. From him I learned that once more a press character had hurt me with the "Establishment." True, I'd been arguing for years that all our carrier battle groups should be nuclear-powered. But all I said, twice, to a twit who called was, "Call CINCLANT on that." Then the reporter quoted me as if I'd just stated my case to him, picking a public fight with CNO Anderson and McNamara. My log says, "I guess my end as a Naval officer is close. Henry Jackson says to relax."

Offsetting that aggravation, Goldwater, a member of the air force reserve, asked if he could fly an F4H Phantom. He was in a war at the time with Nelson Rockefeller for Republican Party leadership, and I said the Republicans would be furious if I let anything happen to him. "Maybe," he laughed, "but Rocky will be happy." So I let him go up with our fighters' skipper, Gerry O'Rourke. They had a nice flight, Gerry said, and back aboard, Barry said he was going to do his next two-week reserve duty aboard the "Big E."

Meantime, and for years afterwards, "spin doctors" got busy correcting the Cuban missile crisis record to pin stars on Kennedy and black crepe on the Pentagon. One proof of the "fix": In mid-May, I discovered that Beakley's "surgical strike" plan—which Bobby Kennedy was claiming the Joint Chiefs had told his brother was "not feasible," anyway—had disappeared from the Joint Chiefs' Operational Plans file. Meantime, on 6 February, my twenty-five-ship task force was under way from Mayport, Norfolk, and Charleston to rendezvous at sea. And on the 8th, bad luck whacked us when a big sea struck the forward elevator, which was down, injuring eight men, washing four overboard and killing two. We were at twenty-eight knots and had had no sea trouble until then. It was a terrible way to start a cruise. That night I wrote a four-line editorial: "McNamara seems hooked on the semantics of offensive versus defensive weapons. He's called the Navy on it, regarding air defense. If I'm off an enemy coast, shooting down his planes to defend my ship, how offensive can I get?"

Another, even more weird exercise he and his "Whiz Kids" were running constantly was what he called "cost-effective analysis," studies on what a proposed new weapon system would do compared to an existing one it was to replace, to reveal if that would be worth the increased (usually) item cost. In effect, he was using arithmetic to make judgment calls on military combat capability; and usually poor ones, at that, because his statistics ignored half the real-world equation, the "people" factor. Yet, in spite of "Management" McNamara, we in "labor" knew that how well a ship or aircraft performs depends not on what the engineering specs say it can do but on how well it's maintained, how skilled the pilot or skipper is, and how committed, even heroic, he is in combat. Synergism between man and machine is what wins battles, not "cost-effectiveness."

Anyway, back on the bridge, my sailing orders, as usual, had included one of those "Unless otherwise directed, do" so-and-so, and of course they never directed me until maybe a week too late. It happened again on 11 February. Behind schedule, moving fast southeast of the Azores, way outside any Atlantic airline routes, suddenly our big radar saw two high-up targets, three hundred miles out. My rule was never give anybody a free ride over my task force, so I sent up two armed F4Hs who reported, a bit later, that they were Soviet Bear bombers. Quickly, I notified CINCLANTFLEET, CINCLANT, the CNO, everybody, that if missiles were hung on the Bears, I would shoot them down. "No!" screamed in over the wire. It seemed everybody knew about those Bears except us poor sailors at sea, once again, intelligence not given to the guy who could use it. They'd flown by Iceland earlier, knew about where we were, and had headed our way. I didn't shoot them down, of course, but they didn't hang around, either.

When I relieved Johnny Hyland in the western Med on 19 February, I was amazed to hear that he hadn't been told on 19 October what the Cuban situation was, and had gone into port on the 23rd. It's a bit scary, realizing Kennedy, worried about Russia invading Turkey, didn't tell Task Force Sixty, a primary counterattack force if Russia had invaded. That, I said, is what comes of trying to run operations from the East Room of the White House. Within a week, I'd filed that worry away when one of our half-dozen navy-secretary guests asked me to be his electronics firm's chairman of the board. I passed. Dressers Industries' offer was far better.

Then death swooped in on us, again. When the Russians overflew us off the Azores, we'd been doing a LoadEx, practicing loading nuclear weapons on our A5-A Vigilantes just recently put aboard the "Big E." Now, on the 25th, an A-5A flew into the ramp and exploded, killing Lieutenants Kruse and Cottle, spreading fire the length of the deck. The firefighters got on it in a hurry, fortunately, to prevent a catastrophe. What hurt most was that Kruse had just become a father for the fifth time, a baby he hadn't yet seen. Next day, Lt. Holter's A5-A also hit the ramp on approach. We did get him into the barricade without much damage, but two crashes in two days had me worried that my VAF-7 group might be a weak outfit. Still, that big (seventy-six feet long, fifty-three-foot wingspan, forty-ton full-loaded) Mach 2 aircraft was a demanding one to fly onto a carrier. It takes a combination of sink rate, power settings, and a fairly high angle of attack over the ramp to avoid hitting it but that also makes the deck seem smaller.

While I was mulling that over, we went up to Naples to off-load Holter's A-5 and two others North American should have kept in its factory. As we did, four Bear bombers swooped down over the *Forrestal* on its way home. They were intercepted on their second pass but not the first—which earned the *Forrestal* skipper a slap on the wrist from the CNO. And I noted in my diary, "This is a strange Russian performance, all of a sudden doing this in both the Atlantic and Pacific. Either they're just training or showing us they can go to sea in a big way. It will have the boys thinking a bit back in Navy OPs."

In early March, I met Lili and her sister in Athens, Greece, and we picked up as we'd left off before the Cuban missile business: sightseeing; parties; "quite a lunch," I noted, "with the King and Queen;" to their AEC laboratory in Depocritus seeing Ambassador Henry Labouisse and his wife, the daughter of Eve Curie of radium fame; then the three of us went to Corinth. I did do some work. On the 23rd, on the *Franklin Roosevelt*, involved in an exercise, I made my three hundredth (it was her twenty-three thousandth) landing on a carrier, then flew back to the "Big E."[1] April stops, most with Lili, included Athens; Palermo; Naples to host the Italian armed forces chiefs of staff; then to Rome for Easter services in St. Peter's Square, to meet Pope John XXIII and see in St. Peter's tomb where we'd begun the carbon dating exercise years ago. Then after a busy week at sea, visits to Aranci and to Cannes to see Paul Gallico and Princess Grace, again. And a brief, mid-month note in my log says, "If I don't get the Pacific ASW slot, I will leave."

Then on 29 April, I was ordered under way quickly from Cannes to the Middle East to "show the flag" over some problem in Jordan, "a threat to its independence," it was said. The "Big E" and my nuclear escort ships were there, on station, almost three days before the rest of my task force arrived, delaying the taking on of fuel, supplies, and ammunition. But as I knew I'd tell the powers-that-be in Washington, while alone out there, I didn't feel at all vulnerable. 'Course, I didn't know whose side I was supposed to be on, either.

While there, I was told Bill Martin would relieve me on 18 May and I was to go all the way to Hawaii to relieve Jimmy Thach as commander, ASW

Force Pacific on 14 June. My hope had been rewarded. On 18 May, the guns fired, I hauled my two-star flag down from the Big E, said good-bye to Paul and Ginny Gallico and Princess Grace, and Lili and I left Nice in a COD S2F Tracker, an uncomfortable airplane. Then in Rabat, we boarded an R5-D, navy designation for the Douglas DC-4, a 250-mile-an-hour mule I knew Lili would not like a bit for it is a long way from Rabat to Norfolk by way of the Azores. At our little house in Atlantic Beach in May, Lili and I worked out the logistics. John, Jr., would drive Lili and her sister to California and they'd ship the car to Hawaii. Meantime, on the 30th, Jenny and I would fly to 'Frisco, then Hawaii.

And on 14 June in Hawaii, I broke out my three-star flag again.[2] I'd wanted this duty for the same reason I'd taken the nuclear-powered task force. I'd put a lot of R&D into ASW and wanted to see firsthand how good it was. Our domain covered a lot of ocean, from Adak, Alaska, to Australia, which is why the Australian chief of naval staff was one of my Task Force commanders. My staff was pretty cosmopolitan, too—an Australian officer, an air force lieutenant colonel, a coast guard lieutenant, my chief of staff a navy captain, my flag lieutenant a young fighter pilot—and our family quarters were paradise, about the best we'd had, ever, a big house on Ford Island with a view of the Arizona Memorial from our front window. And my boss was CINCPACFleet's Adm. U. S. Grant "Oley" Sharp, a tough but friendly guy.

At work, I had an R6-D, navy designation for a Douglas DC-6, fitted with bunks and a steward plus a crew since I'd have to travel to Australia, the Philippines, Singapore, Taiwan, Korea, Japan, and Alaska to coordinate the ASW programs in all those lands. I also had to oversee ASW training in San Diego, preparing to deploy those units to the First Fleet, an ASW subset of the Seventh Fleet. So one of my first trips was there—to discover that my old bluejacket pal, Frankie Williamson, had command of the training unit and, I saw, as my log says, "This Command is a good one, doing a good job." Frank told me that the ASW simulator I was involved in buying while I was in the Pentagon "works well. It was a good investment."

We had the use of three carriers and some fine skippers. The best of them was my old friend, Pete Aurand, who, my diary notes, "has more imagination and drive than all the rest of the group." For the final exam at training command, we'd create a carefully prepared problem, and run an ORI (Operational Readiness Inspection) exercise in Hawaiian waters before they deployed, to see how they operate. Over time, I noted, "Some did better than others. There hasn't been much hard study of how effective we are in the sub-killing business. Our submariners are well trained but lack decent kinds and numbers of weapons. Our nuclear attack submarine carries only 26 torpedoes, same as some Fleet boats in War II. If they get 50% hits, better than was ever done in the war, that's only 13. But when I told my friend Rickover 26 fish make his warship just a low-altitude foxhole, he said weapons are not his department!"

One early decision was to build an underwater firing range off Kauai with control stations to record how effective our submarine force was, firing its weapons. I also arranged with Charlie Weakley, who had the same job in the Atlantic that I had in the Pacific, to have a joint conference each year, the first one in Norfolk, then in Hawaii in 1964, to share information and to submit agreed-upon OR (Operational Requirements) requests to the CNO. One of the first was for a standoff weapon for the P-3. If its crew sights a submarine on their radar, by the time they arrive to drop a homing torpedo, the submarine is long gone. If their radar picks up a surfaced submarine, they can do little about it. The result became, a few years later, the Harpoon with a range of seventy-five-plus miles.

But my notes on our first session, 31 August 1963, did grouse a bit: "Meeting went well. Weakley has a large operation in the Atlantic which always has made out better when it comes to submarines. We have the First Fleet to deploy and a bigger area, harder to cover, but we have only 39 submarines, only 40% nuclear. They all should be nuclear."

A diary entry for 23 September 1963 says I was discouraged about the whole ASW business: "How many Russian submarines carry missiles? Can we cope with a surprise attack? The Soviets have gone into a large program to reduce the noise level of their conventional submarines. Given that, our SOSUS (Sonar Sound Underwater System) is doing reasonably well, getting signals but then we dispatch aircraft with little reward. We know the Soviets are out there and SOSUS should be able to locate them. We must do better."

But by early 1966, all our static sound surveillance systems were in place, in the Atlantic and in the Pacific. AT&T had done all that work for us. The SONUS were lateral arrays on the ocean floor, their sonars at varied heights on the cables in order to penetrate the oceans' thermal layers. We were tracking millions of miles of ocean, detection now measured in hundreds of miles, not yards as the old hunter-killer groups did. At Adak, Alaska, we also had a big array just looking right into Petropawlow, the Russian port, and a ground-based radar to reinforce the surveillance. To identify friend or foe, we had by then a program whose goal was to record the engine sound, the footprint, of every submarine in the world. We had our own submarines' signatures, of course, and all our ASW missions into foreign areas tape-recorded every submarine they heard, most so distinct we could tell not only friend or foe but knew by extrapolation the name or number of a particular boat. The tapes were all highly classified, naturally.

In the Atlantic, through a deep-channel thermal layer, we could detect clear down at Barbados in the West Indies a Soviet submarine as it came around the North Cape out of Murmansk. That data went to stations at Eleuthera in the Bahamas; at Nantucket, Massachusetts; and in Newfoundland where they all listened all the time. Their data also went in real time to an ASW center in Norfolk. Through triangulation, it could then tell right where the submarine was, and, if warranted, sent out the P-3s. In the Pacific from 35 degrees north, we had an Alaska station, one in Washington, and one in San Clemente, California; and their deep-channel detections could give me a triangular fix on a submarine twenty-three hundred miles away. By 1966, we were pretty confident that, if an enemy submarine was at sea, we knew where it was. And with the ASROC (Anti-Submarine Rocket) anti-sub missile, new homing torpedoes, and drone helicopters, we began to give the P-3 a wider reach if it went out to localize and quickly kill the target.

Meanwhile, as all that started, in September 1963, I sent a letter to navy Secretary Fred Korth, as strong as I could make it, urging that our next carrier, to be named the *United States*, be nuclear-powered. I sent a similar letter to Henry Jackson whose Joint Committee on Atomic Energy was to have a hearing on that subject. Then in October, I got a subpoena ordering me to testify on Monday, 26 October. I knew I'd get in trouble if I did. The ship's hold was big enough to house eight nuclear reactors but, like Anderson whom he'd replaced as CNO, if Adm. Dave MacDonald had to let it be a conventional carrier to get it, he would. And I'd learned when he boarded the Big E with Kennedy that McNamara knew nothing about the navy but acted arrogantly as if he was all-wise "management" and we were menial labor. I asked Oley Sharp what he thought I should do and he said, "You've got to go. You've been subpoenaed."

So I went, flying into Washington, D.C., on Sunday night for the hearing, due to begin at 9 A.M., Monday, the 26th. A small crowd was there. On the defense side were McNamara, of course; his director of R&D, Harold Brown from my Livermore Laboratory days; Korth; Paul Nitze, Korth's just-named successor; MacDonald; Rickover, whom I couldn't prove but was sure was who got the committee to send me my subpoena. All my Senate friends were there—committee chairman John Pastore, Henry Jackson, Lyndon Johnson, Margaret Chase Smith. They, especially Mr. McNamara, were amazed when I showed up because, I later was told, McNamara had told them I wasn't available to testify.

As Pastore, Jackson, and the others asked questions and got answers, each, as if rehearsed, would ask McNamara, "Admiral Hayward is the only one of you who has commanded a nuclear task force. Did you ask him about this?" Of course, McNamara had to say, "No," every time. Boy, that was killing me. Then when I was called up to bat, my answers were biting and to the point— and all contrary to the McNamara party line. Afterward, a press type asked if my testimony was due to my not wanting to come back to Washington. "Don't worry," I said, "You won't see me around here ever again." But it went by the board with McNamara who not only opposed nuclear propulsion but also having a lot of carriers, at all. As I wrote in my contribution to the Naval Institute's 1967 edition of their annual *The Naval Review*, "McNamara's reasoning went adrift in arguing for a reduced number of aircraft carriers," and "I'm very disturbed by his consistent hewing to theory in the face of reality and common sense."

After my joust with McNamara iced my career, I thought, I left Hawaii on 3 November to spend most of the month in Japan, my first long visit since I'd gone to Hiroshima in 1945. Part of my job was to help train Japan's Maritime Self Defense Force and tie it into our ASW program, so my first calls were on Burke's friend, Admiral Nakayama, head of Self Defense operations, and Admiral Tamura, their minefield guy. Nakayama promptly invited me to play a round of golf. Adding in the "hot battsu" and twenty-course meal served by the smiling geishas, it took all day to complete, and my log says, "Japan's economy is really booming. Yet, Americans know so little about their culture and they know so little about our people. We all better learn. A lot of our future in the Pacific will rest on our relations with Japan."

After seeing Seventh Fleet commander Tom Moorer and my own ASW Third Fleet boss in Yokosuka, I went to Kanazawa, Kyoto, and Nara to discuss MK-5 mines. Then I went to Shein Maya to see the big eighty-thousand-pound Kawanishi flying boat. The only big seaplane still being built anywhere, its complete boundary layer control was supplied by a separate turbine engine mounted amidships on the upper wing. It could get airborne at only forty-five knots and land nicely even in a rough sea. BuAer was helping Kawanishi, technically, and Kawanishi people said, when I came back, they'd let me fly it. Then, back in Tokyo, under the date, "11/23/63," my diary notes, "Terrible day. My former Big E task force chief of staff, Capt. Harry Cook, and I were down in the Ginza when we heard Jack Kennedy had been assassinated in Dallas. Everyone on the street offered us condolences."

Three weeks earlier, jarred by seeing on our TV news that Buddhist monks were immolating themselves in Saigon, protesting oppression by the Catholic government of Ngo Dinh Diem, a guy we'd put there in the first place, Kennedy had our ambassador, Henry Cabot Lodge, urge a military junta to depose Ngo. The junta had done that, then murdered him. By 25 November, a rumor in Tokyo said that, angered by our "killing Ngo," some South Vietnamese had assassinated Kennedy and Madam Ngo had sent Jackie Kennedy a note, saying, "Now you know how I felt."

Shortly thereafter, the name of Mr. McNamara's fossil-fueled carrier was changed from the *United States* to the *John Fitzgerald Kennedy*, but my track on Vietnam had begun earlier. Some typical notes: 4 September 1963—"Every day's CINCPACFleet morning briefing says southeast Asia looks worse. Where Washington wants us to go out there, or even if they know, is not clear." On 25 September—"Kennedy putting 15,000 Army troops in South Vietnam I believe is unwise. He says they will be merely advisers to South Vietnam's army but because that's a guerrilla war, how can they not get involved? And when they do, what do we do?[3] In November-December —"General Harkins, head of the MAG (Military Advisory Group), is a good man but what's our objective there? Maxwell Taylor, McNamara, if things continue, are going to sink us deeper and deeper into a commitment to Vietnam."

In January 1964—"We've been involved since Truman, pressured by de Gaulle, recognized Vietnam's Bao Dai government, but we were wrong. Ho Chi Minh was our friend. The place is a loser unless we go to all-out war. Half-hearted efforts can lead only to defeat." And in June—"They're trying to run the war from the Pentagon basement. I think McNamara is making an awful mistake. Taylor has taken Lodge's place, Army General Earle Wheeler is now Joint Chiefs Chairman, a General named Westmoreland is replacing Harkins but where's our game plan?"

That month, since my ASW command would do it if ordered, I briefed Wheeler on how we could mine Haiphong harbor. Mostly-Soviet freighters are delivering 85 percent of the war material the North Vietnamese are using, I said, and a minefield at Haiphong would stop that cold. I told him we'd planned this early on and already had the mines in the forward area. We could lay them down in a day, I said, set to go inert whenever he wanted, six months, two years from now. But, he wondered, what if they sank a Russian ship? My answer: Tell the world we're going to mine the place and any ship there will proceed at its own risk. Of course, we didn't get the order to lay the minefield until after Richard Nixon was president. By then, Hanoi had lost every battle but was winning the psychological war to demoralize South Vietnamese and would sour Americans on the whole business so much they'd demand that our politicians pull us out of there.

That same month, Oley Sharp became CINCPAC and ran some very dicey meetings on Vietnam. I realized later why Bruce Krulak—who wanted to go all-out in Vietnam—never became marine corps commandant, and why Oley Sharp—the smartest of all our admirals, I thought—never became CNO nor ended up Joint Chiefs Chairman, as he should have. At one of the closed meetings, I saw Oley in blunt, nasty terms tell McNamara he didn't know what he was doing in Vietnam and, "Westmoreland, you have no strategy to win the war!" In those years, I was very critical of the military on that. Few really spoke up to Johnson and McNamara. The guys who agreed with those two got promoted. Oley was hung out to dry.

My diary was still at it in 1965. In January—"Lots of talk about the Viet Cong attacks on our bases. We may own the day but they own the night, and knowing who is who at night is difficult. We're now MACV (Military Assistance Command, Vietnam), Westmoreland the boss of Army operations. He's supposed to report to Sharp, CINPAC, which the Army objected to strenuously but lost the battle. Fact is, Washington talks to MACV direct, most of the time, expanding U.S. forces in numbers and scope of operations." And in February—"The McNamara-Taylor so-called strategy of flexible response is silly. It lets the enemy control the pace of the war. I agree with LeMay. If we're going to fight, we should bomb them back into the Stone Age." And in April—"Our leaders are adrift. The Tonkin Gulf resolution is a charade and LBJ's we-can-have-both-guns-and-butter routine is a completely dishonest statement."

About then, my command was involved in shipping control in the South China Sea and in being sure that no Soviet submarines got to Yankee Station. As a result, I was in the area a lot and thus flew what became my last navy mission, an ECM (electronic countermeasures) in an A3D off the *Coral Sea*. Its skipper was George Cassell and the ship was part of Carrier Division One commanded by Adm. E. C. "Eddie" Outlaw, my Ops officer when we delivered VC-5 to the Med many years ago. The mission was to tape all the electronic emissions over North Vietnam. Knowing frequencies, pulse width, and repetition rate is the first step to doing countermeasures against enemy electronics. Later, asked how come they let a one-eyed old guy up in that A3D, I said, "Well, with the carrier's ball-light system, it's easy these days to land the plane. And if I got in any real trouble, I had a guy in the back seat to bail me out. Besides, no one seemed to want to tell the Admiral he couldn't fly that airplane."

In June out of Hawaii, Lili and I swung through the Orient together— Manila, Hong Kong, Kowloon, Kyoto, Nara, Nanasawa, Wakura Springs, to

Tokyo by rail, and Yokosuka. Lili loved it. The best part of my taking a demotion in order to command the *Enterprise* was being able to give her the kind of fun time rarely enjoyed even by rich folks. I'd called her my Sea Gull as she met our ship in ports all over the Med, a small compensation for all the years of my being off on a mission, out at sea, leaving her to cope with caring for the children and our quarters at home. This trip seemed a fitting way for us to fade out of the navy.

In August, I was hit hard when Vlad Reichel died suddenly. He'd been like a father to me. And I was hearing only rumors on what the navy had in mind, if anything, for me. Lili and I sailed out of that fog, briefly, to Seattle so she could christen the world's largest hydrofoil, the USS *Plainview*. Then I flew to Dallas to nail down my future with Dresser Industries. Its chief executive, John Lawrence, did that nicely, handing me a check for $10,000, my first month's salary, to take with me. And my log says, "As my career ends, I'm very philosophical. It's been fun, exciting, and who'd have thought I'd ever make Admiral, let alone Vice Admiral."

In late October, the Chief of BuPers, B. J. Semmes, said I could have the Naval War College but added that he couldn't promise it. "A sad, sad commentary," my diary says. "I'm adult, senior Vice Admiral in the Navy! Why play games? Why do they even worry about my leaving?" Then on 3 November, I was in Tokyo when Paul Nitze called personally to tell me the War College job was mine and to call him Friday on the subject. I must say no, I told myself, and on the 5th I wrote, "Today, I quit the Navy! I'm relaxed and ready to go."

But when I told Lili we now had an option besides Dresser, she said, "Stay in the navy if you want. We don't need a lot of money."

And a month later, on 19 December, after I'd mailed back Dresser's $10,000 check, I recorded, "I didn't quit the Navy! Never thought I'd change my mind but Nitze said he'd back anything I wanted to do to make the Naval War College a really first-class graduate school."

On 15 January 1966, as I received a Distinguished Service Medal, Jack Chew relieved me and with all the proper salutes, my flag came down off the *Yorktown*. Since Newport, Rhode Island, is about as far from Hawaii going west as going east, we decided to go west, a three-week journey. We went to Tokyo, then down-country to play golf, to Hong Kong to see more old friends, and to Thailand's tiger country. We reached New Delhi right after Prime Minister Lal Badahur Shastri died and I was told to represent our navy at his funeral. We also saw the Taj Mahal, and the "red fortress" of ancient Indian history; then in Istanbul, the Topkapi museum and the Blue Mosque. I also got a briefing in Istanbul from our naval attaché on what Mc-

Namara and Johnson, principally, were creating in Vietnam, as micromanaged yet irresolute as the Bay of Pigs, ignoring an old adage, "If you're going in, go in hard and fast with a clearly overwhelming military force." Or as Churchill had told Forrestal in 1946, "A gesture of power not fully implemented is almost less effective than no gesture at all."

Next, we saw Rome; Paris, not as much fun with Vlad Reichel gone; Madrid to visit its Prado museum; then the island of Rota for the plodding flight in a piston-engine aircraft to McGuire Air Force Base, New Jersey, and finally an Allegheny Airlines flight to Providence and car ride to Newport. Our quarters at the College, the president's house, was a large, old, well-kept home, overlooking the bay and just three blocks from where my navy career had begun forty-one years earlier. And in late February to ruffles and flourishes, I became Naval War College president. Part of the deal I had with Nitze was that I'd report to him, not BuPers. As one small proof of his sincerity, he let me bring with me from my ASW staff my administrative assistant, Frank Ginn, and public relations guy, Walt Aymond, both excellent at their specialties. Nitze's backing produced the same from Under Secretary R. H. B. "Bob" Baldwin, a former top executive at Morgan Stanley and Nitze's close friend, and from Assistant Secretary for Personnel Chuck Baird.

Established in 1884 in an old "poor house," the Naval War College was the world's first educational institution designed specifically to prepare navy officers for high-level command. And under its first two presidents, Commodore Stephen B. Luce (1884–1886) and Capt. Alfred T. Mahan (1886–1892), it had done so. But subsequent navy operations people never really bought the War College as the place for that mission. They believed officers should go to sea for their education. As a result, in the eighty-two years from 1884 to 1966, the navy had invested only about $11,000 a year, on average, in the War College. A clue to that neglect: Sims Hall, the College Command and Staff school, was built in World War I, and had been my barracks in 1925. The navy's first-priority investment in officer education, all those years, had been the Naval Academy.

Another War College weakness was the faculty. Until I came in, it had been mostly passed-over people, mainly commanders, on their last tour. So were most of the students, mid-ranked officers in the twilight of their careers, unlikely to be appointed to a top-level command or staff position after graduation, let alone make admiral. Many of our World War II leaders had come from the War College: Nimitz, Spruance who was president here after the war, King; but nobody from the War College since Don Felt on the staff in the late 1940s had been selected for admiral. They'd started a run at that problem before I arrived when the vice CNO had approved hiring as the faculty dean a first-rate professional, Dr. Fred Hartman from the University of Florida. He came aboard in July 1966, and we instantly were friends. I relied on him a great deal over time, and gradually, under his leadership, the faculty became primarily civilian, all excellent teachers, expert in their fields. Nitze, Baldwin, and Baird firmly backed that move, too.

Especially helpful was their taking care of the cash flow when I said we must hire more civilian "Chairs" than were allowed, at the time. "We have much too few," I said, "and they're underpaid. We can't compete in the world of academe at these prices."

Initially, however, my agenda had two major projects: build a new college and create a Board of Advisors. As I'd seen at China Lake with CalTech, an advisory board can be very helpful, raising money and pressuring others to help us do what we wanted to do. With Baldwin's help, we enlisted a dozen people. They included Dartmouth College President Fred Dickey to chair the Board; "Containment" strategist George Kennan, who'd become head of Princeton University's Institute for Advanced Studies by this time; and Ed Teller. In addition, we signed up Standard Oil of New Jersey's executive vice president; the president of the Rand Corporation; two retired navy vice admirals; the senior partner at Sullivan and Cromwell; John Nicholas Brown's son, J. Carter Brown, curator at the Mellon Gallery (now the National Gallery of Art); and Columbia University's Maurice "Max" Tauber, a nationally recognized expert on building and managing libraries.

On the new-college project, helped by the chief of staff I'd inherited, Fred Schneider, an experienced, "gung-ho" officer and Ernest King's aide for years,[4] we brought in the architects and told them what we wanted. The upshot was a request for authorization to invest, starting in FY 1968 (1 July 1967 to 30 June 1968), $28 million in construction, $27 million more than had been spent on construction in all its previous existence. To help sell that, I enlisted a "Bonnie and Clyde" team, I called them, two bright captains, Al Pickert and Dan Morgiewicz. They began to crisscross the land with a set of impressive flip charts and a detailed knowledge of the college's history, giving a well-rehearsed spiel to navy field commands and others on why the War College soon would be important not only to the navy but to overall national security and foreign policy. To navy audiences, especially, they also noted that young officers with fine service records soon would improve their chance to make admiral by attending the Naval War College. And of course, strongly in favor of what we were trying to do, my Board of Advisors were disciples, too.

I had a little clout in that department, myself. I invited my friend, Henry Jackson, and Bob Sikes, Congressman from Florida, up to see our mediocre condition and the lack of even minimal maintenance. Of course, Sikes owed me one. When I was in R&D and we wanted to build a Naval Aviation and Space Medical facility, Bethesda Naval Medical Center had wanted it but Sikes wanted it put in his home district, Pensacola, Florida. I'd put it in Florida. He might have been a crook, as Lili said, but he was a good friend of mine and he chaired the House Public Works Committee, which ruled on all appropriations requests for military construction.

However, in just the navy part of the defense construction budget given Congress in 1967, our $28 million was thirtieth on the list, meaning it was not likely to go anywhere. And we'd already taken a hit on the way to Capitol Hill. We'd wanted a new library. Ours had less than one hundred thousand volumes and, to be a top-level graduate school, a college should have at least two million books in its library. But McNamara knocked out the library—and had to dig way down into the tiny details of the navy budget request to do it since the total Defense Department spending request for FY 1968 was approximately $65 billion. I couldn't do anything about that because everybody already was commenting on how "posh" I was making the place. And the navy wanted the money to go to the Naval Academy. Still, that construction budget is a wheeler-dealer kind of thing anyway, and when the navy top brass met with Sikes' committee, he said, bluntly, "If you don't give Chick his money, you won't get any money, at all."

And CNO (as of 1 August 1967) Tom Moorer folded.[5] Jackson and Pastore nursed our request through the Senate. Meantime, in 1966 and 1967, we did a lot more. To the few Chairs the War College had, I added Chairs in economics, naval science, comparative cultures, military management as opposed to command and operations, continuing education, and international law. In cooperation with George Washington University, we also enhanced our foreign affairs and national security Chair so that ours became the first war college authorized by academe to award a master's degree in international affairs.

Since the incoming students were extremely diverse in their professional and education backgrounds, starting that first year, I made all of them take a Graduate Records Exam covering both mathematics and verbal. That upset some of them but the results showed they needed it. Many were very low on the verbal end of the tests. And, I told them, for their thesis, my door was always open and I would discuss any proposed topic with any student who wished to do that; but I made a rule that the aviators couldn't write about aviation and the submariners couldn't write about submarines, but that they must branch out to other subjects and broaden their overall knowledge of the other services. I instituted "President's Hour" once a week at breakfast where I could ask the students any questions I liked and they could ask me anything about policy or what was going on in the college. Initially, I talked mostly about plans for the new college and listened to a lively discussion on Vietnam and the draft business. I really had fun with that one and certainly learned what was bothering them.

With help from Walt Aymond, mainly, I also got invitations to lecture at Harvard, Tufts, Brown, Dartmouth, and many other colleges in order to get their best people to come lecture at the War College—and heard, again, the burning resentment of their students on the unfairness of the draft. We also modernized the Naval War College curriculum and, as we acquired top-drawer faculty, got them accredited to give degrees. We also organized seminars and war games, inviting outside civilians in to participate and learn more about us. One interesting aspect of the war games, I saw, was that the amateurs always talked strategy and tactics while the pros talked logistics, knowing that in a battle, if they didn't, they could be out of ammo and not know it.

And of course, we continued the foreign affairs seminars begun by Arleigh Burke in 1958 over stiff State Department opposition. As he'd decreed, we invited each of, eventually, thirty-nine countries to send just one representative. If we let in two, Arleigh said, after the debates one might go back home and slander the other one to their boss, president, or dictator for whatever he said. So I stuck to the rule, and the Israeli and the Egyptian got along just fine. The whole drill has been a success, great for the United States, especially since we got the best of their naval officers, many becoming head of their own navies. Now, the other war colleges each also have a foreign officers class.

My one regret while I was there was that we had more army and air force and civilian students than navy. The Vietnam war was the main reason. Army and air force individual tours there were only thirteen months, fitting nicely with a school year. Navy people, especially aviators, were tied to combat chores that didn't schedule neatly with the end of a school year. Still, my diary suggests I wasn't totally immersed those years in the War College. On 30 April 1966, I noted, "Nephew Steve Madey goes to the Naval Academy this June. He's a nice lad, will make a good Naval officer." And on 30 May, "John has enrolled at the University of Arizona and its Naval ROTC program, did it himself, a good boost for his confidence." And on 2 May 1967, "Headed a Captain's Selection Board which selected Frank Ginn and Walt Aymond. I'm sure from their records, if I'd not been War College President, they'd probably not have made it. Still, my colleagues tried to promote two guys because they had 20-year 'spotless records.' 'You mean they never argued, never had a difference of opinion?' I asked. 'What kind of man is that?'"

Back at the College, in 1967 and 1968, we broadened the basic curriculum to make it relevant to the navy's operating environment and added wide-ranging elective options. As a result, students received a baseline of instruction in naval warfare and command-staff basics, and a shot at specialized, graduate-level studies in international conflicts, economic aspects of military plans and spending, military "business management" as part of "command," the systems analysis business, theories of war, military roles and missions, and the organization of and budget process in the federal government. As I told a *Wall Street Journal* reporter, "When I got here, there was an obsession with the procedures of military planning. That's strictly a mechanical, not an intellectual exercise. I wanted the War College to be an intellectual experience for every student."[6]

Finally, on the advice of doctors who said the old, one-eyed ex-bluejacket might lose his disability pension otherwise, I retired on 1 September 1968. Except when I talked Red Raborn into relieving me in 1962, my War College tour was the only time my choice was who, in fact, relieved me. A relatively junior rear admiral, Dick Colbert, had just been passed over for some three-star post in Norfolk. He was a bright, talented guy who'd had Arleigh Burke's first foreign-student class at the War College in 1958. So I recommended him to navy Secretary Paul Ignatius for my three-star job and, with Burke's help, he got it. He knew what the War College needed, what we were trying to do, and did a marvelous job seeing to it that what we'd started was completed—including building the big new library.[7] Of course, having a vice admiral in charge, one clearly headed for future top-command posts, didn't hurt at all our wanting people to know how important the Naval War College is. And today, it generally is recognized as one of the finest institutions of its kind in the world.

And my diary for 5 September 1968 reads, "It is over. The ruffles, flourishes, gun salutes echoed across the bay. The President awarded me another Distinguished Service Medal. The aircraft salute as they roared across the College sort of brought a lump to my throat. After 43 years, we own a house in Atlantic Beach and I have the large sum of $25,000 in the bank."

20

Where Are the Leaders?

In his "First Report" on the "National Military Establishment," issued by the U.S. Government Printing Office in December 1948, Secretary of Defense Jim Forrestal noted, "The capacity for making war is not separable from economics any more than it is from diplomacy." Moreover, he added, "An unstable world endangers our security because we no longer are isolated—if we ever were. Advances in science (supersonic aircraft, ICBMs, nuclear submarines with a world-wide range) have wiped out the barriers of time and space, oceans and icecaps. It is not enough for us merely to devise what we hope will be an impregnable defense against new weapons and methods of warfare. As a primary precaution, we must strive to prevent (war's) inception anywhere . . . bring order out of chaos wherever it exists."

My exposure to Fermi, Oppenheimer, Bohr, Lawrence—all those people—had made me realize how little I, like most naval officers, knew that we thought we knew on the technical side. But for a time, we went too far in that direction. For instance, the Naval Academy soon had seven technical fields from which a midshipman could choose a major when MIT had only five. It didn't make sense, not when transportation and communications gains were shrinking a world in political turmoil everywhere, telling us that to cope with it all, we'd need national-security leaders who are broad-gauged. That was largely what drove the expansion and modernizing of the War College curriculum.

What Forrestal had noted and what many of us accepted as self-evident in the late 1960s was what was called, at the time, "the total approach to national strategy." This meant that devising it against possible threats, then implementing it, was not about just science but about economics, politics, psychology, and military power as well. And we have to learn to live with cultures markedly different from ours. A naval officer assigned to a Middle East mission had better know why a Shiite is different from a Suni, for instance.

The same goes for Buddhists and the Japanese culture in the Orient. Our navy education process must produce a well-rounded person.

By 1968, we hadn't yet seen much payback from the new, comprehensive War College syllabus, but we were attracting a much stronger group of students whose service records said they'd likely be given high-level command-management duties, later. Thus, we could anticipate, as I did in a magazine article, "It is ridiculous how people try to isolate today something called the 'military mind.' There's no such thing. If the military are to take their rightful place in the scheme of things, they not only have to be better educated themselves in all areas but also better educate others with whom they deal on just how broad and deep their knowledge has become."[1]

As for my career's impact on my family, the only really heavy load was being always in debt. But the bills got paid, and Lili's parents helped over the years. In fact, it was years before I realized that not all navy wives went home to Mother every summer as Lili did. And at times, as at Inyokern, the Ordnance Lab, and Hawaii, we lived in high-quality quarters rent-free. One big sacrifice was that with constant duty-station transfers, we never had hometown roots like most people, except Pensacola in the children's early years until after Mommie Floss and Daddy Pete died.

Each of our children paid the price at one time or other, being jerked out of one school, angry or sad at having to leave friends and make new ones somewhere else. On the other hand, they were much more mature than I was as a teenager. So what we lost in apples we made up in oranges. The nomadic life may help explain why my son was so rebellious, growing up. At the University of Arizona in his junior year, when, to stay in the navy ROTC program, he had to agree to serve five years after graduation, he left school to enlist in the navy for a two-year tour. That's why he was on the *Barrier,* the same destroyer now on display at the Washington Navy Yard, in the South China Sea during part of the Vietnam war. He turned out fine, spending most of his working life helping General Dynamic's Electric Boat division build submarines—and having as much trouble raising his three boys as Lili and I had raising him.

As for me, Dave Lewis, the chief executive at General Dynamics, offered and I accepted a job as his vice president for international operations with authority to also poke around in its R&D activities. As in the navy, I wasn't too popular, being dumped in there on top of all the guys trying to make it to the top from inside. I operated out of our New York City office but also had an office in Tokyo. To run it, I hired retired Japanese Admiral Hiramatsu who used to shoot at me over Rabaul during World War II. I told him that his being a bad shot was why I hired him, but the fact is, a foreign company just can't do business in Japan unless it has a plant there or a Japanese agent. I also had offices in Canberra, London, Bonn, Athens, Tel Aviv, and Teheran. And as she'd followed my ship when I was in the Med, my "sea gull," Lili, went with me on those international visits. Except she wouldn't go to the Middle East. She stayed in Rome.

Arleigh Burke used to say my navy career was "mangled, not managed." Yet, because I'd flown 150 different aircraft types, built and run laboratories, and managed R&D projects into production during that career was why General Dynamics hired me. To some extent, my new job just meant wearing a business suit instead of a uniform to work. One example, reminiscent of my navy battles, was the air force F-16 fighter we were building. I was shaken, at first, to see that it didn't have a stick in it. The air force would not have gotten the F-16 if air force Secretary Bob Seaman hadn't insisted. His requirements folks would have groaned, "We've never tried fly-by-wire. We can't do that." But with fly-by-wire, the pilot can't get into an unstable position as he could with the old full-span-spoilers business. The onboard computers won't let him.

My first two years with General Dynamics, I earned more money than I'd received in my entire navy career. I worked for them until 1973, then was a consultant to them until 1979. After that, I worked with the Draper Laboratory part-time and with the Hertz Foundation on its PhD program that, because of wise investments, now has more money than when it started. That program pays selectees $20,000 a year, so marriage is not an obstacle to signing on. In return, they have to agree to earn a PhD within three years and work some portion of their subsequent careers for the government. Off an annual applicant list of about a hundred persons, we interview them all and those we don't select we usually offer to the National Science Foundation. Quite a few of our picks, so far, have become Nobel Prize winners. I'm also involved in my fading years—I tell them the body parts are breaking down but my head still works—with the Hope Haven Hospital for Children, Naval War College Foundation, and the Naval Aviation Hall of Honor.[2]

I've had a full life. I've done many things I wanted to do, didn't do a few things I should have, and had to do many things I didn't want to do. All in all, I've got no complaints. Except about the navy, specifically what former navy Secretary Jim Webb calls, very incisively and accurately, "leadership failures at the highest level."

During my navy career, I was disliked, even hated, by some people for my frank, openly stated opinions on everything from what new weapon we needed, to what policy or strategy we ought to employ. So were Ernest King, Arleigh Burke, Deak Parsons, Arthur Radford, Dan Gallery, Tom Moorer,

Tom Connolly, and several other officers in my time. In fact, nearly all the bellwether advances in navy force structure, equipment, and doctrine in that era were instigated by such people, what the so-called "Establishment" called "rebels, disruptive of good order." Today's top navy leaders, and higher-ups clear to the White House, for that matter, act as if being popular with the "politically correct" crowd is more important than being honest.

The evidence is as rife as barnacles on an old garbage scow. Just ten years ago, the navy was shrinking down to 600 ships including 238 surface combatants—100 AAW (antiair warfare), 37 ASW destroyers, 101 frigates. Rather than fight that trend, a succession of CNOs in the 1990s have said they can live with a 300-ship fleet, easy to say since they lack a clear, coherent instruction for its use. Indeed, our highest governmental levels seem indifferent to the need to maintain a strong navy, one that can control the seas, and keep wars far from our shores. Of course, a present day CNO hasn't much leverage on that since he's only chief of the Washington navy staff, not chief of naval operations.

Anyway, it's his successors who will have to live with an eroded naval force, one only the commander in chief, apparently, can restore, but our current CINC, William J. Clinton, seems far more interested in doing "photo-ops" on a carrier. So the trend says that by 2003, the active fleet will be down to under 200 ships including only 112 surface combatants (27 cruisers, 62 destroyers, 23 frigates, no battleships), dragged down by a shipbuilding budget allowing construction of only six or seven new ships a year. By then, it will be impossible to do even today's operations where 49 percent, 172 of 347, of our ships are under way 82 percent of the time. Similarly, though naval aviation's combat-readiness standing has been obscured, willfully, a drumbeat of media reports about crashed planes not replaced, aircraft cannibalized to keep what's left flying, and reserves operating 20 percent shy of authorized (but unfunded) maintenance-crew complements, all say it has the same problem as the fleet. And of course a dwindling resource being run at a wartime pace simply accelerates the rate of decay.

Meantime, all the global war games the War College has played so far have focused on Europe. Yet I think events argue that our next big war, if any, will erupt in the Middle East, and could be a nuclear war, to boot. I recall when Israeli war hero Moshe Dayan had me come to Israel in the early 1950s to lecture on plane waves and shaped-charge explosives—key parts of Fat Man. (He thanked me for my visit but advised me to wear an eye patch over my blind eye like he did over his. "It'll make you look more sexy," he said.) They had the atomic weapons, I'm sure, though he'd not admit it. Finally, I said, "Well, assuming you have them, what will you use them against, Damascus, Baghdad, Cairo?"

He didn't answer, but later, when he was Israeli prime minister, Menachem Begin pretty much said—at least the Arabs thought so—that, if provoked, they'd use it on the Arabs even if that meant pulling the world down around their own ears. The so-called "nuclear family" has grown large since then: Russia, China, the Ukraine, India, Pakistan, North Korea, probably Iran and South Africa, and possibly Iraq. And today, the warhead need not be just nuclear but chemical or bacteriological, a weapon not of war but of terror. Just one nuclear- or anthrax-armed weapon launched from a submarine (a distinctly possible way) could kill thousands, millions in a city like New York, Miami, or Los Angeles. Yet, my country has no antimissile defense in spite of technology proving since the mid-1960s that we could build one. Instead, our leaders procrastinate and rationalize the threat away.

So far, I've heard nothing but weasel words on ICBMs, either. Our leaders insist on chasing the START (Strategic Arms Reduction Treaty) mirage with Russia, which for several reasons Moscow surely won't sign anyway. Further, with no plan to replace our five hundred old Minuteman ICBMs, why not scrap them? Land-based, they make the United States the target and, because of prevailing winds, a nuclear explosion in, say, the Dakotas would drop fallout over the heavily populated northeastern half of the country. We should put our deterrent at sea, treaty or not, fourteen *Trident*-class boats with 24 D-5 six-thousand-mile-range missiles. That, 336, is a lot of missiles.

We have a second big problem in that arena. Our country no longer tests the nuclear warheads nor even makes the tritium initiator, with its 12.5-year half life, any more. We learned the hard way (the pre–World War II magnetic-exploder torpedo is a good example) that weapons not tested in peacetime usually don't work in war. Livermore Laboratory is trying to build a big fusion facility to duplicate a nuclear test. They doubt it will work but they have the billions-plus budget so they're going ahead. 'Course, anyone can test the components but the total critical mass can be tested only by exploding a warhead. It can't just be left in an igloo and pulled out whenever they want to use it. And if a potential enemy doesn't believe it will work, its deterrent value is zero.

Similarly, an evident aversion to anything "military" by the present administration has resulted in a drop since 1989 of 39 percent in the size of the military work force. R&D spending is down 57 percent, procurement 71 percent since this military "down-sizing" began in 1990 during the Bush administration. And defense spending is now 16 percent of the federal budget, down from 39 percent when the Berlin Wall was torn down in 1989. But as a clue to where present-day military priorities are, the Defense Department will spend $28 billion in FY 1999 to comply with "environmental protection" rules and regulations.

And the combat commands still are saddled with the so-called "two wars" concept, a relic from the McNamara era. It required, he said, having men and equipment to fight two Korean-sized wars simultaneously on opposite sides of the globe with enough left over to fight, and presumably win, a "brush-fire" war someplace else. President Bill Clinton's Defense Secretary Les Aspin, a McNamara protege, dropped the brush-fire part, but still, when today's Pentagon leaders imply we have that capability, they are being disingenuous. To deploy, today, a force the size President George Bush sent into the Persian Gulf War would require, for instance, every combat division the army has, worldwide—assuming they're all at full strength, which almost none are.

The navy soon, in ships, aircraft, and submarines, will be what it was in 1941 when the Japanese hit Pearl Harbor, saying "Americans have no stomach for war." In 1938, in spite of our navy's weak size, morale was pretty good because nobody expected much from us. However, as today's navy fades, just in the past few years, navy and other forces have been deployed to Haiti, Somalia, Bosnia, off Africa, and several times to the Persian Gulf, which is becoming sort of a permanent duty station, all paid for largely by cutting back on O&M (Operations and Maintenance) and training expenses that are supposed to maintain combat readiness. On the other hand, does NATO really need more U.S. troops in Germany than Britain has in its entire army?

The people who man the ships, fly the aircraft, and deploy the submarines, along with the competent technical enlisted personnel who support them, are all that really count. Yet the general situation has changed compared to what we had in the 1930s or even the 1960s. The personnel are smarter than we were at that particular stage in our careers. The pay and benefits, inflation-adjusted, are as poor for enlisted personnel and junior officers as they were in the 1930s, but the complaint I hear most often is being away from their families. True, in my career, I might be gone a month or two but I wasn't deployed, as most of them are now, for six months, a brief return home, then off again. Since most will have to start a new career at the end of twenty years anyway, they ask themselves, "Why wait? Start it now so I can be in a stable home, helping my children grow up."

They know they have to prepare for it. The navy's a great outfit but when it drops the gate, you're on your own. And as long as the Washington leaders remain oblivious to that reality in the fleet and persists in conducting its social experiments, the victimized yet competent essential personnel are going to leave. Consequently, I was not surprised, just dismayed, when marine combat veteran and former navy Secretary Jim Webb reported that, in 1995 alone, 53 percent of naval aviation's post-command commanders, those due to be promoted to captain, resigned rather than continue their careers, twice the worst number ever suffered in any previous year.

The Tailhook inquisition by Congress and "politically correct" civilian authority in the navy has just beat down officers who would step up to fight for the navy's people. And of course, the feminists' carping, such as that by former Congresswoman Pat Schroeder, about women in combat has had a traumatic impact. There is a valued place for women in the navy but not in combat. We have a female vice admiral now, excellent because she's in education. But when I'm urged to agree to women in combat, I think of the hell that was Guadalcanal. And perish the thought. Israel tried women in combat and quickly stopped. Too many of their men were being killed trying to protect the women instead of attacking the enemy.

And of course, put a bunch of young guys and gals together for a while and there's going to be sex, period! I recall, shortly after all this women-in-combat took hold, that the *Eisenhower* carrier came back from six months at sea with about 15 percent of the women pregnant. So the captain decides to check all women aboard—and it hits the fan, "Invasion of Privacy!" and all that stuff, which just drives morale into the deep-six. So should the captain go ahead with the testing? And if a woman's pregnant, does he off-load her? And how does he off-load her?

He should do it the way Rear Adm. Riley Mixon handled the drug-abuse business when he had command of the carrier *Oriskany* in the 1970s. He didn't wait for any Pentagon directives to tell him what, if any, the accepted procedure was to avoid a suit over "invasion of privacy." His first priority was to maintain discipline and a firm hold on his chain of command. So he told his crew, when they come aboard after a shore leave, they'd be tested for drug use, and anyone testing positive would be brought before the mast, right away.

I was lured into that awful mess myself briefly in 1973, when Captain Eastling invited me to the south Atlantic to ride the *Franklin Roosevelt* in for the last time on her way to the scrap heap. As I boarded her, the ship was a cluttered mess, the smell of marijuana smoke everywhere. The chaplain's cabin had been ripped off a couple times. They had three guys before the mast and the captain did practically nothing to them. I went down to the chief's mess where a couple of my crew from the old *Point Cruz* were still aboard the *FDR* and asked them what gives with this ship. And they said, "Don't ask. We just do what we're told."

What I learned told me that breakdown in discipline was partly the ship's captain's fault, partly her chaplain's, but mostly CNO Elmo Zumwalt's. I sent him a blistering letter, reciting my opinion of his "Z-Grams" to "All Hands," which, I said, were telling enlisted men, especially, and black (the commonly used label now is African-American) enlisted men, in effect, that their feelings were more important than discipline and unit cohesion. Later, after our first mutiny ever aboard a U.S. Navy ship, the *Kitty Hawk,* I wrote an essay, published in 1974 in a nationally distributed magazine, saying, in part, "As leader of the permissive approach to discipline, Adm. Zumwalt earns the discredit for what happened. True, he inherited a racial problem born during Mr. McNamara's reign when the 'policy' was, a black without a high school education was worth more than a white with one. Is this 'equal opportunity' for the man who, unable to pass through any of the Navy's technical schools, ends up doing menial chores? Is he held down because he's black, or because he simply isn't qualified?

"But the real question is, since the Navy's mission is control of the seas, why would such a goal-oriented organization pursue such a policy? In sum, the seeds of mutiny were sown before Zumwalt became CNO. My criticism is of what he did when it happened. He and the Secretary of the Navy assembled all the flag officers in the Washington area and dressed them down as if it was their fault. The Media was delighted, of course. Indeed, to many of the officers, aware of how complex the problem was long before the *Kitty Hawk* incident, it seemed the whole meeting was held simply for its public grandstand value. One wonders what would have happened if they'd had to fight a war at sea, right then."

A defiant "Black Power," symbolized by sticking a clenched fist in the air, rumbled across the land at the time, in shipboard "sit-down" strikes, in city-street riots, not to advance some noble cause, really, but as an excuse to loot stores. The Zumwalt-type answer made no sense to me, even today. How can racial "integration" be achieved through segregation? Far better, I thought, for navy morale and combat readiness, was to go "one-on-one" like the time when, with navy Undersecretary Bob Baldwin's help, I got Cdr. Sam Gravely command of a destroyer—not because he was black but because he was bright, capable, and I thought could become, as he later did, our first African-American admiral.[3] The "black equality" business is just an interesting statistic, all it deserves to be if the goal is, as it ought to be, building and maintaining the strongest navy in the world.

Adm. Tom Hayward was the guy who put the "squash" on drugs in the navy, starting about 1975. (My dad once told me all the Haywards in the

United States are related, but I never bothered to check that out.) Today, get caught using drugs in the navy and it's goodbye, right now, sayonara. But with today's vaguely undefined "discrimination" against women in combat and "sexual harassment," it's really become the history of the drug-use days revisited. Mike Boorda and I began to correspond, mostly on that subject, when he was in the personnel business and a good deal after he became CNO. We had in common a strong bond. We'd both begun our careers as bluejackets. I'd made it to vice admiral. He'd become the first sailor to make it all the way to CNO without having graduated from the Naval Academy.

In 1991, he sent a plane down to Atlantic Beach to take me up to Norfolk to join Arleigh Burke for the christening of the first *Burke*-class destroyer, making Arleigh one of the few navy men to have a class of ships named in his honor while he was still alive. Before the ceremony began, Mike said, "I guess you and Admiral Burke never expected to see me here." We both laughed and I said, "That's right."

That first ship had awesome firepower, surface-to-air missiles, Tomahawk and Harpoon surface-to-surface missiles, Phalanx rapid-fire guns, and torpedoes, but Arleigh didn't like it because it didn't have a helipad on the fantail. Even after that was corrected, it still had no hangar back there to store a helicopter in bad weather. Yet that helicopter is what can bring to bear the full potential of the destroyer's firepower. A helo's synthetic-aperture radar gives a destroyer the ability to see out at the far end of a two-hundred-mile radius around the ship. Moreover, what the helicopter sees can be piped back to the ship's skipper on the bridge so he sees in real time what the helo sees, not a blip on a radar screen. He can identify targets by type, and fix precise coordinates. And now, with satellite data links a practical fact, a Pentagon command post also can see what the skipper sees—which can cause a problem. If some kibitzer in the back room starts trying to run the battle, should the skipper shut off the kibitzer's radar or not?

Anyway, when Arleigh Burke died on 1 January 1996, Mike sent a plane down again, to take me up to Annapolis for the funeral, gave me the assistance of an aide and everything. He put me in the front row at the service, between Ross Perot and President Bill Clinton (who delivered a very nice eulogy.) Afterward, one of my grandsons, a Naval Academy first classman at the time, said, "You've got some funny friends, grandpa."[4]

Mike Boorda was under heavy fire by then. In one case, a female pilot with "only two down-checks" got killed, trying to land on the *Lincoln*. As her forty-ton airplane started its approach, she overshot the line. Her fatal mistake was trying to maneuver back into line, a desperation move, no need for it. A good pilot would have gone up and around again. Then they tried to blame the crash on engine failure, a cover-up. The cardinal rule since the navy first got airplanes is that a full, honest investigation is done on all crashes, not to punish someone but to see if anything needs to be done to improve safety in that inherently dangerous occupation. Then they grounded the plane's back-seat guy. That was Boorda's mistake.

On the heels of that tragedy, a Navy Board sent Vice CNO Stan Arthur a report saying another female pilot, with seven "down-checks" from seven different instructors, was going to kill herself, or somebody else, if she wasn't grounded, permanently. Arthur, a five-hundred-mission Vietnam combat veteran who also commanded in the Persian Gulf War the largest task force we'd assembled since World War II, endorsed the report. So she wrote to Sen. David Durenberger from Minnesota, claiming she was sexually harassed, discriminated against because she was female. Then Mike nominated Arthur to be CINPAC and Durenberger said, "No, I won't approve it." Mike had said nothing as Congress ruined a lot of innocent navy pilots' careers over Tailhook. Now, he was torn between doing what was right and what the "feckless," one magazine called him, Navy Secretary John Dalton wanted him to do: ram the "women in combat" crusade down navy line officers' throats. And Mike made another mistake. He flew out to Chicago to try to influence the woman, offer her a job. He was crazy to do it. All she did was tell him what he must do to get women "accepted equally with men in navy combat roles," then repeat that at a press conference. My advice to him was, "Put your four stars on the table, fight it."

But he didn't. And Arthur had to retire. That was a crime. Then on 15 May, I got a letter from Mike in which he said he'd mishandled the female business and "worse, how I let down my good friend, Stan Arthur" by not defending him. In the margin of his neatly typed letter, he scribbled in his own hand, "If I thought it would help the Navy, I'd kill myself."

I was shocked when I read that but the way he phrased it, I assumed he wasn't serious. But on the 16th, he committed suicide, a tragic end for a fine person. Though Mike was Jewish, I know, if my mentor, Father John Brady, had been around, that wouldn't have happened.

On a much broader scale, I think World War III is developing right now, in the Middle East, Africa, the nation-states once part of the Soviet Asian empire, wherever our system is interfacing with an opposed, dictatorial system. Besides a rise in terrorism, in the Islamic jihad business, some sixty border and civil wars are being fought, right now, in Asia, Africa, Eastern Europe, and South America. In December 1998, Russian Prime Minister Yevgeny Primakov, from New Delhi, no less, called for a "strategic alliance" of Russia, China, and India, a "troika," Nikita Khrushchev probably would have called it. Were that to happen, it would ally three of the world's largest countries, all armed with nuclear weapons, two of them exporting that technology not only to each other but to other enemies of ours. Each already is negotiating bilateral treaties with each of the others "to bring greater stability" not just there but "to the world."

And now North Korea, China, India, Pakistan, and Iran all have ballistic and cruise missiles—in spite of our intelligence community (also now decimated by budget cuts) insisting, recently, that it would be fifteen years before they had those missiles. Moreover, I can see only a high risk to our security in President William Clinton's allowing American companies to sell militarily useful technology to China; trying to cede China our Long Beach, California, naval base; shunning Taiwan; and, on China's demand, banning our navy from international waters between the two.

In sum, our irresolute leaders are chasing minor, outdated issues while we really face an ominous political, scientific, engineering, economic, and military challenge we must meet. Yet we're not devising plans, weapons, or communications to face, let alone fight it. We paid a terrible price from 1941 to 1943 for that dereliction of military and civilian duty in the 1930s. Are we destined to repeat that awful mistake? I'm afraid we are. Only this time may be the last time.

Appendix

Thumbnail Biography of Marine Corps Brigadier General Richard W. Hayward, 1906–1989

In 1988, Chick Hayward's elder brother, Dick, told the author of *Military Leaders since World War II*, published by Facts on File, Inc., "We had to get Jack into the Naval Academy to keep him out of jail."

Dick Hayward's own military career was as colorful, as heroic, in many respects as his younger brother's. Richard Wright Hayward V was born 8 January 1906 in New York City; he attended Oakdale Military Academy on Long Island and schools in the Washington, D.C., area. He was attending Cornell University, Ithaca, New York, when, in 1926, he left to enlist in the United States Marine Corps.

In 1927, President Calvin Coolidge sent Hayward's marine unit into Nicaragua to end a revolution started by guerrilla "general" Cesar Augusto Sandino. (In the early 1900s, U.S. business investment in Central America was booming. Most noted—some said "notorious"—was Boston-based United Fruit Company. At the time, it dominated the area's agricultural (mostly fruit and sugar) industry, and whole governments of some "Banana Republics" as well, in order to protect its investments. Even before 1927, U.S. presidents had used "Gunboat Diplomacy" to do the same. It's still used today, mostly for political, not economic, reasons.) For his leadership and "bravery under fire" in Nicaragua, Dick Hayward received a battlefield promotion to second lieutenant.

From 1928 to 1943, he was in the Philippines and China on shore duty and/or commanded the marine unit on ships assigned to our Asiatic Fleet. His one near-fatal crisis in those years was aboard the USS *Fulton* on antipiracy patrol in the South China Sea in March 1934, when an explosion and subsequent fire destroyed it. All hands were rescued by the Royal Navy's HMS *Wishart*, which became Lord Louis Mountbatten's flagship in World War II until the German Luftwaffe sank it.

On the eve of World War II, he was one of the first officers to volunteer for the marine parachute program, becoming a parachute battalion commander, then regimental executive officer until both the paratroop combat units and the program were disbanded. His Pacific-war combat record, most of it in command of the Eighth Marine Regiment's First Battalion, began in New Zealand and went from there to combat on Guadalcanal, New Georgia, Saipan, Tinian, Okinawa, and finally to occupation duty in Kyushu, Japan. During the last half of the 1940s, he was a student, then on the faculty, at the Army Command and General Staff College, Fort Leavenworth, Kansas; an analyst in the Office of Naval Intelligence, then chief staff officer to the

Joint (army, navy, air force) Secretaries, then senior marine corps aide to the secretary of the navy, all the latter three assignments being in Washington, D.C.

In February 1951, he took command of the First Marine Division's Fifth Marine Regiment in South Korea; and led them north from Pusan until, at the Yalu River, North Korea's northern border, a giant Chinese Communist counterattack (China called it "The Fifth Phase Offensive") surprised the allied advance and turned it into a rapid retreat. Of the Chinese spearhead hitting U.S. Marines on the Korean peninsula's west side, the April 24, 1951, New York *Herald Tribune* reported, "The massive Chinese attack on UN (United Nations) lines was opened by Red buglers at 3 A.M. blowing 'Open the Door, Richard,' [a pop music hit in the United States that year] in front of the 5th Marines [suggesting China's spy work was excellent]. Colonel Richard Hayward of New York, Commanding Officer, slammed the door and kept it shut."

That assessment was a bit rosy. Still, on 31 May, Hayward was told to seize control of a narrow mountain pass north of Mundong, Korea, to protect the First Marines' retreat south from Chosin Reservoir. In a fierce rain under intense enemy mortar, artillery, and small arms fire on treacherous, mud-choked terrain, he did just that until relieved on 2 June. For "incredible courage under fire, intrepid actions and consummate devotion to duty," says the citation, in part, he was awarded, by order of Gen. Matthew Ridgway, Eighth Army commander at the time, the Army Distinguished Service Cross, its highest award for valor. (Quickly after that, he flew to Hawaii to explain to his wife, Elizabeth, that top *Herald Tribune* war correspondent Marguerite Higgins' stories of "How I spent the night in a foxhole with Dick Hayward" were "just journalistic hyperbole, not what you might think, at all.")

After 1951, he was deputy chief of staff, Fleet Marine Force, Pacific; a Naval War College graduate, then in its Advanced Study Group on Strategy and Seapower; served on the staff of NATO's SACLANT (Supreme Allied Commander, Atlantic) and on the staff of the Joint Chiefs of Staff in the Pentagon. In March 1958, because of his combat decorations (Army Distinguished Service Cross, Legion of Merit with combat "V", Bronze Star with combat "V", Air Medal with two gold stars, Purple Heart with one gold star, and Nicaraguan Medal of Merit with one Silver Star), he received a retirement promotion to brigadier general.

Next, he was Washington-based manager of international operations for Aerojet General chief executive Dan Kimball, working with NATO governments and industries under a U.S. policy of sharing with NATO the results of American research in space and weapons technology. Then he went to McLaughlin Research, Inc., a pioneer in using computers to control military and aviation-industry supply-logistics. In 1976, he and his brother (spelled "Chic" in the Marine Corps History Division's write-up of Dick's career) formed a consulting firm on international and U.S. government affairs.

By then, he and marine corps brother-in-law Dan Pollock, noting that the Continental Congress had voted on 10 November 1775 to raise two battalions of marines, making the marine corps "our Nation's first armed force," had needled Admiral Hayward's being in "the junior Service." Admiral Hayward returned the needle by reminding them that John Paul Jones, for one, already had an officer's commission in "the Navy" by 1775, and that the corps did not become permanent, and a part of the navy, until 1798. Chick Hayward usually also added, "Anyway, you guys obviously caused so much trouble the navy had to create a Chaplain's Corps on 28 November 1775" just eighteen days after Congress created the Marine Corps.

Dick Hayward died of heart failure on 2 February 1989, after a long illness, and was buried with full military honors in Arlington National Cemetery. At the time, an obituary noted, "In addition to his wife, he is survived by two daughters, Elizabeth Bray and Cynthia Davis; a son, Richard W. Hayward VI; two sisters, Elinor [actually baptized Eleanor, see chapter 1] and Marjorie Madey; a brother, 'Chic' [meaning "Chick," but it was misspelled in the press release sent to the paper]; and three grandchildren."

Notes

Chapter 1. *The Dropout Years*

1. The gunboat's exploits are recorded in *Gunner Aboard the Yankee: From the Diary* by Henry Harrison Lewis, published in 1898 by Doubleday & McClure.
2. The Glenn H. Curtiss JN-4D ("JN" for "Jenny") biplane with a ninety-horsepower engine weighed 2,130 pounds full-loaded and had a top speed of seventy-five mph. With two open cockpits atop the fuselage in tandem, it trained 95 percent of World War I's Canadian and American pilots.
3. On 2 July 1919, R-34 (a clone of the huge airships from Zeppelin, Germany), commanded by Major G. H. Scott, with a crew of twenty-eight (and one stowaway), left East Fortune, Scotland, on a 108-hour, twelve-minute flight to Hazelhurst. Its return took 75 hours, three minutes. It was the first roundtrip flight over an ocean; first intercontinental flight to go east to west; and first North Atlantic crossing by an airship.
4. James Harold Doolittle was an army pilot instructor in World War I, a daredevil air-race celebrity by 1935, and a national hero when, on 18 April 1942, as a lieutenant colonel (he later became a lieutenant general), he led a surprise raid on Tokyo by sixteen carrier-based army air corps B-25 bombers.

    Casey Jones—an early U.S. airmail pilot—and J. Wathen helped build the Ed Link trainer, a closed "box" that simulated instrument flying. To market it, Jones then formed the JWL company with Wathen and Link as partners. It was the first and for years the best such device in the industry.
5. "Yes, we have no bananas" became the title of a song by Frank Silver and Irving Cohn, performed first in 1922 by Silver's "Music Masters" and by Eddie Cantor in a revue, "Make It," by The Pied Pipers in a 1948 film, *Luxury Liner*.
6. In March 1918, Germany launched its last major offensive along a fifty-mile line northeast of Paris, driving to within fifty-six miles of Paris by May. On 6 June, 27,500 U.S. Marines counterattacked a small bastion, Belleau Wood, in the German line. After seven days' fighting against dug-in machine guns and artillery, costing them 55 percent casualties, the marines had taken that stronghold for good.

    On 6 July in the Second Battle of the Marne, the Germans threw all their dwindling reserves into a final desperate thrust. It failed, largely because of the marine

corps stand on the Allied right flank. Nine days later, with some 270,000 Americans now at the front, a huge attack using all of Germany's forces was crushed within three days. Said the Kaiser's chancellor, "On the 18th, even the most optimistic among us knew all was lost. The history of (the war) was played out in three days."
7. Admiral Hayward has noted, "We ex-enlisted men, and proud of it, tend to forget over the years the many first-rate officers who put the care and success of enlisted folks first on their agenda, people like Admirals Holloway and Smedburg and many others, all down the line."
8. In one text, *Literature and Life: Book One,* the stanza reads: "Oh sleep! it is a gentle thing,/ Beloved from Pole to Pole!/ To Mary Queen the praise be given!/ She sent the gentle sleep from Heaven,/ That slid into my soul." A tough question, maybe, but as with many other "classics," reading that poem was routine for high schools up to about 1965.

Chapter 2. *Up-Checks and Down-Checks*

1. His service record says: "Enlisted 29 June 1925; Honorably discharged 12 July 1926; Midshipman, 13 July 1926"—not exactly when it all happened.
2. On 1 June 1813, at the start of the Battle of Lake Erie, Captain James Lawrence of the U.S. frigate *Chesapeake* was being carried below, mortally wounded, when he said, "Tell the men to fire and not to give up the ship. Fight her 'til she sinks." They didn't. Instead, they soon surrendered the battered ship to the British frigate *Shannon.*
    At the time, Oliver Hazard Perry, veteran of the Barbary Wars, was there to command a just-built fleet of ten warships. Told what Lawrence had said, Perry put a new flag, with "Don't Give Up the Ship" sewn on it, for his flagship. On 10 September, made of green wood, manned by novice sailors, his ships hit head-on a veteran British fleet. After the battle, Perry sent General (later President) William Henry Harrison his now-famed message: "We have met the enemy and they are ours."
3. Known in Germany as the Battle of the Skagerak, called by World War I historians, "The greatest naval battle in history," it was fought 31 May–1 June 1916 off Jutland Peninsula by a German battle cruiser and the German High Seas Fleet under Admiral Reinhard Sheer and Vice Admiral Franz von Hipper, respectively, against a British cruiser and the British Grand Fleet under Vice Admiral David Beatty and Admiral John R. Jellicoe. Before the Germans withdrew, they'd lost 11 of their 110 ships. The British lost 14 of their 149 but gained undisputed control of the North Sea.
4. William Sowden Sims, Commander, American Naval Forces, Europe from 1917 to 1919, later criticized Navy Department policies severely. In 1920, he and coauthor Burton Hendrick won a Pulitzer Prize for *Victory at Sea.* Sims died in 1936. Ironically, so did army air corps Gen. "Billy" Mitchell (see endnote chapter 2, n. 5).
5. Brig. Gen. William "Billy" Mitchell (1879–1936) was one of the most decorated heroes of the army air forces he had organized and led in World War I. As army air service assistant chief in 1921–25, he publicly accused the Navy and War Departments of "almost treasonable administration of the national defense" for "neg-

lect of Airpower." He repeatedly predicted a Japanese attack on the United States, and urged creation of an Air Force Department equal to the other services and "unified" in a single civilian-led Defense Department. Within eleven years after he died, his prediction and proposal both had happened.

6. Langley (1834–1906), with many scientific successes, suffered the irony of being best known for a single headlined failure. His eminence in "aerospace" and as secretary of the Smithsonian Institution led Congress to give him $50,000 to build "a man-carrying airplane" he called "Aerodrome." Powered by a five-cylinder engine, it was shot off a Potomac River houseboat on 7 October 1903, tangled in its catapult, and crashed. It crashed again on 6 December. As he fled into retirement, derided as a "fraud," a "dreamer," Wilbur and Orville Wright did eleven days later at Kitty Hawk, North Carolina, what he'd tried to do.

Chapter 4. *Life on the Covered Wagon*

1. Academy class of 1916, Arthur William Radford served on a battleship in World War I, entered flight training in 1920, was chief of naval aviation training, then an air-combat unit commander in World War II. Vice chief of naval operations in 1949, in the Korean War he was commander in chief, Pacific, and also sent military advisers to help France fight the Vietminh in Indochina. In 1957, he retired after serving four years as chairman of the joint chiefs of staff.

2. An army revolt in Spanish Morocco in 1936 migrated, victoriously for the rebel "Insurgents," to garrison towns in Spain. Nazi Germany's air and naval forces and, over time, fifty thousand to seventy thousand Italian army "volunteers" aided Gen. Francisco Franco's rebels. Russia backed the Loyalists, the seated government. The war ended in March 1939, when Madrid and Valencia fell to the Insurgents who named Franco Spain's dictator. The war's cost: seven hundred thousand people killed in battle; thirty thousand executed or assassinated; fifteen thousand killed in air raids, mostly Spanish civilians by German *Stukas*.

3. The building of the carriers *Wasp* and *Hornet* also was approved in 1934, but the *Wasp* keel was not laid until 1936, the *Hornet*'s in 1939. They were not launched until 1939 and 1940, respectively.

Chapter 5. *The Mid-1930s: Getting on the Step*

1. Born about 1540, Drake was master of his own ship, the seventy-five-foot-long *Golden Hind*, by his early twenties. First Englishman, second European (after Ferdinand Magellan) to circle the globe (1577–1580), he was called the "Scourge of the Spanish Main" for plundering Spanish towns and ships in the Americas during his voyage. Returning to England in 1580 with tons of stolen bullion, he was knighted by Queen Elizabeth I. He became an English hero when, in Cadiz, he destroyed thirty-three Spanish ships preparing to attack Britain, then helped defeat the Spanish Armada in 1588. After that, until his death, he was a privateer.

2. World War I air ace "Eddie" Rickenbacker, who built Eastern Airlines into a commercial giant, nearly died by not doing a "square-circle." In 1942, searching for possible military-base sites in the South Pacific, he left Hawaii for Canton Island, twelve hundred miles to the south, got lost, and kept going south until, out of fuel,

he had to ditch his aircraft. Luckily, he was found—three hundred miles off course, halfway to the Fijis. On a raft twenty-four days (he wrote a book about it), he, his pilot, and six crewmen were rescued on 11 and 12 November.
3. As for its namesake, the Polaris fleet ballistic missile, developed from 1956 to 1960, the project director, Rear Adm. "Red" Raborn, said, "We thought our toughest engineering task would be igniting it underwater. Turned out, it was making the Polaris realize which way was 'Up' after it had popped out of its tube."
4. And where, on the base golf course, "Chick" Hayward met army air corps Lt. Bernard A. "Benny" Shriever, a low-handicap golfer as was he. Their career paths would cross again twenty years later.
5. Coco Solo base commander Capt. John Sidney "Slew" McCain graduated from the Naval Academy in 1906. Chief of BuAer, then DCNO for Air (OP-05) early in World War II, he was promoted to vice admiral and given command of a task force in August 1943. His planes, in battles at Peleliu, Leyte Gulf, the Philippine Sea, Mindoro, Luzon, Formosa, the Ryukyus, and off Japan, once sank forty-nine Japanese ships in a single day and from 10 July to 14 August 1945, destroyed three thousand enemy planes on the ground. He died shortly after the Japanese surrender.

Adm. John McCain, Jr., was serving in London in the 1960s when he was moved to CINCPAC, Hawaii, hoping, the navy said, that would encourage North Vietnam to release his son, a navy pilot being held prisoner in the infamous "Hanoi Hilton." Hanoi did offer to do it but Slew's grandson said, "Not until everyone else is released, too." That John McCain is now a United States senator from Arizona.

Chapter 6. *Peace in Our Time?*

1. Under the heading, "Graduate School . . . University of Pennsylvania, Moore School and Temple University," records at the Naval Historical Foundation say that from 1937 to 1940, Hayward completed courses in theoretical physics, electronics, very high frequency currents, mathematical analysis, elementary foundry work, experimental atomic physics, applied gyrodynamics, procedures in experimental physics, and magnetism. Many subjects obviously were in the "atomic business," as Admiral Hayward calls it, but others clearly related to his work at the factory.
2. Officially, a nautical mile—the length of a minute of arc on a great circle of the Earth—equals 1.15 United States land-miles but, since the Earth is not a perfect sphere, the "true" length of a nautical mile changes as one moves from the equator to the poles.
3. According to the *Reader's Digest Family Encyclopedia of American History*, "The Naval Observatory is the only astronomical observatory in the U.S. that determines time, and its determinations are accurate to one-tenth of a millionth of a second."

Chapter 7. *Preparing to Fight while Fighting*

1. Arnold, known for "involvement in even the smallest details of Army Air Force activity," was taught to fly by the Wright brothers in 1911. Named chief of the

air corps in 1938 when it had less than twenty-one thousand people and four thousand aircraft, he was still its chief when it was renamed the army air forces in 1941. By late 1944, he wore five stars and his air force was 2.3 million people, nearly two-hundred thousand planes.
2. While fighting, marines, with captured Japanese equipment, finished the airstrip within a week, naming it Henderson Field to honor marine Maj. Lofton R. Henderson, killed at Midway attacking enemy ships with the only four dive-bombers he had left. From Henderson, starting 20 August, navy aircraft took control of the whole Solomon Islands area.
3. In January 1944, Mitscher took command of Carrier Division Three, then its successor Carrier Task Force Fifty-eight to destroy the best of what was left by then of the Japanese fleet and carrier aircraft in the battles of the Philippine Sea, Leyte Gulf, and Okinawa. In 1945, he was given at his request command of the Eighth, then the Atlantic Fleet where he served until he died in February 1947.
4. Hayward earned a Distinguished Flying Cross for that 5 October raid, the citation saying in part, "His inspiring leadership and indomitable fighting spirit . . . at great personal risk, were responsible in large measure for the (mission's) outstanding success. . . ."
5. In the New York Yankees–St. Louis Cardinals 1943 World Series, Game One was played on 5 October in Yankee Stadium. The Yankees' starting pitcher was seven-year veteran Spud Chandler. In the sixth inning, the Yankees scored two unearned runs on a St. Louis error to break a two-two tie to record the game's final score, and went on to win the World Series, four games to one.
6. Whoever transcribed his 23 June 1944 debriefing to the CNO's top people in Washington typed 160 "miles per hour," which, of course, is not what he said.
7. Marines began flying the F4U out of Cactus in February 1943, the navy its carrier version a few months later. With six 50-caliber machine guns or four 20-mm cannon plus two 1,000-pound bombs or eight rockets under its distinctive gull wing, its top speed was more than four hundred mph. A pilot adage, "Don't dogfight a Zero"—too swift, too nimble—died with arrival of the F4U. Rated the best carrier-based aircraft of its type in the war, its kill-to-loss ratio was eleven to one.

Chapter 8. *From Air Combat to Atomic Bombs*

1. That "Gold Star in lieu of a second Distinguished Flying Cross" actually was for, reads the citation in part, "Heroism and extraordinary achievement. . .8 November 1943 to 24 February 1944 . . . [in] many [search/combat] missions . . . launching a smashing . . . attack on Kapingamarangi Island to score four direct hits on an enemy bivouac, sinking several supply barges on three other occasions and spotting three . . . cargo vessels [at] Kavieng, enabling our forces to destroy them. . . ."
2. In 1944, Hayward aired his gripe on "inconsistent, late" awarding of medals clear up to the navy CNO. He was campaigning for all VB crews, of course, not himself, since in the war he received the Distinguished Flying Cross four times (once by the army), five Air Medals, the Silver Star, Bronze Star, Legion of Merit with Combat "V," and two Purple Hearts.
3. For 1–23 April "heroism, extraordinary achievement in aerial . . . operations against Japanese forces," Hayward received a "Gold Star in lieu of his third Dis-

tinguished Flying Cross." During VB-106's nine months in combat, it sighted 406 enemy ships, attacked 104, sank 43, damaged 54. In the air war, it shot down 15 aircraft in aerial combat plus 22 "probables," damaged 9, and destroyed 5 "Rufes" on the ground.

4. A hint to the stress in those months is in the citation for Commander Hayward's Legion of Merit with Combat "V," received July 1944. It says, in part, " . . . [from] 25 March 1944 to 1 June, [VB-106 flew] 305 long-range, armed reconnaissance and strike [missions, sinking 26 enemy cargo vessels and barges, damaging 23 more and shooting down 12 enemy aircraft] in aerial combat."
5. As later chapters in this book document, quantum gains in technology would result, and Hayward would be part of the navy leadership that did in the 1950s what he had urged in 1944.
6. The memoir reference is to Charles Furey, *Going Back: A Navy Airman in the Pacific War* (Annapolis, Md.: Naval Institute Press, 1997).
7. There are several kinds of electroscopes today, to (a) measure the intensity of an object's radiation; (b) detect the presence of an electric charge on a body and tell if the charge is positive or negative.

Chapter 9. *A Sailor's Start into the Nuclear Age*

1. Peenemünde stumbled briefly in 1944, a year after mass production began on V-1 and -2 missiles, when SS leader Heinrich Himmler, trying to take control of Peenemünde, threw in jail its top scientist, Wernher von Braun. Hitler personally had to order von Braun's release.
2. J. D. Gerrard-Gough and Albert B. Christman said, in the monograph *The Grand Experiment at Inyokern* (Washington, D.C.: Government Printing Office, 1978), "It takes a special kind of personality to function and deliver under such a schizoid command structure and (Hayward) was such a personality. He became known as a very human character who not only was predisposed to buck the system when it slowed down progress, but also was a grand master at that particular science."
3. Vincent Davis, *The Politics of Innovation: Patterns in Navy Cases,* vol. 4, monograph 3 (Denver, Colo.: University of Denver, 1966–67), 25n.
4. His service record lists CalTech courses in exterior and interior ballistics of rockets, chemistry of explosives and solid propellants, aerodynamics, optics, microtime physics, strength of materials, and explosive casting techniques. The University of New Mexico and Los Alamos courses were contemporary physics, physics of the atmosphere, wave mechanics, critical assemblies, and uranium/plutonium/nuclear processes.

Chapter 10. *A-Bombs and Turf Wars*

1. Some people credit Army General and one-time Secretary of State George C. Marshall as the source of, "There's no limit to the good a man can accomplish," etc. Military "officialese" habitually uses "accomplish" instead of "do," suggesting that Marshall's version was not spontaneous. Chick Hayward said Parsons made the "can do" remark to him several times, but he never heard of Marshall saying

it. "On the other hand," he said, "maybe somebody else said it first and they both just read that guy's book."
2. An 1873 Naval Academy graduate, Albert Abraham Michelson (1852–1931) left the navy to study advanced physics in Europe, then returned to the academy to teach. (It had one opening in physics, one in math. Following navy tradition, he and a friend flipped a coin to decide who would do which. Michelson lost and had to teach physics.) There, using his improvement on a device invented by French physicist Jean Foucault, he made one of the first precise measurements of the speed of light in a vacuum: 186,262 miles per second.

His 1907 Nobel Prize in physics was for work done in 1887 with American physicist Edward Williams Morley. With an "interferometer" he had invented to record the Earth's speed as it moved through "luminiferous ether"—which scientists assumed, back then, filled all space beyond the Earth's atmosphere—they proved that "ether" did not exist. "Their experiment also implied," said Hayward, "that, paradoxically, the speed of light does not change as one moves." Sending physicists everywhere "back to basics," that led to new postulates, including Einstein's special (published in 1905, relating size and speed of matter to energy) and general (published in 1916, defining the corollaries of time to space) theories of relativity.

Chapter 13. *This Is Not a Boat, Hayward*

1. Hayward's service record says, in part: ". . .Temporary appointment as Captain terminated, 1 January 1948; (promoted to) Captain (T), 1 August 1948; Captain, 1 July 1951, to rank from 1 August 1948." Said Hayward, "Obviously, they predated the event."
2. In the 1950s, Kimball became chief executive of Aerojet General, a supplier to the military of solid propellant fuel, among other things, and Gen. Dick Hayward's employer for a time. (See appendix.)
3. In theory, DOD was the acronym for Department of Defense, OSD for the Office of the Secretary of Defense. But until about 1965, most army, navy, and air force officers referred to just the Office of the Secretary of Defense as "the DOD." As Forrestal said in his first annual report as secretary of defense, ". . . military professionals thinking in terms of the Service to which they have devoted their entire adult lives is to be expected. [But] true unification [will not be achieved until] generals and admirals, ensigns and lieutenants, soldiers, sailors, airmen and civilians all learn we are working together for a common cause."
4. For the result of his audience with Pope Pius XII, see chapter 14.

Chapter 14. *Turf Wars: Some Angry, Some Deadly*

1. General Nichols told Chick that at a meeting in Washington, D.C., in October 1996, where he and Hayward had been invited in order to help the Energy Department nuclear-weapons people fill in some blanks. (The nuclear folks had lost all records of what happened in the atomic business between the Bikini tests and the creation of the AEC in 1947.)
2. Army Signal Intelligence began to amass intercepted Soviet cables in 1943, a top-

secret "Verona project" the National Security Agency continued until 1980. The Verona Papers went public in 1996 and *The New York Times* snorted, "[The cable reprints are in] fragments . . . uncorroborated, easily misunderstood. . . . One [even] suggests pre-eminent (columnist) Walter Lippmann met regularly with a Soviet spy [which was] 'not possible.'" But he did. "The Verona cables," noted *The American Spectator,* "identify his secretary, May Price, as the Soviet agent."

3. Also called radiocarbon because it is radioactive, carbon 14 is part of the carbon dioxide created by cosmic rays bombarding nitrogen in the atmosphere. Absorbed by green plants, it enters animals through the food chain. Over time, it reverts to nitrogen, a loss replenished as living organisms breathe and take in food. At death, the resupply stops. University of Chicago physicist Willard F. Libby, in about 1946, isolated the carbon 14 isotope, fixed its half-life at 5,730 (give or take 40) years, proved that it decayed to nitrogen at a slow, steady rate, and devised a way to carbon-date fossils. Though the analytic process is complex, the math is simple: Count the carbon 14 atoms still there; proportion that to 5,730. In other words, if a bone died with a million and has only five hundred thousand carbon 14 atoms left, it died 5,730 years ago.

Chapter 15. *The George Cruz Ascom Story*

1. Hayward's helicopter was a Sikorsky, the first twin-rotor one, which could carry up to six people. It was used mainly as a plane guard, replacing destroyers "which didn't want the job, anyway," he says. "It was the helicopter's first major navy role, really. Until then, a carrier was required to have a destroyer plane-guard nearby when conducting flight operations."
2. A story, "1,000 Men and a Baby" by Lawrence Elliott, was in the December 1994 *Reader's Digest.* A write-up, "George," in the 7 December 1997 *People* magazine was tied into the TV movie shown that night. On the latter, the navy refused to cooperate at first, telling CBS, "We've never had any captains who act that way." But they became very helpful after Hayward called the navy CNO, "Mike" Boorda, who also had begun his career as a bluejacket. "What got Mike on board," Hayward thinks, "was my saying the movie tells people how fine the American sailor is—a speech I gave a lot, back in the 1950s."

Chapter 16. *On Making Admiral*

1. An aside to "Oppie's hanging"—in the January 1998 issue of *Chronicles* magazine, a twenty-three-year CIA operations officer, Lawrence B. Sule, revealed, in part, the following: Harry Hopkins, an advisor so close to FDR that he was called "assistant president," and head of our Lend-Lease program to Russia, ordered some uranium sent to Russia, according to Leslie Groves, when "the Soviets weren't even supposed to know of our atomic-bomb development, much less be given our uranium for their own program." Air corps transport officer Maj. George R. Jordan confirmed it after the war with copies of the shipping manifests. "No wonder," the article says, "the Soviets considered Hopkins a more important agent than even Alger Hiss," or Oppenheimer, certainly.

2. An OP-03 himself, in 1957 Admiral Hayward was instrumental in the F4H being designed to carry a radar operator. At first, pilots strongly opposed putting a second man in their "little" (51,700 pounds of takeoff weight) airplane. "I can't imagine," he answered, "how just one man can fly a Mach 2 aircraft [in the Phantom, 1,415 mph at thirty thousand feet] and operate its AI radar at the same time."

Chapter 17. *The Pentagon: A Place to Pass By*

1. Admiral Hayward's service record at the Naval Historical Center says, "Jan 1957 - Office of the Chief of Naval Operations, Navy Dept. (Director, Strategic Plans Division)," which obviously is incorrect.
2. Flying above seventy thousand feet to avoid leaving a vapor trail, and too high for Soviet fighters to reach, that Lockheed U-2 had 540-mph speed, a visual range of three hundred miles (to the horizon), and could detect enemy radar and radio signals. Kelly Johnson's Lockheed "Skunk Works" went from design in 1953 to first flight in 1955, an amazing record even in that era when design-to-first-flight time averaged about four to five years.
3. When Thomas Gates became secretary of defense in 1959, he began to go to Joint Chiefs meetings regularly. "That," he said, "cleared up lots of papers that had been around quite a while. In one case, in twenty seconds I decided the navy would be responsible for military activity related to Africa. That freed up thirty-two position papers on a minor subject. You can imagine how many they had on complicated questions." On dissent generally, he said, "I was suspicious as hell if I got a single, non-dissenting position out of the Chiefs, and God help the country if we ever have a single strategy."
4. He'd met Sarazen in 1923 and 1924 when Jesuit Father Farley was teaching teenager Hayward to play golf. Sarazen renewed the friendship "after all these years," said Hayward, after the admiral went to OP-03.
5. In September, the Senate voted, "Hayward, J. T., Rear Admiral, to rank from 1 August 1957."
6. For more on Admiral Boorda, see this book's final chapter.
7. Burke's view: "The man in charge can be a big help to any organization by deciding what projects may be suitable, which should go first; but to do it wisely, he must pick the brains of the technical people."
8. At the time, 1959, as commander of Task Group ALFA charged with building and deploying an ASW defense, Thach said, "Just twelve Soviet missile-carrying submarines deployed off our shore could knock out 70 percent of our industry, road, and railway networks."

Chapter 18. *From R&D to Cuba*

1. In "The Oxcart Cometh—And Goeth at Mach 3.2," author William Burrows says, "[Engine] problems . . . would have broken the program financially had the Navy's $38 million not bailed it out." See *Air & Space* (Feb–Mar. 1999), 71.
2. After that, the next carrier authorized, in 1967, commissioned in 1975, was the *Nimitz*, first in what is to be, if the plan is completed, a fleet of ten nuclear-

powered carriers. Its two convection-cooled A1-G reactors have fuel enough to run for fifteen years. Compared to even the *Enterprise,* it carries more aircraft, fuel, and "bullets." As important in an era of threatened chemical warfare, said Hayward, "It's so much better protection. The engine room is completely airtight, hasn't any funnels going up through the deck."

3. In 1991, Tom Moorer, CNO, 1967–70; then Joint Chiefs Chairman, 1970–74, agreeing with Hayward, said, "Nobody in this century has made as great an impact on the navy as Admiral Burke did."

4. For *Military Leaders since World War II,* Vice Adm. (ret'd.) William I. Martin, an expert in night operations off carriers, told the author he hadn't worked for Admiral Hayward but did see him occasionally and "I don't think he ever ran into a science or engineering problem he didn't understand. The rest of us navy officers knew how to fly airplanes, drive ships. He knew the technology. Several times, I saw him in front of a group and, scribbling with his chalk, fill two or three blackboards with all kinds of formulas and equations. There may have been two or three PhDs in the group who understood what he was talking about. None of the rest of us did."

5. The detail: a vice admiral since 25 April 1959, he reverted to rear admiral, effective 9 March 1962.

6. Of the *Thresher,* in a diary entry for 30 December 1963, the admiral said, "The accident report says little about the boat's HY-80 grade steel or the fact that, as steel is made tougher, it gets more brittle; but I believe the failure of the safety system caused the accident, anyway. It hadn't sunk due to flooding. It had imploded. That was the signal we got on the SOSUS (Sonar Sound Underwater System). It was at 1200 feet when it lost power in a hurry, thus losing the ability to blow its tanks. As a result, it dropped below crush depth in a hurry. Rickover won't admit it's a design problem but I'm sure they focused on preventing a nuclear accident and failed to realize an instrument failure can knock out the safety system. If that happens at depth, it also can result in disaster."

Chapter 19. *From Cuba to the Naval War College*

1. The jet engine was changing the whole equation about then, Admiral Hayward has pointed out. Most of his 13,023 navy flying hours were in piston-engine aircraft, his longest being a twenty-six-hour nonstop flight from San Diego to Panama in a PBY. Today, a carrier aircraft mission rarely flies more than two hours, and today's navy pilots are likely to have as many carrier landings in a career as he had flying hours.

2. Specifically, says the record: first promoted to vice admiral on 25 April 1959, reverted to rear admiral effective 9 March 1962; vice admiral, 13 June 1963, to rank from 25 April 1959.

3. In *Street Without Joy,* published in 1965, French author Bernard Fall, writing from Saigon in early 1964, said, "Americans are dying in Vietnam, dying in American uniforms. And they die fighting."

4. In getting reacquainted, Schneider asked Hayward if he'd noticed, in King's portrait at the Naval Institute, that the ribbon for his Distinguished Service Medal is located above his Navy Cross, then told Hayward why. Bitterness had erupted

in World War I when every combat-ship commander was given a Navy Cross, even those who'd not seen combat. King was furious, Schneider said, "And when I told him, 'Admiral, you're wearing your medals wrong. Your Navy Cross is supposed to be above your DSM,'" King snarled, "Like hell I've got them on wrong. I did something for my DSM and it's going to stay there."
5. During the preceding five years, Moorer had been Seventh Fleet Commander, CINCPAC, and CINCLANT/Supreme Allied Commander, Atlantic. Clearly he was being steered toward being MacDonald's CNO successor. So, when Defense Deputy Secretary Cyrus Vance asked Hayward, "If Mr. McNamara offers, will you accept a nomination to be CNO?" Chick said, "No, for two reasons: It's Tom Moorer's turn and I could not live in the management environment your Mr. McNamara has created."
6. In the paper's 28 June 1968 issue, the article added, under a headline, "School for Admirals: An Intellectual Officer Tries to Widen Outlook of Future Navy Leaders," that the College's ten-month course "traditionally concentrated on battle planning, fleet maneuver and control, logistics and weapon systems. But now a guest lecturer blasts the U.S. presence in Vietnam. A professor tells his class the State Department's rationale for the Cuban quarantine is all wet. A picture of Ho Chi Minh hangs prominently in an administrative office. And a favored new text is Che Guevara's on Guerrilla Warfare."
7. Sums up the authors of the book, *Sailors and Scholars*, in part, "Colbert's mission was not to revolutionize the College; Hayward already had done that. The Colbert presidency was deliberately low-key with respect to the evolution of the academic program which he believed was then in excellent shape. [So] his primary attention was given to enriching the existing overall program with better student housing, creating the Naval War College Foundation, and establishing the College as a center for international naval discussions through symposia, conferences and courses; work with foreign navies on common solutions to common problems; [focus Hayward's core curricula study] of U.S. naval capabilities to engage in world-wide operations in support of national policies supplemented by war games to confirm or disprove those assessments."

Chapter 20. *Where Are the Leaders?*

1. Quoted from Admiral Hayward's essay, "The Second-Class Military Advisor, His Cause and Cure," in the November 1968 issue of *Armed Forces Management* magazine, p. 68.
2. Into which Admiral Hayward himself was inducted, as was Sen. John Glenn, in May 1998.
3. A Naval War College graduate, Gravely retired as a vice admiral in command of Hayward's old ASW Pacific task force, after which Hayward got him elected to the Draper Laboratories board of directors.
4. One of Hayward's twenty-two grandchildren (eleven male, eleven female). He also has eighteen great-grandchildren (nine male, nine female).

Bibliography

Chapman, Capt. William C. (USN, ret'd). "The Bay of Pigs: The View from Prifly" (Primary Flight Control), paper given to the Ninth Naval History Symposium. Annapolis, Md.: U.S. Naval Academy, 20 October 1989.

Christman, Al. *Target Hiroshima: Deak Parsons and the Creation of the Atomic Bomb.* Annapolis, Md.: Naval Institute Press, 1998.

Davis, Vincent. *The Politics of Innovation: Patterns in Navy Cases,* vol. 4, monograph 3. Denver, Colo.: University of Denver, 1966–1967.

Elliott, Lawrence. "1,000 Men and a Baby," *Reader's Digest* (December 1994).

Fall, Bernard B. *Street Without Joy: Insurgency in Indochina, 1946–1963.* Harrisburg, Pa.: The Stackpole Company, January 1963.

Furey, Charles. *Going Back: A Navy Airman in the Pacific War.* Annapolis, Md.: Naval Institute Press, 1997.

Gallico, Paul. "The Dropout Who Made It to the Top," *Reader's Digest* (November 1966).

Gerrard-Gough, J. D. and Albert B. Christman. Monograph: *The Grand Experiment at Inyokern.* Washington, D.C.: Government Printing Office, 1978.

Hammond, Paul Y. "Super Carriers and B-36 Bombers: Appropriations, Strategy and Politics" in *American Civil-Military Decisions.* Birmingham: University of Alabama Press, 1963.

Hattendorf, John B. *Sailors and Scholars: A Centennial History of the U.S. Naval War College.* Newport, R.I.: Naval War College Press, 1984.

Hayward, Vice Adm. John Tucker. "Chaplains Corps" for the Oral History Program. Annapolis, Md.: Oral History Department, U.S. Naval Institute, 1984.

———. "Chick," an Oral History. Annapolis, Md.: Oral History Department, U.S. Naval Institute, first draft done in 1984.

Knott, Richard C. *A Heritage of Wings: An Illustrated History of Navy Aviation.* Annapolis, Md.: Naval Institute Press, 1997.

McMaster, H. R. *Dereliction of Duty: Lyndon Johnson, Robert McNamara, the Joint Chiefs of Staff, and the Lies that Led to Vietnam.* New York: HarperCollins, 1997.

Miller, Edward S. *War Plan Orange: The U.S. Strategy to Defeat Japan, 1897–1945.* Annapolis, Md.: Naval Institute Press, 1991.

Parrish, Thomas, ed. with Brig. Gen. S. L. A. Marshall as chief consultant-editor. *Encyclopedia of World War II.* New York: Simon & Schuster, 1978.

Preliminary History of NASA: 1963–1969. Monograph authored by several National Aviation and Space Administration executives of that era. Washington, D.C.: NASA, 15 January 1969. On file at the Lyndon B. Johnson Library, Austin, Tex.

Rearden, Steven L. *History of the Office of the Secretary of Defense,* vol. 1, *The Formative Years 1947–1950.* Washington, D.C.: Historical Office, Office of the Secretary of Defense, 1984.

Scarborough, Capt. William E. (USN ret'd.). "The North American Savage," in *The Hook* (fall 1989 [p. 28] and winter 1989 [p. 16]). San Diego, Cal.: The Tailhook Association.

"Special Section: The Atomic Bomb—Making It Possible, Making It Happen, Its Technology Legacy." *American Heritage of Invention & Technology.* New York: American Heritage division of Forbes, Inc. (summer 1995).

Swanborough, Gordon and Peter M. Bowers. *United States Naval Aircraft since 1911.* Annapolis, Md.: Naval Institute Press, 1968, 1976, 1990.

Taylor, Theodore. *The Magnificent Mitscher.* Annapolis, Md.: U.S. Naval Institute Press, 1991.

Index

Abelson, Philip, 120
Ablard, James, 229
Acheson, Dean, 198–99
aircraft: for A-bomb deployment, 167–69; for China Lake, 125–26; debate over, 14; development of, 53, 78, 81–82; Morrow Air Board on, 32–33
aircraft carriers, 15, 166–67, 170–71
AJ-1 (North American Aviation [NAA] A-2As), 167, 168, 182, 188, 195, 196, 197
Allen, Ross, 84
Alvarez, Luis, 116
Anderson, Carl, 115, 127
Anderson, George, 258–59, 264, 266, 272
Antisubmarine Warfare (ASW) Force Pacific, 277–80, 282
Argonne, 45
Arkansas, 15–16, 148
Ark Royal, 213–14
Armed Forces Special Weapons Project (AFSWP), Sandia, N. Mex., 156, 157, 158
Armitage, Jack, 126
Arnold, Henry H. "Hap," 86, 140
Arthur, Stan, 299
Ashworth, Frederick L. "Dick," 120, 128, 135, 139–40, 183; VC-5 squadron and, 169, 171
Aspin, Les, 294
Astoria, 88
ASW Force Pacific, 277–80, 282
Atlantique ASW (antisubmarine warfare) aircraft, 257
atomic bomb(s): assembly of, 160–61, 173; Chicago Pile One and, 128; Japan and, 138–40, 142–45, 146; limitation treaties for, 205; manufacturing, 158–59; Mark 57, 231–32; to Mediterranean, 196; military or civilian control of, 146–47, 153–54; research, 128–33; *versus* small tactical weapons, 201; stockpile security for, 174–75; testing, 134, 164, 203, 294; testing "Mike," 206, 207–8; testing on Bikini, 147–49. *See also* Fat Man; Little Boy
Atomic Energy Commission (AEC), 153, 158, 160, 161, 198; discussions of role for, 206–7; Fermi Medal for Oppenheimer by, 228; General Advisory Commission of, 153, 161; Military Applications Division of, 197; Military Liaison Committee of, 159, 175; research project assignments and, 202; storage/security monitoring by, 173, 174–75
Aurand, Pete, 172, 204, 247, 266, 278
Australian Coast Watchers, 88, 90
Aymond, Walt, 285, 288

B-29 bomber, 135, 136, 187–88
Babcock, J. V., 19, 22
Badger, Admiral, 156
Badger, Henry, 177
Bahm, George, 61
Baird, Chuck, 285
Baker, "Bos'n," 35
Baker, Jim, 53, 55
Balao, 100
Baldwin, R. H. B. "Bob," 171, 285, 297
Bales, Ray, 94, 96, 106, 108
ballistic missile program, 246–48
Barlow, Lt., 94
Barrett, Edward R., 195
Barrier, 291
Batista, Fulgencio

319

Batista, Juan, 64
Battle Fleet exercises (1939), 73
Beakley, Wallace M. "Miles," 191, 192, 268; Cuban missile crisis and, 269, 270–71, 272, 275
Begin, Menachem, 294
Bennett, Eddie, 4
Bennett, Ralph, 228–29
Berliner Joyce scout plane, 46
Bettys, Japanese, 104–5
"Bevo" (alcoholic beverages oath) sheet, 36, 42
Birch, Tommy, 94
Blackburn, Tommy, 98
Blackfish, 100
blacks in the military, 297
Blandy, William H. P. "Spike," 118–19, 148, 190
Bleil, Dave, 229
Bloch, Rene, 257, 262, 268
Bloch, Yolande, 268
BM-1 dive-bombers, 53
Boak, Captain, 99
Bock's Car, 139–40
Boe, Bob, 210
Boeing F4-B, 50
Boeing NB-2 biplane, 28
Boeing XPBB-1 Sea Ranger, 86
Bohr, Neils, 116
Boise, 73
Boorda, Jeremy "Mike," 245, 298–99
Borden, Bill, 200
Bowen, Harold "Ike," 116, 120, 125, 148
Bradbury, Norris, 120, 128, 201, 202; as guest on *FDR*, 233, 235
Bradley, Omar, 162, 187, 191–92, 194, 198–99
Brady, John J., 8–9, 11, 22, 101
Brereton, Lewis, 140
Briggs, Ellis, 222, 238
Briscoe, Bob, 222, 223
Britain, World War II and, 75, 76–77, 84
Brode, Wallace, 153
Brooks, Louise, 20, 28, 29
Brooks, Overton, 254
Brown, Cat, 244–45
Brown, Chuck, 182
Brown, Harold, 202, 280
Brown, J. Carter, 286
Brush, Fuller, 225, 226
Buchanan, Charles A., 241
budgeting, naval, 256–57

Buhl Bullpup midwing monoplane, 27–28
Bulkeley, John, 213
Burck, Gail and Roberta, 155
Bureau of Aeronautics (BuAer), 84, 118–19
Bureau of Naval Weapons, 253
Bureau of Ordnance (BuOrd), 118–19, 141
Bureau of Personnel (BuPers), 104, 114
Bureau of Supplies and Accounts, 104
Burgess, Johnny, 51
Burke, Arleigh, 164, 190, 213, 260, 298; CIA Cuban invasion and, 261–62; dawn attack on Kavieng harbor and, 105–6; Hayward and, 231, 234, 251; leadership of, 292–93; Naval War College and, 288; on nuclear engines for ships, 258–59; on R&D, 244; retirement of, 264; Revolt of the Admirals and, 185; on technology revolution, 241
Burroughs, Sherman Everett "Ev," 115, 116, 118–19, 128, 138
Busch, Roland C., 216–17
Bush, Vannevar, 115–16, 120, 137
Byrd, Harry, 256
Byrnes, James, 155
Byrnes, Pat, 147

California Institute of Technology, 115, 122
Campbell, Clarence, 274
Campbell, William R. "Soupy," 181
Canberra, 88
Canterbury, Monte, 157
Carbonella, Agua de, 60
Carney, Robert B. "Mick," 210, 222, 223
Carpenter, "Gillie," 43, 45
Carpenter, Steve, 171
Cassell, George, 283
Castro, Fidel, 259–60
Central Intelligence Agency, 242, 260
Chadwick, George, 130
Chapman, Bill, 262
chemical diffusion plant, 131
Chew, Jack, 284
China Lake, Calif., 115, 116, 118–33; aircraft for, 125–26; atomic weapons research at, 129; classified nature of, 127–28, 132; Fat Man development, 130–33; Fat Man testing at, 134; Hayward's education at, 123; Office of Scientific Research and Development (OSRD) and, 118–20, 122; Parsons and, 120–21; personnel responsibilities at, 118–19; rocket-building at, 126; smart bomb de-

velopment, 125; solid-fuel propellant testing at, 124–25
Christmas, Lee, 25
chronometers, World War II, 77–78
Churchill, Winston, 205
Cincinnati, 48
Civil Aeronautics Administration, 78–79
Clagett, Tom, 233
Clark, J. J. "Jocko," 43, 162, 213–14, 215, 217, 222
Clark, Johnny, 261
Clifford, Clark, 155
Clifton, Joe, 31, 98
Clinton, Bill, 203, 298
Clipper, Sam, 36
Coker, George, 24
Colbert, Dick, 288
Cold War, 151–53, 176–77, 203, 213; changes in Russia and, 209–10; Kennan's containment policy on, 178; Middle East, 239–40; Poland/Hungary revolts, 240; Soviet atomic testing and, 187–88; space race, 249–50, 252, 266–67; Truman Doctrine and, 154–55, 179
Collins, J. Lawton, 160, 191
Combs, Commodore, 107, 109
Commonwealth, 5–6
Compton, Arthur, 128
Connecticut, 3
Connolly, Tom, 157, 171, 292–93
Conover, Al, 182
Consolidated P2Y-2 flying boats, 53
Consolidated P2Y-3 flying boats, 55
Consolidated PBY patrol-bomber, 82, 86, 90
Constellation, 7
Contocook, 23
Convair B-24D Liberator bombers, 86–87, 101, 103, 114; accounting for extra, 110, 113; bombing ships from, 97; escape hatches for, 109; in hurricanes, 105; modifications to, 87, 93, 95; problems with, 96–97, 111–12, 114; speed variations among, 96
Convair B-36 bomber, 185–86, 189
Cook, Harry, 281
Cooksey, Don, 201
Coral Sea, 166, 172, 183, 283; task force tour and, 244; VC-5 on, 193, 194, 197
Cornwell, Zemp, 34–35
Covered Wagon. See *Langley*
Crommelin, John, 35, 99, 113, 135, 189, 191
Crutchley, V. A. C., 88

Cruz Ascom, George, 219–23, 224–25
Cuba, 259–62, 268–71, 272–74, 275
Cunningham, Al, 245

Dahlgren Proving Ground, 118
Dalton, John, 299
Daniels, Josephus, 9
Dare, Jim, 171
Dassault, Marcel, 257
Davis, Burton, 234
Davis, "Dagwood," 94, 95, 97, 107–8, 109
Davis, David, 87
Davis, Leverett, 116
Davis, W. B. "Little Bill," 196
Dayan, Moshe, 293–94
Dean, Gordon, 200, 201, 202, 207
Decker, Edward A., 195
Decker, George, 226
Defense Reorganization Act (1958), 252
Denebrink, "Rinky-Dink," 165
Denfeld, Louis E., 165, 168, 190–91
Dennison, Robert, 266, 270, 275
de Poix, Vince, 127
Detroit, 8
Development of Substitute Materials (DSM) Project, 128
DeWolfe, "Sock," 43
Dickey, Fred, 286
Doak, Robert, 24
Doan, "Carp," 53
Dobey, Commander, 171
Doolittle, Jimmy, 37
Dorgan, Thomas A. "Tad," 4
Dorland, Gil, 157, 173
Douglas, Catherine, 71
Douglas A3D-1 Skywarrior, 169–70, 203–4
Douglas DC-2 transport, 62
Douglas DC-6, 278
Douglas P2D2 twin-float torpedo bombers, 53
Douglas SBD scout/dive-bombers, 53
down-checks, 33; on female pilots, 298–99
Draper, Stark, 80
Draper Laboratories, 292
Dresser Industries, 276, 284
drugs in the navy, 296–98
Dryden, Hugh L., 254
Duke D'Aosta, 71
Duke De Savioa, 71
Dulles, Allen, 259–60
Dulles, John Foster, 260
Dumpstrey, Herbert, 101

Duncan, Donald "Wu," 43, 124, 163
Dunlap, Sam, 62
Durenberger, David, 299

earthquake, Long Beach, Calif. (10 March 1933), 39
Eaton, William, 60
Ebbe, Gordon, 94
Ecklemeyer, Ed, 63, 64–65
Eckstrom, "Swede," 51
Einstein, Albert, 82, 128
Eisenhower, 296
Eisenhower, Dwight, 203, 208, 209, 243, 249–50
Elliott, Richard, 26
Enola Gay, 138–39
Enterprise, 83, 85, 90, 258–59, 265–66
Enthoven, Alan, 263
Essex, 261

F4B-2 biplane, open-cockpit, 55
F4U-4 fighter, 137
F4U Corsairs, 98
Fairless, Clyde, Jr., 196
Farley, John, 5
Fat Man, 129, 130–33, 137–38, 146; Nagasaki and, 139–40; successors to, 160; testing, 134, 136
Fechtler, William, 203
Feklisov, Alexander, 199–200
Felt, Don, 230–31
Felton, Luke, 51
Fermi, Enrico, 116, 120, 122, 128
Fife, Commodore, 100
fission, description of, 129–30
Flag, Allan, 43
Flately, Jimmy, 180
Fleet Problems, 46–49, 56, 74
Fletcher, F. J., 85
Floberg, Jack, 194
Florida, 16
Forrestal, 230, 258–59
Forrestal, James, 108, 141, 160, 183; National Military Establishment report (1948), 290; suicide of, 186; Truman Doctrine and, 154
Foster, Johnny, 202
Fowler, W. A. "Willy," 115
France, World War II and, 75
Franke, William B., 253
Franke Board, 253
Franklin Delano Roosevelt, 166, 172, 184, 196, 277; around South America, 236–37,
238–39; clean-up of, 233; discipline breakdown on, 296–97; Hayward and, 232; mechanical trouble for, 233–35
Friede, Dick, 210
Froman, Carol, 164
Fuchs, Klaus, 200
Furnas, Clifford, 250
Fye, Paul, 229

Gaither, H. Rowan, Jr., 247
Gaither Report, 247–48
Galapagos Island, 56
Galbraith, John Kenneth, 145
Gallery, Dan V., 174, 176, 181, 204; leadership of, 187, 292–93
Gallico, Ginny, 267, 278
Gallico, Paul, 4, 267, 277, 278
Garcia, "Beppo," 43
Gates, Tom, 243, 248, 249, 251, 266
Gaulin, Vic, 51
Gaulle, Charles de, 262
Gauyner, Ray, 167
Gavin, Jim, 262
General Advisory Commission, AEC, 153, 161
General Dynamics, 163, 226, 291–92
George Washington, 248
Germany, technology of, 53, 124
Ghormley, Robert L., 89
Gilpatric, Roswell, 262, 265–66
Ginn, Frank, 285, 288
Glennan, T. Keith, 254
Goddard, Robert, 123–24
Gold, Harry, 200
Goldthwaite, Bob, 212
Goldwater, Barry, 275
Gouin, Marcel, 32
Grace, Princess of Monaco, 267, 277, 278
Gravely, Sam, 297
Gray, Gordon, 227
Great Depression, 17, 36–37, 38
Greenglass, David, 200
Grenfell, Admiral, 259
Griffin, Virgil, 43, 44
Griggs, David, 200
Gross, Courtland, 248
Groves, Leslie R., 115, 121–22, 128, 146; Aurand and, 162–63; experimental freedom for, 131–32; on honors for atomic bomb, 139; Joint Chiefs of Staff (JCS) and, 134; retirement of, 164; Sykes and, 147; on transporting Little Boy, 130
Grumman: A2F/A-6 Intruder, 274; atomic-

powered aircraft plan of, 176; W2F/E-2 Hawkeye, 274–75
Guantánamo, Cuba, 3, 21–22, 49, 72–73
Gunn, Ross, 120
gun size, debate over, 14–15

Hafstad, Larry, 121
Hagen, John, 250
Hagerty, Jim, 249
Hall, Theodore, 200
Hallen, Herman, 43
Halsey, William F. "Bull," 31, 89–90, 91, 106, 136
Hammond, Paul, 177
Hampton Roads Naval Air Station, Va., 18–19
Hancock, 204
Hanford, Wash., nuclear reactor, 128–29, 131, 158
Hardy, Lou, 210
Harkins, Paul D., 216
Harlow, Chief Quartermaster, 7–8
Harper, Kelly, 171, 188
Harris, Hunter, 180
Harrison, Lloyd, 64
Hart, "Pumpkin," 51
Hartman, Fred, 286
Hartman, Greg, 164, 229, 230
Hawkins, Jack, 261
Hayward, Billy, 3
Hayward, Charles Brian, 1, 2–3, 7, 29–30, 55, 65–66; during Depression, 36–37, 65
Hayward, Charles Brian, Jr., 65
Hayward, Eleanor (Elinor), 1. *See also* Kitchingman, Eleanor; Pollock, Eleanor and Dan
Hayward, Jennifer, 92, 228, 233, 267, 274, 278
Hayward, John Tucker: admiral selection and, 232, 235, 236–38, 243; AJ-1 and, 182, 188, 195, 196, 197; ASW (Antisubmarine Warfare) Force Pacific and, 277–80; atomic bomb deployment test flying by, 135, 167–68, 183, 184; on atomic bombs for Korean War, 208–9; B-24D test flights by, 87; birthdate of, 1, 17; boot training for, 6–8; Brady influence on, 8–9, 11; Buhl Bullpup monoplane of, 27–28; captains *Point Cruz,* 210–11; at China Lake, 121–33; Cuban missile crisis and, 269–74, 275; damage assessment in Japan, 142–45; early life of, 2–4; enlistment in Navy, 1, 5–6; eye problem of, 237, 239, 240, 252–53; first flight of, 3; Fleet Problems and, 46–49, 56, 74; generals'

nuclear processes class by, 162–63; in Japan, 281; Joint Strategic Planning Committee (JSPC) and, 242–43; judge-advocate duty of, 60–61; Korean War orphan and, 219–23, 224–25; Korean War POWs and, 214–15, 216–17, 218; marriage of, 36–37; Naval Academy and, 9–11, 12–18; at Naval Air Factory (NAF), Philadelphia, 63–64, 66, 74–76, 82–83; at Naval Ordnance Laboratory (NOL), White Oak, Md., 226, 228; Naval War College and, 29, 284–89, 292; nicknamed "Chick," 6; night and bad-weather flying and, 51, 78–79; Panama posting and, 51–52, 56–58; Pensacola flight-training for, 31–35; pilot hours for, 45–46; PTA work by, 150–51; Space Committee hearings and, 250–51; Sykes and, 141–42, 156, 242; Truman and, 209; as VB-106 commander, 92–96; VC-5 squadron and, 169–74, 175–76, 180–81; on Vietnam, 282, 283; Whiting's tour of *Langley* and, 43–44; World War II combat debriefing, 111–15; in World War II Pacific theater, 96–103, 104–11; World War II supply-support by, 89. *See also* honors and awards; research and development (R&D)
Hayward, John Tucker, Jr., 150, 176, 228, 233; character of, 265, 291; education of, 288; in Europe, 267; holidays with, 274; on *Independence,* 268; move to Hawaii and, 278
Hayward, Leila Marion, 56, 59, 182, 206
Hayward, Lili, 48, 49, 115, 181; on air force service, 163; in Asia, 283–84; China Lake and, 121, 138, 150, 154; death and, 65, 206; in Europe, 267, 268, 277–78; in Florida, 61, 224; holidays with, 274; illness of, 176, 177, 230, 231; on John T. Hayward's navy career, 192–93, 284; Long Beach posting and, 73; marriage of, 36–37; at Naval Ordnance Laboratory (NOL), 228; Panama duty and, 52, 55, 59; Philadelphia post and, 64, 66, 82–83; in Virginia, 242; World War II and, 87, 88, 92
Hayward, Marjorie, 1, 28, 65, 66
Hayward, Mary Shelley, 45, 52, 53, 182; as mother, 231; Panama duty and, 55, 59
Hayward, Richard Wright, 1, 67, 87, 189, 301–3
Hayward, Rosa Valdetarro, 1, 2, 29–30, 55, 234; during Depression, 36–37, 65, 66

Hayward, Tom, 297–98
Hayward, Victoria, 83, 228, 233
Hazard, "Hap," 43, 44
Healey, Denis, 263
Hean, "Red," 156–57
Hearst, William Randolph, 191
Heinemann, Ed, 170
Helena, 121
Henderson Field, Guadalcanal, 90
Hertz Foundation, 292
Hickman, Clarence, 123
Higley, Frank, 51
Hindenberg, 63
Hiramatsu, Admiral, 291–92
Hitler, Adolf, 62–63, 75, 83–84
Holloway, Jim, 253
Holt, Frank, 94, 107, 110
Holt, Walter, 50
Honduras, peace-keeping in, 23–25
Honolulu, 73
honors and awards: in 1962, 264; for atomic bomb, 139; Bronze Star, 92; Congressional Medal for Life-Saving, 24; Distinguished Flying Cross, 105–6; Distinguished Service Medal, 284, 288; Letter of Commendation, 147; Navy's inconsistency in, 105, 112, 114, 139; Order of the British Empire, 77; Southern Cross (Brazil), 70–71
Hooper, Ed, 232
Hope Haven Hospital for Children, 292
Hornet, 85, 89, 90
Hornet (II), 106
Hovde, Fred, 116, 119
Howerton, "Sunshine," 43, 45
Howland, Jack, 233
"How Long Is a Mile?" (article), 57
Hull, Cordell, 83, 164
Hussey, George, 146–47, 153, 156
Hyer, "Daddy Pete," 36, 52, 115, 181, 229; Haywards and, 202, 291; prosperity of, 65
Hyer, Leila Marion "Lili," 29, 35–36. *See also* Hayward, Lili
Hyer, Mary, 38
Hyer, "Mommie Floss," 36, 51, 52, 115, 181; death of, 206; Haywards and, 202, 291; red silk pajamas from, 46
Hyer, "Waddy," 36–37
Hyland, Johnny, 275, 276

Independence, 148, 258, 266, 267–68
Indianapolis, 137, 138, 140

Ingersoll, Royal, 127–28
Institute of Aeronautical Sciences, 59
Inyokern, Calif., 150. *See also* Naval Ordnance Test Station (NOTS), Inyokern, Calif.
Italy, World War II and, 75

Jaap, Joe, 171, 245
Jackson, Henry "Scoop," 202, 234, 249, 250; Hayward and, 253, 268, 275; Joint Committee on Atomic Energy and, 280–81; Naval War College and, 286–87
Jackson, J. P. Orton, 7
James, Jules, 64, 65
Japan: atomic bombing of, 138–40; damage assessment in, 142–45; Maritime Self Defense Force, 281; U.S. military intelligence on, 145–46; World War II and, 84–85
Jepson, Morris R., 138–39
Jette, Eric, 116, 129
John Paul XXIII, 277
Johnson, D. P. "Dippy," 53
Johnson, "Dub," 144
Johnson, "Johnny," 94, 107, 108
Johnson, Kelly, 258
Johnson, Louis, 183, 184–85, 191–92; 1951 budget request of, 193; Convair and, 189; Matthews and, 187
Johnson, Lyndon, 250–51, 280
Joint Chiefs of Staff, 178–79, 184–85, 262, 269
Joint Strategic Planning Committee, 242–43
Jordanoff, Assa, 27
Joy, C. Turner, 181, 225
Jutland, Battle of, 14

Kaltenborn, H. V., 99
Keating, Kenneth, 269
Keenan, Hugh, 221, 223, 224–25
Kellogg-Briand Pact (1928), 16
Kelly, "Jocko," 50–51
Kelly, Merwin, 158
Kemp, "Pluvie," 14
Kennan, George F., 178, 286
Kennedy, Bobby, 275
Kennedy, Jack: Cuba and, 259–60, 261, 269, 271; death of, 281; Hayward and, 264; visits *Enterprise*, 265–66
Kennedy, Joseph, 260–61, 262
Kenney, W. John, 185
Killian, James, 247

Kimball, Dan, 187, 248
Kim Il Sung, 218
Kimmel, Husband, 73, 74
King, Ernest J., 74, 128, 136, 292–93; Hayward's Interview and, 113–15; *Indianapolis* and, 140–41; naval aviation and, 31, 50, 112; on order SPO-220, 98; in Panama, 61–62
Kistiakowski, George, 116, 123, 131
Kitchingman, Eleanor, 37
Kitchingman, Ray, 37
Kitty Hawk, 297
Knowland, William F., 217
Knox, Frank, 108, 117, 119
Koga, Mineichi, 101, 107
Korean War, 193–94, 205–6; atomic bombs consideration for, 208–9; effect of, 224; POWs from, 214–15, 216–17, 218; truce in, 212–13
Korth, Fred, 265–66, 268, 275, 280
Krause, Ernie, 202
Kriendler, Bob, 97–98
Krulak, Bruce, 283
Kurzweg, Herman, 229

Labousisse, Henry, 277
Langley, 15, 41–54; Fleet Problem (1933) and, 46–48; Hayward ordered to join, 37–38; night-flying from, 50–51; takeoffs from, 41–42
Langley, Samuel Pierpont, 15
Lanning, Admiral, 43–44
Lapwing, 56
Latin America, U.S. in, 25
Lauritsen, Charles, 115, 119, 126, 147, 175; missile development and, 123, 124
Lawrence, Ernest, 116, 122, 123; cyclotron and, 131, 132; as guest on *FDR*, 233, 234, 235–36, 238; Livermore Laboratory and, 201, 202; task force tour and, 244; VC-5 briefing for, 184
Lawrence, John, 284
leadership, 290–300; development of, 292; Establishment and, 293; Forrestal on, 290; Middle East and, 293–94; military downsizing and, 294–95; needs in future for, 299–300; in nuclear armaments, 294; with personnel, 295–96; social problems in military and, 296–99; total approach to, 290–91
Lee, Sanjo, 216
LeMay, Curtis, 136, 138, 139, 160, 262

Lewis, Dave, 291
Lexington, 15, 50, 83, 85; Fleet Problems (1934), 48–49; night-flying from, 50–51; takeoffs from, 41
Libby, Ruthvan, 234, 241
Lightfoot, Bosun's Mate First Class, 70
Lincoln, 298–99
Lindvall, Fred, 116
Little Boy, 129, 130, 131, 138–39, 160
Livermore Laboratory, 200–202, 294
Lockwood, Charley, 111
Loening Amphibians, 45
London Naval Conference of 1930, 16, 30
Long Beach, 249, 258, 265
"Long Lance" torpedo, 90–91
Los Alamos, N. Mex., 128–29, 131
Lovett, Robert, 247
Lyon, R. S. D. "Ross," 43
Lyons, Evelyn, 13

MacArthur, Douglas, 102, 106, 136; in Japan, 142, 144; Korean War and, 194–95, 198–99
MacDonald, David, 258–59, 280
Maday, Steve, 288
Madsen, "Pinky," 143–44
Mahan, Admiral, 104
malaria, 107, 115
Maloney, Guy, 25
Manhattan Project, 115
Marcus, Groome, 69
Marine Corps Aeronautical Company, 30
Mark 52 mine, 229
Marlow, Don, 229
Marshall, George, 195, 198–99
Marshall Plan, 155, 178
Martell, Charlie, 171, 245, 253
Martin, Bill, 277
Martin, Harold M. "Beauty," 222
Mason, Charlie, 91
Massey, Lance, 35
Mathews, "Matty," 43
Matthews, Francis P., 186–87, 189–90
McCain, J. S., 61, 62
McCloy, John J., 247
McDonnell F-4 Phantom fighter, 257
McElroy, Neil, 251
McLean, Bill, 127
McMahon, Brien, 153
McNamara, Robert, 248, 262–63, 265–66, 272, 276, 280
McNeil, Wilfred, 187, 248

McVay, Charles B., III, 140
McWhorter, Captain, 60
Michels, Ken, 200
Michelson, Albert, 147
Michelson Laboratory, 125, 147, 154, 175
Middle East, Israel and wars in, 175, 293–94, 299–300
Midway, 166, 172, 180, 191
Mikawa, Gunichi, 88
Military Applications Division, AEC, 197
Military Liaison Committee, AEC, 159, 175
Missileer project, 263
Missouri, 144, 155
Mitchell, Ben, 94, 97, 104, 107, 108
Mitchell, Billy, 14
Mitscher, Marc A. "Pete," 49, 88–89, 136, 137; on aircraft for carriers, 167; China Lake and, 115; in Pacific theater, 91; on regulations, 141
Mixon, Riley, 296
Moffett, Bill, 23, 35, 87
Moffett, William A., 23, 31
Monroe, Jack, 157
Monsanto's Mound Lab, Portsmouth, Ohio, 129, 130
Moorer, Tom, 144–45, 157, 191, 281, 287, 292–93
Morgiewicz, Dan, 286
Morrison, Steve, 171
Morrow Air Board, 32–33, 53
Morton, "Mushmouth," 99
Moses, Captain, 126
Muesel, Bob and Emil, 224–25
Mumma, Al, 259
Murphy, John N. "Mother," 167
Murray, Tom, 201, 205
Mussolini, Benito, 75

Naff, Henry, 19
National Advisory Committee on Aeronautics (NACA), 30
National Air and Space Administration (NASA), 254
National Defense Research Committee, 115–16
National Security Council, 155–56, 177, 262
Nautical Almanac, 77
Naval Academy, 9, 12–18; first class year, 17; graduation from, 17; plebe year at, 12–14; Prep School, 9–11; second year, 15; summer cruises, 14, 15–16; technical education at, 290; youngster year, 16

Naval Aircraft Factory (NAF), Philadelphia, 63, 74–75; research and development at, 76–80
Naval Aviation Hall of Honor, 292
Naval Conference of 1935, 62
Naval Observatory, U.S., 77
Naval Ordnance Laboratory (NOL), White Oak, Md., 226, 228
Naval Ordnance Test Station (NOTS), Inyokern, Calif., 115–17; military or civilian control of, 146–47. *See also* China Lake, Calif.
Naval Research Laboratory (NRL), 118, 201–2
Naval Science and Education program, 245
Naval War College, 29, 284–89; Foundation, 292
navigation: celestial, 58–59, 77, 78; by dead reckoning, 100, 113
Navy: air force–navy feud over aviation, 165, 166–67, 179–80, 181, 182–83, 189–90; air force–navy feud over nuclear weapons, 200–201; Articles for the Regulation of, 60; Bureau system in, 104; Revolt of the Admirals, 173, 185–86; roles-and-missions debate and, 174; shrinkage of, 293
Navy Courts and Boards, 59–60
Navy Postgraduate (PG) School, 59
Navy Tactical Data System, 274–75
Nehru, Jawaharlal, 215
Nelson, Bill, 102–3
Nevada, 14, 83, 148
Nichols, Brum, 18
Nichols, Ken, 161, 164, 199
night-flying, 50–51, 78–79, 112
Nimitz, Chester W., 89, 100, 102, 110–11, 136, 167
Nitze, Paul, 144–45, 280, 284
Nixon, Richard, 222
Northampton, 90
North Atlantic Treaty Organization (NATO), 155, 176, 187, 203, 209, 262, 295
nuclear power, 120, 258. *See also* atomic bomb(s)

Oak Ridge, Tenn., 128–29, 130, 131–32
O'Beirne, Emmett, 31, 196
O'Beirne, Frank, 274
Office of Scientific Research and Development (OSRD), 115–16, 118–20, 120–21, 122, 128

Ofstie, Ralph A., 43, 144–45, 168, 183, 190
O'Hare, Butch, 100
O'Higgins (Chilean navy flagship), 237
O'Neil, Butch, 210
Operation Crossroads, 147–49
Operation Ivy, 206, 207–8
Operation Platform, 215, 217, 218–19
Oppenheimer, J. Robert, 116, 122, 200–201; atomic weapons research of, 129; cancellation of security clearance, 227–28; Fat Man tests at Los Alamos, 136; Livermore Lab and, 203; on personnel at China Lake, 119; on quantum mechanics, 123; Site S at Los Alamos and, 131; on test data *versus* information, 125
Oriskany, 197, 296
O'Rourke, Gerry, 275
Ostrum, Whitey, 99
Outlaw, E. E. "Eddie," 171, 194, 197, 283

P2V aircraft, 167, 168–69, 183, 196–97; carrier-landing testing for, 188–89; on *Midway*, 191–92
P-3 Orion, 255, 256, 279
P-59 jet, 123
Pacific Aircraft Battle Force (AIRBATFOR), 42
Pact of Paris (1928), 16
Panama, 48–49, 52–53, 56–58
Pandell, George, 38
Parent-Teachers Association, Inyokern, Calif., 150–51
Parish, Judy, 52, 55
Parish, "Rags," 43, 47, 52, 53, 55, 59
Parsons, William S. "Deak," 80, 116–17, 128, 137; atomic weapons research of, 129; bomb-assembly and, 138–39, 161; China Lake personnel and, 120; death of, 227; on deterrence of atomic bombs, 146; experiments and, 132; Goddard and, 124; leadership of, 292–93; on personnel at China Lake, 119; proximity fuse and, 121; radar and, 120–21; on smart bombs, 125; on solid-fuel propellants, 118, 124; Sykes and, 147; VC-5 squadron and, 169, 171
Pastore, John, 280–81, 287
Pate, Oscar, 51
Pate, Randolph McCall, 214
Patella, Sam, 94, 95, 104
Patten, Bob, 49
Payne, Tommy, 144
Pearl Harbor, Japanese attack, 83

Pearson, Drew, 174, 175
Pearson, Robert, 119–20
Penny, William, 142, 144, 205
Pensacola flight-training, 18, 28, 31–35
Perot, Ross, 298
Petain, Henri Philippe, 75
Peto, 102–3
Philadelphia, 63–64, 66
Phillips, Claire, 23
Phoenix, 66, 67–68, 69–73, 74
Phoenix technology, 263
Pickert, Al, 286
pigeons, carrier, 42
Pirie, Bob, 258, 263
Pius XII, 204–5
Plainview, USS, 284
PM-2s, 53, 55
Point Cruz, 209, 210–11, 212, 225, 233
Polaris FBM program, 248–49
Polignac, Louis de, 267
Pollock, Eleanor and Dan, 87, 236
Pool, Bob, 244
Powers, William J., 220
Practical Aeronautics, 3
Pratt, Admiral, 72
Pratt and Whitney, 30, 257–58
Price, J. D., 181
Pride, Mel, 43, 181, 233
Primakov, Yevgeny, 299
Project Vista, 201
proximity fuse, 121
Purdon, Dave, 171, 194

Quarles, Donald, 158, 250
Quesada, Pete, 214
Quincy, 88
Quinn, Allen, 19, 22

Raborn, William Francis "Red," 248, 264
radar, 120–21
Radford, Arthur W., 45, 52, 165, 190, 292–93; heads Joint Chiefs, 210; naval aircraft and, 168–69; navy nuclear-strike capability and, 166, 182–83
Ragan, Lt. Cdr., 94
Ragg, Tony, 76
Rainer, Prince of Monaco, 267
Ranger, 49, 51, 258
Rankine, J. W., 69–70, 72
Read, A. C., 43
Reichel, Vlad, 79, 80, 82, 244, 284; as guest on *FDR*, 233, 235, 238

Reina Mercedes, 12
Renard, "Chick," 38
Renard, Kit, 38
research and development (R&D): antisubmarine warfare, 255; for atomic bomb, 128–33; budgeting for, 244–45, 254–55; at China Lake, 118–20; competition in, 251–52; Defense Reorganization Act (1958) and, 252; Franke Board and, 253–54; military or civilian control of, 146–47, 153–54; at NAF (Naval Aircraft Factory), 76–80; at NRL (Naval Research Laboratory), 118, 201–2; at OSRD (Office of Scientific Research and Development), 115–16; Pentagon, 243–44
Revolt of the Admirals, 173, 185–86
Reynolds, Wally, 201
Rhee, Syngman, 215, 225
Rhine River Patrol, 257
Rhodes, "Dusty," 43, 180
Rhodes, Warner, 147
Richardson, J. O., 74
Richmond, 19–28; Fleet Problems (1934) and, 48–49; fuel efficiency on, 19–20; at Guantánamo, Cuba, 21–22; to Newport, R.I., 25–26; peace-keeping in Honduras, 23–25; "tare" and "baker" drills from, 21; testing NB-2 biplane from, 28
Rickover, Hyman, 120, 198, 255, 258, 259, 280
Ridgway, Matthew, 198, 226
Riehlman Committee's Public Law 313, 157–58
Riley, Edward O'Neill, 210, 217, 219–23, 224
Riley, Herb, 29, 35
Ritter, Doctor, 56
Rivero, "Rivets," 161, 171, 245
Robinson, T. Douglas, 7
rockets, 1944 development of, 126
Romberger, Bill, 171
Roosevelt, Franklin Delano, 39, 74, 75–76, 134, 137; Einstein and, 128; on *Philadelphia*, 64
Rose, Ensign, 10, 14
Rosenberg, Ethel and Julius, 199–200
Royall, Kenneth, 181
Royer, John, 37
Ruddy, Joe, 31, 36
Rude, Lt. Cdr., 94
Rule, Abie, 59
Ruppert, Jacob, 4
Rusk, Dean, 261

Russell, Jim, 164, 262, 263
Russell, Richard, 273

Sacramento, 25
Sage, Bruce, 124, 153
Salem, 244–45
Sanchez, Mike, 29
Sandia, N. Mex., Armed Forces Special Weapons Project (AFSWP) at, 156, 157, 158
San Roman, Jose, 261
Saratoga, 15, 49–50, 83, 148, 258; night-flying from, 50–51; takeoffs from, 41; in World War II Pacific theater, 98
Saunders, Bill, 84
Saunders, "Hitch," 49
Savannah River, Ga., nuclear reactor, 158–59
Scarborough, William, 171–72
Schaefer, Nev, 157
Schenck, 61
Schneider, Fred, 286
Schoeffel, Malcolm, 147, 153–54
Schreffler, Roger, 109
Schriever, Benny, 251
Schroeder, Pat, 296
Schulten, Lawrence, 10
Seaman, Allen, 94, 100, 102, 106, 107–9
Seaman, Bob, 292
Sears, Harry, 93, 98, 104, 105, 230; crew fatigue and, 107, 110
Sease, Hugh, 43
Semmes, B. J., 284
Settle, Tex, 37
Shadburn, J. E., 191, 194, 197
Sharop, Grant "Oley," 278, 280, 283
Shastri, Lal Badahur, 284
Shenandoah, 8
Sherman, Forrest, 43, 111, 113, 191, 192; death of, 203; Truman Doctrine and, 154
Shoeffel, Malcolm, 119
Short, Giles E., 50
Shoup, David, 261
Shryock, Bill, 171
Sides, John H. "Savvy," 190
Sikes, Bob, 286–87
Simpson, Bill, 193
Sims, William, 14, 19, 53
Singh, Gurk-Bakh, 217
Skate, 102
smart bombs, 125
Smith, Don, 43
Smith, Freddie, 62
Smith, Herschel, 73

Smith, Levering, 121, 248
Smith, Margaret Chase, 250, 280
Smith, Page, 268
Smythe, Harry, 205, 228
Snead, Bill, 93, 101
Snyder, C. P. "Peck," 17
solid-fuel propellants, 118, 124–25, 247–48
Somers, Dick, 108–9
SON-1 seaplanes, 63, 64, 69–70
Soviet Bear bombers, 276, 277
Soviet Union, former, 299–300
SPO-220 order, 98, 99–100, 102, 112–13, 114
Spaatz, Carl "Toohey," 139
Spanish Civil War, 54
Sprague, Thomas L. "Ziggy," 55, 163
Spruance, Ray, 144
Stennis, John, 265–66
Stone, Mac, 235
Stout, Thomas, 24
Strach, Dore, 56
Strategic Arms Reduction Treaty (START), 294
Strauss, Lewis, 228
Stroh, Red, 270, 274
Strong, Bernie, 171
Stroop, P. D., 230, 235, 253
Stump, Felix, 43, 180, 183, 194, 234
submarines, 53–54, 259
Sullivan, John L., 174, 177, 185
Surface, Hank, 94, 98
Swan, 56
Sweeney, Charles W., 139–40
swimming, 15, 16, 24, 25–26
Switzer, Wendell, 157
Sykes, James Bennett, 138, 141, 151, 157, 238; Hayward efficiency report by, 156, 242; on military or civilian control of atomic research, 146–47, 153–54
Symington, Stuart, 163, 174, 186, 189, 191–92, 250
Szilard, Leo, 128

Tailhook investigation, 296
Tangier, 107
Tauber, Maurice "Max," 286
Taylor, Jimmy, 84
Taylor, Maxwell D., 160, 215, 216, 222
Teller, Edward, 116, 122, 131, 164, 198, 242; Livermore Lab and, 202; Naval War College and, 286; task force tour and, 244–45
Temple University, 75
test and evaluation (T&E), 254

TF-30 engine, 263
Thach, Jimmy, 190, 255, 277–78
Thompson, L. T. E. "Tommy," 116, 130, 147, 153
Thorat, S. P., 217
1,000 Men and a Baby (movie), 221
Thurlow, Tommy, 76, 78–79
Thurmond, Strom, 269
Tibbets, Paul W., Jr., 132, 138–39
Tinian (island in Marianas), 135, 136, 137, 138
Tinker, Frank, 54
Tojo, Hideki, 89, 136
Tolman, Richard C., 115, 116
Tooley, Bill, 220
Training and Tactical Command (TTC), air force, 161
Trapnell, Fred "Trap," 189, 190, 194
Treaty of 1921–22, 15
Truman, Harry S., 137, 198–99, 207; military reorganization under, 152–53, 155–56; military spending under, 177–79, 183–84
Tuck, Fred, 94, 98, 107, 108
Turtle, "Aunt Jenny" and Bill, 36
Tuve, Mel, 121
Tydings, Millard, 187

Unification Act (1947), 152, 155–56, 187
United States, 166, 185, 279
University of California, Berkeley, 201
University of Pennsylvania, 75
University of Portland, 123
up-checks, 33
Updegraf, William, 47, 48
Uranium Advisory Committee, 128
Uremovich, Al, 142
Utah, 16, 83

Vandegrift, Alexander, 87
Vanderbilt, Muriel Church, 20
van Voorhis, Bruce, 73, 97
Vargas, Getúlio Dornelles, 70–71
variable time (VT) fuse, 121
VB-101 squadron, 87, 90
VB-104 squadron, 93, 104–5, 106
VB-106 squadron, 92; air combat in Pacific by, 104–11; Gilbert/Marshalls reconnaissance mission, 94–95; Hawaii training for, 93–94; Hayward fires chaplain of, 101; leadership in, 95–96; Nimitz greets, 110–11; preparedness of, 92–93; Wake Island mission, 94

VB-115 squadron, 110
VB-116 squadron, 106
VC-5 composite squadron, 169; carrier modifications for, 172; demands on Hayward of, 172–73, 180–81; hardware for, 169–71; home base for, 175–76; in Mediterranean, 196–97, 203; move to Norfolk by, 193; personnel for, 171–72; *Roosevelt* training for, 195
VC-6 composite squadron, 195
VD-3 photoreconnaissance squadron, 94
Very Long Range Search (VLRS) missions, 97, 100–101, 103, 104, 108
VF-1 fighter squadron, 43
VF-2 fighter squadron, 43
VF-3 fighter squadron, 43
VF-6 fighter squadron, 43
Vietnam War, 282–83, 284–85
Villanova University, 260–61
Vincennes, 88
Vinson, Carl, 186, 189, 190, 191, 265–66
Vogelsang, Walt, 105
von Neumann, Clarie, 122
von Neumann, John, 116, 122, 130, 246
Vorhees, "Mac," 51
VP-2 patrol squadron, 53
VP-3 patrol squadron, 53
VP-5 patrol squadron, 53
VS-1B scouting squadron, 43, 44–45
VS-2 scouting squadron, 43
VS-3 scouting squadron, 43

Wahoo, 99
Walker, Tom, 157
Ward, Corky, 272
Warner, Art, 153
War Plan Orange, 84–85
Warren, Stafford, 138–40
Washington Naval Conference (1921–22), 15
Watson, Carl, 116
Weakley, Charlie, 279
Weapons Systems Evaluation Group (WSEG), 185–86
Webb, Dr., 65
Webb, Jim, 292, 296
Webb, Lee, 43

Wehle, John, 78–79
Wendover Air Base, Utah, 129
Weonah, 3
Westmoreland, General, 282, 283
Wheeler, Earle, 282
White, "Bill," 43
White, Ferris, 236
Whiting, Kenneth, 43–44, 49
Wiesner, Jerome, 247
Wilbourne, Willie, 232
Wilbur, Dwight, 8
Wilkins, Lieutenant, 60
Williamson, Frankie, 245, 278
Wilson, Roscoe "Bim," 132, 200
Winchell, Walter, 175
women in combat, 296
Woodhull, Roger, 167
Woods, Ensign, 10
World War II: air combat in Pacific, 104–11; aircraft-submarine communications in, 100; first defeat in, 87–88; outbreak of, 75; Pacific theater, 96–103, 135–37; precursors to, 39–40, 54, 67, 73–74; stopping, 146; U.S. enters, 83–84
Worth, Cedric, 189
Wright, Gerald, 244
Wright, Whitney, 233, 234
Wright "Whirlwind" engines, 30
Wyoming, 8

Yamamoto, Isoroku, 90, 91–92
Yankee, 3
Yankees, New York, 4–5
Year of Decisions (Truman), 137
Year of Trial and Hope (Truman), 198
York, Herb, 202, 244
Yorktown, 85, 106, 284
Young, Dave, 180
Young, J. S. E., 35

Zar, Marcus, 71
Zeroes, Japanese, 86, 104
Zogbaum, Rufus, 32
Zuckert, Eugene, 207
Zumwalt, Elmo, 297
Zuni Indians, 46

The Naval Institute Press is the book-publishing arm of the U.S. Naval Institute, a private, nonprofit, membership society for sea service professionals and others who share an interest in naval and maritime affairs. Established in 1873 at the U.S. Naval Academy in Annapolis, Maryland, where its offices remain today, the Naval Institute has members worldwide.

Members of the Naval Institute support the education programs of the society and receive the influential monthly magazine *Proceedings* and discounts on fine nautical prints and on ship and aircraft photos. They also have access to the transcripts of the Institute's Oral History Program and get discounted admission to any of the Institute-sponsored seminars offered around the country. Discounts are also available to the colorful bimonthly magazine *Naval History*.

The Naval Institute's book-publishing program, begun in 1898 with basic guides to naval practices, has broadened its scope in recent years to include books of more general interest. Now the Naval Institute Press publishes about one hundred titles each year, ranging from how-to books on boating and navigation to battle histories, biographies, ship and aircraft guides, and novels. Institute members receive discounts of 20 to 50 percent on the Press's more than eight hundred books in print.

Full-time students are eligible for special half-price membership rates. Life memberships are also available.

For a free catalog describing Naval Institute Press books currently available, and for further information about joining the U.S. Naval Institute, please write to:

Membership Department
U.S. Naval Institute
291 Wood Road
Annapolis, MD 21402-5034
Telephone: (800) 233-8764
Fax: (410) 269-7940
Web address: www.usni.org